RETROFITTING OF COMMERCIAL, INSTITUTIONAL, AND INDUSTRIAL BUILDINGS FOR ENERGY CONSERVATION

RETROFITTING OF COMMERCIAL, INSTITUTIONAL, AND INDUSTRIAL BUILDINGS FOR ENERGY CONSERVATION

Edited by
MILTON MECKLER, P.E.

VAN NOSTRAND REINHOLD COMPANY
NEW YORK   CINCINNATI   TORONTO   LONDON   MELBOURNE

Copyright © 1984 by Van Nostrand Reinhold Company Inc.

Library of Congress Catalog Card Number: 83-1148
ISBN: 0-442-26226-4

All rights reserved. No part of this work covered by the copyright hereon may be reproduced or used in any form or by any means—graphic, electronic, or mechanical, including photocopying, recording, taping, or information storage and retrieval systems—without permission of the publisher.

Manufactured in the United States of America

Published by Van Nostrand Reinhold Company Inc.
135 West 50th Street
New York, New York 10020

Van Nostrand Reinhold Company Limited
Molly Millars Lane
Wokingham, Berkshire RG11 2PY, England

Van Nostrand Reinhold
480 Latrobe Street
Melbourne, Victoria 3000, Australia

Macmillan of Canada
Division of Gage Publishing Limited
164 Commander Boulevard
Agincourt, Ontario MIS 3C76, Canada

15 14 13 12 11 10 9 8 7 6 5 4 3 2 1

Library of Congress Cataloging in Publication Data

Main entry title:

Retrofitting of commercial, institutional, and
  industrial building for energy conservation.

  Includes index.
  1. Buildings–Energy conservation. I. Meckler,
Milton.
TJ163.5.B84R47   1983   696   83-1148
ISBN 0-442-26226-4

Contributors

JAMES M. ARCHIBALD, Vice President, Smith Environmental Corporation, Chapter 15.

MORRIS BACKER, P.E., Senior Vice President, Bovay Engineers, Inc., Chapter 16.

LARRY BICKLE, PH.D., P.E., Bickle/CM, Inc., Chapter 4.

JAMES T. BRODIE, President, Creative Power Management, Chapter 26.

FRANK T. CARROLL, Product Line Manager, Environmental Control Division, American Air Filter Company, an Allis-Chalmers Company, Chapter 17.

KENNETH M. CLARK, P.E., Burns & McDonnell, Chapter 5.

DALE S. COOPER, P.E., Dale Cooper Consulting Engineers, Inc., Chapter 18.

EUGENE E. COOPER, PH.D, Mechanical Systems Division, Mechanical and Electrical Engineering Department, Civil Engineering Laboratory, Chapter 24.

WILLIAM S. FLEMING, W.S. Fleming and Associates, Inc., Chapter 23.

MICHAEL S. GOODMAN, W.S. Fleming and Associates, Inc., Chapter 23.

HARRY T. GORDON, R.A., Burt Hill Kosar Rittelmann Associates, Chapter 22.

JOHN J. HALAS, P.E., Edward A. Sears Associates, P.E., Chapter 12.

KENNETH S. HARMON, JR., P.E., President, Texas Energy Engineers, Inc., Chapter 9.

JOHN H. HIRT, Hirt Combustion Engineers, Chapter 14.

FREDERICK H. KOHLOSS, Frederick H. Kohloss and Associates, Inc., Chapter 20.

ROBERT C. LEMAY, R.C. LeMay Associates, Inc., Chapter 13.

MILTON MECKLER, P.E., President, The Meckler Group, Chapters 1, 2, 8, 10 and 25.

BILLU MERRARO, Vice President, The Meckler Group, Chapter 21.

PAUL W. O'CALLAGHAN, School of Mechanical Engineering, Cranfield Institute of Technology, Chapter 6.

J. BOB ROBERSON, Southern California Edison Company, Chapter 3.

DONNA L. RYBISKI, Public Affairs Department, Tenneco Inc., Chapter 8.

GIDEON SHAVIT, PH.D., Honeywell, Inc., Chapter 11.

L. G. SPIELVOGEL, P.E., Lawrence G. Spielvogel, Inc., Chapter 7.

FRITZ A. TRAUGOTT, Robson & Woese, Inc., Chapter 19.

J. PHILLIP UPTON, P.E., Department Manager, Bovay Engineers, Inc., Chapter 16.

Foreword

Those of us who have long been concerned about this country's energy future seem to be winning the first and, no doubt, most important battle in ensuring an adequate, affordable supply of energy. We have the public's ear. Expensive home heating oil, the threat to the continued availability of those products (such as plastics) that are made from petrochemicals, the flood of dollars overseas, pollution and the environmental question, the prices at the gas pump—all of these front-page issues are driving home the fact that the ground rules of our sophisticated, highly mobile society are changing. We cannot play the game as recklessly as we have in the past.

The question, therefore, no longer turns on whether or not there is a continuing energy crisis. The question now is how will we respond? Will we be forced into designing for a future of *less,* or will we be called upon, indeed challenged, to use our creative minds to design for a future of *better?*

To plan intelligently for the tomorrow that we know will be radically different from today requires that we have the best, most up-to-date information available. In the area of energy conservation, there is no lack of such information. What we need are tools to bring this information within reach.

This handbook is such a tool. For those who design and build structures as well as those who operate these structures, this book makes accessible information in the developing field of retrofitting for energy conservation. It provides this information in an especially persuasive manner because not only is the material that appears in these pages at the cutting edge of current theory and practice, it is presented to the reader incrementally; one step leads logically to the next. Beginning with a theoretical framework—that is, the concept of energy-conscious design—this handbook proceeds through data gathering, the analysis of data, and the constraints of present codes, and from there to action, management, and evaluation. More than a comprehensive reference book, these pages outline a methodology for the important art of retrofitting.

Why important? Because we have learned that a carefully conceived retrofit program can cut significantly the energy consumed by our present building stock. In fact, the building in which I am writing this, the American Institute of Architects national headquarters in Washington, D.C., is just completing an extensive retrofit that is projected to cut its consumption of energy by 50%.

The authors of the chapters that follow provide the means to learn about what has already been done and to build upon that knowledge. In rising to the challenge implicit in this handbook, in making the best, most energy-efficient use of today's buildings, we will better serve the public, our ultimate client, in the years to come.

Charles R. Ince, Jr.
President
AIA Foundation

Washington, D.C.

Preface

RETROFITTING OF COMMERCIAL, INSTITUTIONAL, AND INDUSTRIAL BUILDINGS FOR ENERGY CONSERVATION is designed for use by building owners, managers and operators, investors, architects and design engineers, contractors and plant engineers. It broadly covers commercial, institutional and industrial buildings and other energy consuming facilities. Actual hands-on and operating experience in surveying, auditing, planning, design, construction, and the maintenance of energy conserving equipment and systems, which have been employed and retrofitted within existing operations, are emphasized.

While the term "retrofitting" can and does have different meanings for the various engineering disciplines as employed in construction we mean: "The modification, rearrangement, removal or addition to existing (building or process) energy consuming systems of components and or controls so that the energy use profile of systems operations, utilizing the existing energy demand, is substantially reduced without sacrifice of prior comfort and productivity standards for the maintained environment."

This book has been arranged to provide guidance to all levels of engineering and construction managers engaged in reducing energy consumption through well-planned and cost effective retrofitting programs.

This book cuts across a broad range of topics. Energy analysis of and reconciliation with historical data and the projection of results of proposed new energy conserving measures, including modeling and computer simulations, and evaluation of utility rates, are covered. Life cycle and economic analysis through the physical surveying and auditing of operations and maintenance procedures are also dealt with, as are budgetary, design, construction, pre- and post-construction surveys.

The information is presented in an easy-to-use format so that it can be applied directly.

There are twenty-six chapters arranged in nine sections which include the following classifications: Energy Management and Retrofitting, Preparing for Retrofit, Retrofitting for Energy Conservation—Approaches and Results, and Retrofitting for Cogeneration.

The need to provide a practical approach and methodology for the design profession and construction industry with proven methods employing both conventionally fueled and alternative energy sources constitutes the principal objective of this book. With an expected increase in energy cost and the inflation pressures that will surely result in the years ahead, most building and industrial facilities engineers and managers must begin immediately to explore all potential energy conservation opportunities. In many cases, they can take advantage of available state and federal tax incentives through creative and directed design efforts.

Conservation of existing HVAC systems, elimination of energy waste, improved insulation, and installation of building automation systems to monitor and control building operations are some of the options now available. But what do we do first? Often viable energy conservation measures can be adopted at little or no capital cost. Minor control adjustments or modifications frequently result in major energy savings. Yet their potential impact is often undiscovered until a studied review of historical building information employing simulation modelling, by computer or other suitable manual methods, are employed.

While this book places special emphasis on practical measures that can be implemented immediately, it also deals fairly with innovative techniques and challenging new approaches.

This book has been written by practicing energy engineer consultants and designers selected for experience in areas of their special expertise. They have been given complete freedom to express their ideas and have given of their time to share their knowledge and experience with all of us. I would like to acknowledge my gratitude and appreciation for their work. Their contributions should be appreciated by all.

In presenting RETROFITTING OF COMMERCIAL, INSTITUTIONAL, AND INDUSTRIAL BUILDINGS FOR ENERGY CONSERVATION, I wish those of you engaged in reducing our nation's dependence on foreign oil, in promoting improved utilization of our domestic energy supplies and in rejuvenating our nation's existing building stock every success, and believe that the information provided in this book should help further these goals.

Milton Meckler, P.E. C.Mfg.E.
Editor

Contents

Foreword, *Charles R. Ince, Jr.* / vii
Preface / ix

PART 1. PROJECTIONS AND PERFORMANCE: THE ENERGY MANAGEMENT CHALLENGE / 1

1. Energy Management: The Realities and the Dilemmas, *Milton Meckler* / 3
2. Economic Analysis of Energy Retrofit Projects, *Milton Meckler* / 24
3. Utility Rates and Retrofit Economics, *J. Bob Roberson* / 33

PART 2. ENERGY AUDITS AND SURVEYS / 41

4. Energy Audits and Energy Management, *Larry W. Bickle* / 43
5. Managing the Energy Use Survey and Audit, *Kenneth M. Clark* / 48
6. Industrial Energy Management, *Paul W. O'Callaghan* / 59

PART 3. ENERGY ANALYSIS TECHNIQUES AND STRATEGIES / 65

7. Estimating Energy Consumption in New and Existing Buildings, *L. G. Spielvogel* / 67
8. Schoolhouse Energy Efficiency Demonstration: An Energy Audit Program, *Donna L. Rybiski and Milton Meckler* / 89
9. HVAC System Analysis, *Kermit S. Harmon, Jr.* / 172

PART 4. COMMERCIAL RETROFITTING CASE STUDIES / 187

10. A Computer Simulation Approach to HVAC System Retrofitting, *Milton Meckler* / 189
11. Centralized versus Distributed Fan Systems in High-Rise Buildings, *Gideon Shavit* / 205

PART 5. INSTITUTIONAL RETROFITTING AND CASE STUDIES / 215

12. Institutional Rehabilitation: An Opportunity for Significant Energy Savings, *John J. Halas* / 217

PART 6. RETROFITTING FOR INDUSTRY / 237

13. Industrial Energy Observations and Opportunities, *Robert C. LeMay* / 239
14. Energy Conservation in the Industrial Combustion Field, *John H. Hirt* / 247
15. Retrofit Economics and the Utilities, *James M. Archibald* / 253

16. Case Study of a Commercial Retrofit Project: East Texas State University, Commerce, Texas, *Morris Backer and J. Phillip Upton* / 265
17. Case Study of a School Building Retrofit, *Frank T. Carroll* / 298
18. Retrofitting the Astrodome for Energy Conservation, *Dale S. Cooper* / 304
19. Heat Reclaim System Project for a Chemical Laboratory, *Fritz A. Traugott* / 315

PART 7. SOLAR RETROFITTING POTENTIALS / 327

20. Engineering Solar Retrofit Projects, *Frederick H. Kohloss* / 329
21. Case Study of an Energy and Solar Heat Study, *Billu Merraro* / 341
22. A School Solar Retrofit Case Study, *Harry T. Gordon* / 350

PART 8. RECONCILING ACTUAL RETROFIT PERFORMANCE / 355

23. Reconciling Actual Retrofit Performance with Projections, *William S. Fleming and Michael S. Goodman* / 357

PART 9. RETROFITTING AND POWER GENERATION / 365

24. Cogeneration and Retrofit Opportunities, *Eugene E. Cooper* / 367
25. Employing Solar Thermal Hybrid HVAC/Power Generation Systems in Buildings and Community Energy Systems, *Milton Meckler* / 385
26. Industrial Cogeneration, *James T. Brodie/* / 399

Index / 407

RETROFITTING OF COMMERCIAL, INSTITUTIONAL, AND INDUSTRIAL BUILDINGS FOR ENERGY CONSERVATION

Part 1

Projections and Performance: The Energy Management Challenge

1. Energy Management: The Realities and the Dilemmas

Milton Meckler, P.E.
President, The Meckler Group
Encino, California

	Page
I. Introduction	3
II. Energy Management Objectives	4
III. Fostering Commitment and Cooperation	4
IV. Energy Utilization Index (EUI)	5
V. Developing the EUI	6
VI. Conducting the Building Survey	8
VII. Energy Management Program Implementation	9
VIII. Monitoring Your Program for Effectiveness	12
IX. Mini and Micro Computer Impacts	13
X. Data Transmission	14
XI. Computer-Based Data Acquisition	14
XII. Networks and Distributed Data Processing	17
XIII. Comparing and Selecting Processors	19
References	22

I. INTRODUCTION

The value of air conditioning equipment installed in commercial buildings in 1981 was estimated at 2.3 billion dollars [1]. Of this total, approximately 70% was for new construction, and 30% was associated with modernization, upgrading, and retrofit projects. If we look at the estimated breakdown by building type of air conditioning systems installed since 1970, we find office buildings at the high end. They require roughly twice the HVAC equipment of hotels and motels at the low end.

Building managers, who must operate and manage many building types, suddenly find themselves thrust into an area in which they have little training or understanding: energy management. Until the 1973 oil embargo, energy expense represented a minor cost in planning most building ventures, so managers had minimal interest in managing energy costs as a separate line item, apart from general operating and maintenance costs. In attempting to develop effective HVAC systems for new and existing buildings, mechanical and electrical engineers coined some new terms, such as "energy management program" and "energy audit." These terms have not caught the attention of the management community. Unfortunately, building managers and owners, with a few exceptions, tend to have little direct experience and few skills in this area, and often do not recognize energy use trends as something they can control [2].

From the building owner's viewpoint, energy conservation efforts must pay off on the bottom line. Are those of us involved in reducing building energy consumption sufficiently aware of this? Are we focusing on long-term, high-capital opportunities and perhaps ignoring short-term, low-cost options that can generate immediate cash flow?

What makes energy conservation often seem like a muddle is that involved principals, be they architects, tenants, designers, owners, building managers, operators, or contractors, often relate to the same problem differently. As a result, communication is often cloudy, and cues are misread. Once the obvious opportunities, such as the elimination of unnecessary lighting, shortened building or HVAC system daily operating hours, and reduced weekend schedules, have been explored, sophisticated engineering methods are required. At this point, computer analysis often comes into play. This is also the point where tenant and employee irritation is most likely to surface.

Wasteful attitudes still prevail, since tenants usually bear added energy operating costs through prior lease escalation commitments. Yet tenants often pass increases in energy-related costs on to the consumer. Tenant resistance to automatic escalation clauses is growing, and building owners are now taking some initiative to avoid further tenant disenchantment with runaway expenses.

Reliance on the many "idea lists" floating around or on a direct vendor approach often results in a shopping list of gadgets being added to existing systems. This can mask the major building energy wasters. Should a building owner be satisified with an 18 to 20% reduction in energy consumption compared with 1973 base-year operations? It's hard to say. What about future utility costs that will continue to escalate? In most cases, an 18 to 20% reduction over 1973 consumption levels approaches the same dollar cost in 1978 dollars as the 1973 utility bills.

II. ENERGY MANAGEMENT OBJECTIVES

Although we will not deal with policy matters until later, a lack of building management competence in dealing with energy-related decisions can, if allowed to continue, produce disastrous consequences. Businesses may fail, and essential resources may be misused. Real-location problems may require painful choices in priorities for regional energy managers. The proper fusion of appropriate engineering disciplines, energy management technology, and basic building management objectives makes good economic sense. However, a mechanism for a broader exchange between building management and engineering groups is still lacking. Perhaps the energy management techniques presented here will help building managers develop the new skills and attitudes essential for more profitable buildings that also conserve energy.

Turning over energy responsibility to an employee without defining specific energy reduction goals through a comprehensive energy audit is a most common and unfortunate practice. It is often thought that energy conservation is related to some design or operating problem that can be remedied. Actually, energy management, with emphasis placed on the word management, requires a careful plan of attack, performed by an individual whose performance can be measured against results. Investing in energy reduction is no different from investing in real property. Aside from the deep national concern to preserve our vital resources, energy conservation must be *managed by objectives.* Reliance on operators or managers without adequate skills can prove more costly than seeking outside help from competent professionals.

III. FOSTERING COMMITMENT AND COOPERATION

In the development of an energy management program for a given building, it is important to obtain the commitment and cooperation of all people whose participation in the program is deemed essential. Maximum effectiveness is usually achieved by assigning to one person the responsibility for managing the entire program. In addition to being accountable for results, that individual must have the authority to solicit support from subordinates when necessary. Otherwise, fragmentation of authority can result in confusion in the lines of authority and communications, which are vital to the success of the program.

Generally, the key factors in any building energy management plan come from the following list: owner, manager, maintenance and operating personnel, clerical staff, tenants, or other occupants.

Without the owner's active cooperation, little is likely to happen. While an owner will usually encourage any attempt to conserve energy usage and cost, it is important that an owner set aside adequate monies for conducting an energy audit of the type subsequently described. An owner must be convinced that an energy audit represents a wise investment, and that it may lead to conservation measures that require little or no capital cost.

Where records exist, it may be helpful to review the energy consumption of the owner's building and compare it with others of its general type.

The building manager must also be committed to the program. The manager is often placed in a difficult position between the building owner, demanding results, and the operating and maintenance personnel, who may not know how to deliver those results. The manager's ability to generate enthusiasm for the owner's expectations and the operator's concerns is necessary to maintain cooperative interest among the parties most affected.

The cooperation of operating and maintenance personnel is essential, particularly during the early stages of the program. These workers will be required to implement any of the changes needed, and some employees tend to regard initial energy management efforts as a questioning of their own capabilities. This is particularly true when an outside consultant is brought in to conduct the initial survey. To overcome hostility or evasiveness, it is often desirable to adopt the attitude that energy management involves "doing things differently." At this stage, employee confidence is gained by focusing on the profitability of various proposed conservation measures and on the need for continued employee participation in seeking new energy conservation opportunities.

In most cases, the individual in charge of the overall program will require clerical assistance to develop and assemble the base year and other historical data necessary for evaluating various alternative energy-conserving strategies. Special efforts may be necessary to find the data required and to verify their accuracy. Care in assigning these tasks to responsible individuals is also essential for program success. When such employee assignments are made, the importance of individual contributions should be stressed. Where possible, select the same personnel used during the implementation phase. It may be helpful to point out early that they will also be collecting subsequent operational data on a monthly basis. Thus personal interest is intensified through an awareness of a further responsibility for monitoring of program results.

Although tenant participation in the initial phases of an energy management program is minimal, valuable operational and occupancy data can be obtained through interviews during the survey phase. Tenants should be made aware of the program's objectives and how they can benefit from its success. Good public relations can pay off handsomely, particularly when tenant operational modifications are requested later.

Let's next proceed to establish an energy conservation goal.

IV. ENERGY UTILIZATION INDEX (EUI)

Effective management requires that all tasks have goals, which are tangible ways to measure results in terms of performance and aid in any redirection of efforts. A study of 18 commercial office buildings in Philadelphia [3] demonstrated the value of energy budgets based on comparison of an energy utilization index (EUI), defined as the consumption of Btu's per gross conditioned space, or cubic feet (cubic meters) per year. With the EUI, a building owner or manager can determine the amount of total energy his building *should* be using to be considered *efficient*. He can then implement appropriate energy conservation measures (ECM) that result in reductions of the annual energy consumption consistent with the listed EUI values for comparable buildings and climates. In the Philadelphia study [3], investigators found that four factors could explain 78% of all variance in EUI's computed on a square foot (square meter) basis of the 18 office buildings sampled. After one computes the EUI's on a cubic foot (cubic meter) basis, the following nine factors can explain 95% of the variance of the EUI's reported:

1. Computer room operating hours and refrigeration capacity
2. Ventilation rate
3. Hours of cooling system operation
4. Type of HVAC system installed
5. Gross wall thermal transmission
6. HVAC system (heating) operating hours
7. Maintenance and systems control practices
8. Building gross conditioning volume
9. Lighting power density

Our ability to identify these factors should eventually enable prediction of EUI's for most general building types.

The data necessary to compute the EUI for a given building are presented in the energy management form illustrated in Figure 1-1. Although this form seems a bit complex at first glance, it was developed to present the necessary information in as straightforward and uncomplicated a manner as possible. This same form can also be used in:

1. Comparing EUI's of two or more buildings of similar design for relative efficiencies.
2. Evaluating energy consumption patterns of a given building for comparable periods in previous years to uncover irregularities in utility billings, estimating procedures, etc.
3. Determining whether the building is receiving the lowest possible utility rate consistent with its operation.
4. Projecting the future cost of energy use in the building by matching the latest utility rate base to the energy consumed in prior years.
5. Comparing variances in EUI's from period to period to get a quick idea of the efficiency of the various building operating systems.

Often, an impending mechanical problem shows up through actual comparison of EUI data, rather than through casual observation of equipment operation.

Notice that the recommended energy management form illustrated in Figure 1-1 is based on a 12-month period, preferably the one prior to undertaking the proposed ECM, although, as a practical matter, any 12-month period can be used.

V. DEVELOPING THE EUI

The process of establishing an energy conservation goal begins by first establishing base-year data. The same procedures for recording this data monthly apply to all subsequent years. Non-base data should also be recorded monthly so that the EUI information is always current [4].

Once the energy management form is completed as described above, we can set some initial or interim goals. We might try for a 10% reduction in annual energy consumption, for example. The goals should be realistic and based on some mix of the following: business objectives, economics, financial and personal capabilities (and resources), existing building operational and maintenance constraints, existing energy requirements, and estimated potentials for conservation. Bear in mind that goals often vary from year to year, and that the annual energy reduction percentage selected must take into account all potentially viable energy conservation options.

We must examine more closely, at this point, the management part of the energy management program. We must recognize that the ultimate program goal is the attainment of the highest possible degree of building energy efficiency. We must also ensure maintenance of that level through continuing scrutiny of the data collected. When comparing energy consumption on a year-to-year basis, we must make adjustments for climatic factors and building use factors that reflect changes in occupied building hours or patterns of use. Means must be established to evaluate the impact of such factors in comparison with the

Building ______________________ ______________ Year

Month	Heating Deg. Days	Cooling Deg. Days	Electricity						Purchased			Steam			Fuel							Total Energy Cost
															Oil			Check Gas ☐ Coal ☐ Other ☐			Fuel/ Deg. Days	
			KWH	KWH/ Deg. Days	KW Demand		Cost		M (lbs.)	M (lbs.)/ Deg. Days	Mos/Hr	Demand	Cost		Quant. (Gal.)	Cost		Quant.	Cost			
					Actual	Billed	Total	Per Ut.			Actual	Billed	Total	Per Unit		Total	Per Unit		Total	Per Unit		
1	2	3	4	5	6	7	8	9	10	11	12	13	14	15	16	17	18	19	20	21	22	23
Jan.																						
Feb.																						
March																						
1st Quarter																						
April																						
May																						
June																						
2nd Quarter																						
July																						
Aug.																						
Sept.																						
3rd Quarter																						
Oct.																						
Nov.																						
Dec.																						
4th Quarter																						
Total Per Year																						

Building Data

Gross Area $(ft)^2$ ______________________

Gen Notes ______________________

____

____

Annual Energy Consumption in BTU's

Quantity	Conversion Fee	BTU/Yr
1. ________ kwh Electricity	3413	________
2. ________ Steam (M) lbs Purchased Steam	1,000,000.	________
3. ________ MCF Natural Gas	1,030,000.	________
4. ________ Gallons Oil	#2 138,700 #6 140,700	________
5. ________ Other Fuel	________	________
	6. Total	________

Energy Utilization Index

$$EUI = \frac{\text{Total Energy Consumption BTU's/yr}}{\text{Gross Area } (ft)^2}$$

$$= \underline{\qquad\qquad} \text{ BTU's/ft}^2\text{/yr}$$

Figure 1-1. Energy management form. (From Ref. 4.)

base year. To do this properly we must rely on internal records as well as utility records. Some possible suggestions include:

1. Obtain data on local heating and cooling factors. A possible source is the National Climatic Center, Asheville, North Carolina.
2. Analyze the relationship between heating and cooling degree days (and other available weather data), and energy consumption in your building [5]. Take into account any major changes in building operating systems or maintenance practices.
3. Obtain records of tenant usages, occupancy, and any special equipment, such as heat-producing machinery, and record any changes in total building refrigeration capacity, hours of cooling system operation, HVAC system heating hours, maintained indoor temperatures and humidities, the type of HVAC system installed, ventilation rates, lighting, maintenance or systems control practices, computer room(s) operating hours, and associated refrigeration capacities.

VI. CONDUCTING THE BUILDING SURVEY

One key in any energy management program is the thoroughness with which the building survey phase is conducted. It is important that those assigned to this task be aware of the specific types of information they must gather. They must be prepared to present such findings in an unbiased manner. It is usually advisable to select as team leader a trusted and knowledgeable employee or registered engineering consultant with extensive experience in mechanical and electrical building systems. This leader should be knowledgeable enough to recognize probable sources of wasted energy and suggest possible remedies.

We start with an up-to-date set of records, drawings, and specifications describing all architectural, mechanical, and electrical work in place. The first step is to become familiar with construction details, facilities, and machinery. Often information is unavailable or incomplete; it may become necessary to develop single line or layout drawings of existing mechanical and electrical systems by piecing together information obtained during a brief walk-through with information available from owner's records. It is important also to obtain current information on prevailing operating and maintenance practices. Valuable sources of such information are operations and maintenance (O & M) manuals normally supplied by the original equipment manufacturers. Such manuals may also have been developed by the original designer or installing contractor. Likewise, we should become familiar with the utility rate schedules and those building codes and regulations that relate to the possible economic or design effects of a planned building modernization program. Refer to Figure 1-2 for a form developed by NEMA [4] for use in assembling the necessary building information during the building survey phase.

Once one is familiar with the various building systems and the type, class, and age of equipment likely to be found, a thorough walk-through survey is in order. By establishing early contact with building tenants, the way is often paved for expediting tenant approval of subsequent visits to their premises.

Although one needs only a pen or pencil, ruler, and pad, it is helpful to have a tape recorder, camera, and instruments such as light meters, velometers, and psychrometers on hand.

Often the trained eye will spot opportunities to reduce lighting levels, heat losses or gains, and infiltration, and to improve operating systems and tenant practices contributing to excessive energy consumption [6, 7]. Obtaining unbiased opinions from in-house personnel, particularly from the operating and maintenance staff, can be difficult; for this reason, employing an outside professional may prove worthwhile. Experienced outside personnel are generally sensitive to in-house employee attitudes. They develop tactful interview strategies to determine the extent to which employees perform their duties and what their biases are regarding possible energy conservation measures. Building management should continually assure the operating and maintenance cluding estimates of the payback periods based on life cycle costing, return on investment, etc.,

DATE OF CHANGE	ENERGY CONSERVATION MESURES IMPLEMENTED
	BUILDING ENVELOPE : _____
	VENTILATION : _____
	HEATING & AIR CONDITIONING : _____
	LIGHTING : _____
	ELEVATORS & ESCALATORS : _____
	DOMESTIC HOT WATER : _____
	OPERATION & MAINTENANCE : _____

Figure 1-2. Energy conservation log. (From Ref. 4.)

staff that the purpose of the survey is not to find fault with them but to obtain an objective and independent view of building operations.

The survey results should be reflected in a simple but thorough report detailing those building system features that are energy wasters, and those non-energy and human systems that contribute to excessive building energy consumption. A description of various alternative energy conservation measures (ECM's) is usually provided to remedy the problems identified. When such ECM's involve capital expenditures, a cost estimate, is used to rank the various choices for subsequent management decisions [5]. When significant capital outlays or operating interruptions are anticipated, strong justification may be required for recommendations. All relevant facts must be considered in detail so that management is apprised of any potential major work delays or employee complaints that may result. Care must also be taken to demonstrate that all recommendations can be undertaken in a manner consistent with the comfort and needs of the building's occupants.

VII. ENERGY MANAGEMENT PROGRAM IMPLEMENTATION

When the survey phase is completed, a competent analysis should be undertaken to identify those ECM's that might materially reduce building energy consumption. Let us assume hypothetically that various ECM's have been recommended, some of which would affect tenant operations. Building management has been presented with a report setting forth a set of the prioritized, recommended ECM's, and a logical description of their technical and economic feasibility. At this point, the building owner/manager must evaluate this proposal on the same basis as any new business venture. He must consider the various proposed ECM's in terms of:

1. Allocating manpower and financial resources.
2. Developing policies and procedures to be followed by the operating, maintenance, and management staff.
3. Assigning specific duties or roles, and obtaining the cooperation of in-house employees and tenants.
4. Establishing any necessary changes in maintenance procedures or schedules.
5. Preparing necessary contract documents detailing systems and equipment that must be repaired, modified, or replaced.
6. Estimating the time needed to accomplish this work and to coordinate any construction with the affected building occupants and employees.

At this critical stage, employee and tenant cooperation should be actively encouraged. Because problems and frustrations generally crop up, a positive attitude is essential. Building management must recognize that some problems uncovered during the survey phase may be symptomatic of shortcomings in specific system capabilities, or related operating and maintenance procedures. Having determined what to do about specific system or equipment problems, the owner/manager must provide the operating personnel with specific instructions.

Guidance is often available from equipment manufacturers, publications and magazines, continuing education programs sponsored by local colleges, engineering societies, or local equipment suppliers. Specific training [8] may be provided by qualified consultants or contractors, or operating and maintenance manuals.

Now, the tenants, who have essentially been spectators to this point, must be asked to participate directly in ways they might normally resist. For example, consider carefully a tenant's possible reaction to the following requests:

1. Lower the thermostat during the heating seasons, and raise it during the cooling seasons.
2. Have employees use the stairwell for any one-flight trips.
3. Turn off (some) electric water fountains.

Any such presentation must be tailored to each tenant. It is often helpful first to examine the different leases operative to determine building management and tenant rights and who must bear the costs of any proposed changes. With this information as background, it is advisable next to meet with tenant representatives. Present them with specific data involving energy and operating costs for the building, including relevant impacts on their leased area.

It is generally helpful to begin your discussion with a brief explanation of any proposed building operating changes. Explain why they have been selected to participate, and review their impact on energy and operating costs. Explain when changes are likely to take place. By identifying each change and explaining what results are anticipated, tenant cooperation can be improved. In some situations, certain lease changes on items of minor importance must be negotiated. In dealing with major energy consumption items, however, a firm stand must be maintained. Personal contact at this stage is essential; attempting to explain the program by letter can be disastrous.

Once the various tenant representatives agree to cooperate, offer to assist their office managers in explaining the proposed conservation measures to their co-workers. Often such joint meetings allow tenant employees to air grievances. This may help them avoid future misunderstandings and tensions. It is advisable to bring some materials for distribution to tenant employees. A memorandum of the

TO: Employees, Zebra Tint Manufacturing Co
FROM: John Doe, Cosmos Management Company
RE: "Operation Energy Savings"

As you know, we are in the midst of a national program to conserve energy. The goal is to make our country independent in terms of foreign energy supplies. Because of this fact, and due to the skyrocketing costs of energy, Cosmos Management Co. has embarked on its own program called "Operation Energy Savings". To make it a success, we need your cooperation and assistance.

Here are some of the specific things we'll be doing which will affect you in one way or another.

a. Elevators: We have six elevators in the building. Two of them will be stopping at even-numbered floors only, two at odd-numbered floors only, and two will not be affected. On occasion you will have to walk up or down a flight of stairs. It's an inconvenience, we know, but at least it will help us get some of the exercise we all need.

b. Lighting: We will be removing lamps from certain lighting fixtures; moving others, and disconnecting some decorative or non-essential fixtures completely. Although the lighting levels will not be as high as they were before, the quality of lighting provided will be perfectly adequate for the type work being performed in the various affected areas.

c. Humidity: Because of changes being made in our air conditioning system, there may be times when it may be a bit humid inside. We don't expect that it will happen very often or for long periods of time.

d. Temperature: We plan to be lowering the temperature to 68°F. during the winter and raising it to 78°F. during the summer. To compensate, wear a sweater or some other warm clothing during the heating season, also we suggest you consider wearing light clothing during the cooling season.

We'll probably be making some other changes too, and we will let you know about them beforehand. In the meantime, there are a few things that you can do that will help us out a great deal. We are also placing signs around to remind you, too, so we hope you will cooperate.

1. Turn off electrical equipment when not in use, for example, desk lamps, typewriters, or coffee pots. We realize that some equipment can only be turned off at the end of the day. Please be sure you do turn them off, particularly copying machines.

2. Turn on lights only when they're needed and the space is occupied. Turn them off when they're not needed. While life of fluorescent lamps is reduced when they're turned off and on more frequently, the energy they use when they're on is wasted and doesn't make leaving them on worthwhile.

3. Be sure that you close all operable windows and doors leading to unconditioned areas whenever the building's heating, ventilating and air conditioning systems are working.

4. Whenever you can, group your trips together so you don't have to use the elevators as much. Try whenever possible to do four things on one floor while you're there rather than making four separate trips. Whenever possible use the stairs rather than the elevator.

5. Please do not attempt to change your thermostat. It has been carefully adjusted and calibrated.

6. Please report any leaky faucets, piping, valves etc., to the maintenance department.

7. Please close the venetian blinds or drapes when the air conditioning is running during the day. During the colder weather, though, open the blinds and drapes on the sunny side of the building during the day. We can all use the heat and besides, its free.

8. Send us your ideas and suggestions. We'll all benefit by saving energy.

We will keep you informed regularly of our progress. We're trying to reach a goal of 30% reduction in energy use this year, and with your help and indulgence we are hopeful of achieving it.

Figure 1-3. Representative energy conservation memorandum for building tenants. (From Ref. 4.)

DID YOU REMEMBER TO TURN OFF THE LIGHTS?

PLEASE BE SURE DOOR IS CLOSED SECURELY.

PLEASE TURN OFF THIS LIGHT WHEN ROOM IS NOT IN USE.

DO YOU REALLY HAVE TO USE THE ELEVATOR, OR CAN YOU USE THE EXERCISE?

PLEASE DO NOT TAMPER WITH CONTROLS. WE ALL NEED TO SAVE ENERGY.

Figure 1-4. Some representative slogans/signs used in energy conservation programs. (From Ref. 4.)

type presented in Figure 1-3 may be useful. The placement of signs, such as those in Figure 1-4, in strategic locations often serves as a helpful reminder.

VIII. MONITORING YOUR PROGRAM FOR EFFECTIVENESS

Program results must be measured in terms of the quality of implementation and effectiveness.

Has the installing contractor completed his work in a satisfactory manner and in accordance with agreed-upon schedules? Are maintenance personnel aware of all changes in their preventive maintenance schedules? Are such changes being made in the most effective manner? These changes are best recorded in a log of the type illustrated in Figure 1-2. It is also helpful to keep on file duplicate copies of all maintenance manuals and schedules, cost and inspection records, and so on.

In monitoring building energy consumption, it may be useful to refer to the information developed earlier on the energy management form, illustrated in Figure 1-1. This information may be applied to the entire system or its major subsystems for spot checking of metered data, pertaining to lighting, elevators, HVAC systems, and so forth. Such checks often involve the placement of meters on the principal electrical feeders to the various subsystems.

In larger buildings or older buildings where several subsystems are served from a single feeder, an empirical survey, performed by a qualified person, often establishes needed data on connected loads, usage, load factors, and so on. Such a survey evaluates energy consumption levels for key subsystems. Whether obtained by actual metering or by an empirical survey, such data can serve as the basis for an energy budget for the various tenants. Once this is established, a written modification is prepared for each lease agreement, allowing the tenant to consume a designated amount of energy. This amount is specified as part of the normal rental payment. Any excess, however, is to be paid for directly by the tenant. By means of subsystems monitoring, building management can readily identify excessive tenant energy subsystem usage. This kind of arrangement provides a strong economic incentive for tenant cooperation.

In some cases, data may already be available for checking. In one instance, metered data on all electric HVAC systems in a number of smaller office buildings located in Philadelphia revealed an overall annual energy consumption profile as follows:

Space cooling	20%
Space heating	14%
Base electrical load (including lighting, elevators, etc.)	66%

Such values vary from building to building and often depend on building envelope characteristics, HVAC systems installed, and use and operation.

Monitoring must remain a continuing process, involving the exploration of further improvements from time to time. Where recorded data suggest such improvements are possible, energy conservation measures that earlier may not have proved cost-effective should be reevaluated. Program goals should also be revised when new insights or operational circumstances justify the change.

Ideally, monitoring should be a routine building management function. It can thereby be made more visible through periodic meetings with building maintenance personnel. At such times, accomplishments, disappointments, future plans, and new ideas can be discussed. Similar exchanges with tenants are also encouraged to stimulate those often self-sacrificing efforts required to achieve still further reductions in building energy usage.

A key area in the monitoring and management of various energy conservation schemes involves the application of cost-effective multiplexing and related processing techniques, which will be described in the subsequent sections of this chapter.

IX. MINI AND MICRO COMPUTER IMPACTS

The minicomputer has taken much of the limelight recently in the field of data processing. Market estimates [9] suggest that it may become the most prominent data processing element of this century. It has been projected that by 1987 total sales may exceed $30 billion. More than 60 companies now manufacture microcomputers and minicomputers, exclusive of peripheral, software, and other system companies now entering the market.

The minicomputer is little more than a small, low-cost (as compared with larger, main-frame versions), general-purpose computer. The microcomputer, on the other hand, is a low-cost minicomputer. Its name relates both to its physical size and its limitations on word size, storage capacity, registers, instruction repertoire, and software format. Basic minicomputer costs range from $15,000 to $500,000 for a complete system. Costs for a microcomputer system range from $300 for a microcomputer with no memory, to $900 to $15,000 for a microcomputer system with a 4K, 8K or 16K memory. Table 1 presents a comparative summary of mini versus micro computer system formats.

The traditional computer optimization approach favors developing a number of programs for the same central processing unit (CPU). Other approaches employ a number of CPU's in some cost-effective distributed fashion. A microprocessor, which is the central part of a computer (the CPU), has now been realized on one single component chip. When this single chip is combined with memory and input/output circuits of a general nature, the assembly is called a microcomputer. It comes in the form of a printed circuit card. A microcomputer system performs a specific task and consists of a microcomputer plus interfaces to sensors, associated instruments, and programs.

Hardware costs for processing power have become negligible compared with software costs. Microprogramming is one way to realize the instruction repertoire of a computer or to implement a particular sequential control circuit. The technique has been used since the very first computers were constructed. A microcomputer may be microprogrammable, but often it is not. As a result, energy software systems designers [10], when selecting from among 4-, 8-, and 16-bit micro or mini processors, for example, must carefully take into account the required speed and complexity of the calculation. Intercommunication needs between various data processors and sensors plus data handling and storage capacity must also be taken into account.

Microprocessors are divided into three

Table 1-1. Characteristics of Current Mini versus Micro Computer System Formats.

	MINICOMPUTER	MICROCOMPUTER
Storage mode	Magnetic core and semiconductor	Semiconductor
Capacity of storage	130K bits	32K bits
Size of word	16 bytes	4 and 8 bytes
Memory cycle time	2 to 5 msec	5 to 10 msec
Secondary storage mode	Disc, tape	Very limited
Control modes	ROM hardware and microprogramming	ROM microprogramming
Composition	Integrated control, I/O, and memory	Integrated control
Available software	Assembler FORTRAN, COBOL, BASIC	Assembler

Source: Ref. 9.

groups, according to the word length of the central arithmetic element:

4-bit calculating or instrument-type microprocessors
8-bit general-purpose microprocessors
16-bit minicomputer-like microprocessors

The 4-bit microprocessor is mainly used to handle very small processing functions, like those performed by a calculator and a small on-line instrument. The 8-bit general-purpose microprocessor is the most widely used type of microcomputing systems. The 16-bit microprocessor resembles a lower-cost minicomputer. It lacks the 8-bit microprocessor's advantages because it has a greater need for byte-oriented data-interchange.

X. DATA TRANSMISSION

In a multiplex system, two or more signals are combined so that they can be transmitted together over one physical cable or radio link. Data transmission capacity frequently determines the speed of the various interconnecting communication lines. It is often desirable to provide communication links with as wide a band width as possible and then divide the band width among many users. In this way, many separate signals can be multiplexed together to travel as one signal over a wide band width. In multiplexed systems [11] several printers, card readers, or other input/output devices operate simultaneously. The bits sent out to them or received from them are intermixed as they travel along a given single channel. The bits are transmitted at a rate that is slow compared to the multiplexer's inherent scanning speed, so that it is capable of overlapping its handling of many data links. Multiplexing is the key to cost reduction in automated energy management systems. The backbone of a truly effective multiplexed system is an effective digital transmission system.

XI. COMPUTER-BASED DATA ACQUISITION

There has been dramatic growth in the size and versatility of building automatic systems in recent years [7]. Digital transmission (multiplex connected) systems provide numerous advantages over prior generation analog (point-to-point connected) systems. (See Figure 1-6.) HVAC, security, fire protection, and safety functions can be combined initially with other tasks and subsystems, such as public address and traffic control. Demand limiting features can be added when the need arises, or as the system becomes cost-effective. Thus, the basic initial system can be relatively low in cost, and, as its effectiveness is demonstrated, additional displays, sensors, controls, and computer capabilities can be added.

Refer to Figure 1-5 for a relative cost trend comparison of hardwired versus multiplexed transmission. Digital transmission systems have significant advantages over analog transmission systems. Digital signals may be transmitted virtually any distance with high accuracy. Multiple types of transmission media are available including wire, microwave, telephone, and radio.

Noise (errors) introduced into the digital transmission system may be quickly and economically detected and corrected. The response or sampling rate and accuracy of digital signal transmission systems can be made virtually independent of the transmission distance. As transmission distances increase, analog transmission with high accuracy becomes more difficult. The greater the transmission distance, the greater the amount of noise that is introduced into the system. Removal of the noise by filtering is time-consuming, difficult, and expensive. It also tends to be individualized for each installation. As the amount of filtering increases, the response time (speed) and accuracy of the system decrease.

Remote digitizing and digital transmission of analog signals greatly simplify the costly calibration and maintenance of a system. Conventional span and zero adjustments are eliminated. Since the analog signals are converted to a digital format at the remote points, the transmission system is not involved in any calibration requirements. In other words, when there is a questionable sensor reading, it is localized to the specific sensor element or transducer, and isolated from the transmission cable.

ENERGY MANAGEMENT: THE REALITIES AND THE DILEMMAS 15

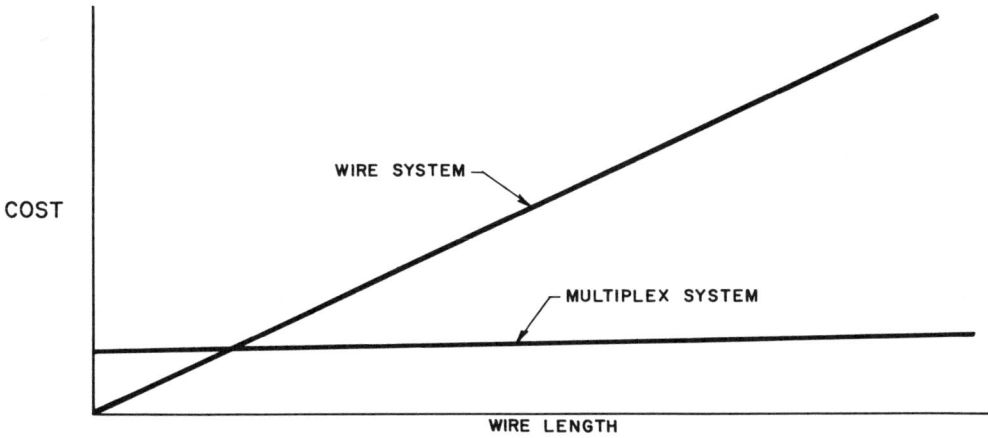

Figure 1-5. Cost comparison for wiring versus multiplexing. *(Courtesy of American Multiplex Systems, Fullerton, California.)*

The technique of digital transmission of digitized analog data makes it simple to provide signal isolation between the remote points and the central control console, often a severe and expensive problem in analog transmission systems.

When remote analog inputs such as temperature and outputs such as set point adjustments or commands are digitized remotely, only one simple data transmission system is required between the remote points and the central equipment. This single system transmits digitized analog values as well as discrete (binary) ON/OFF commands and status moni-

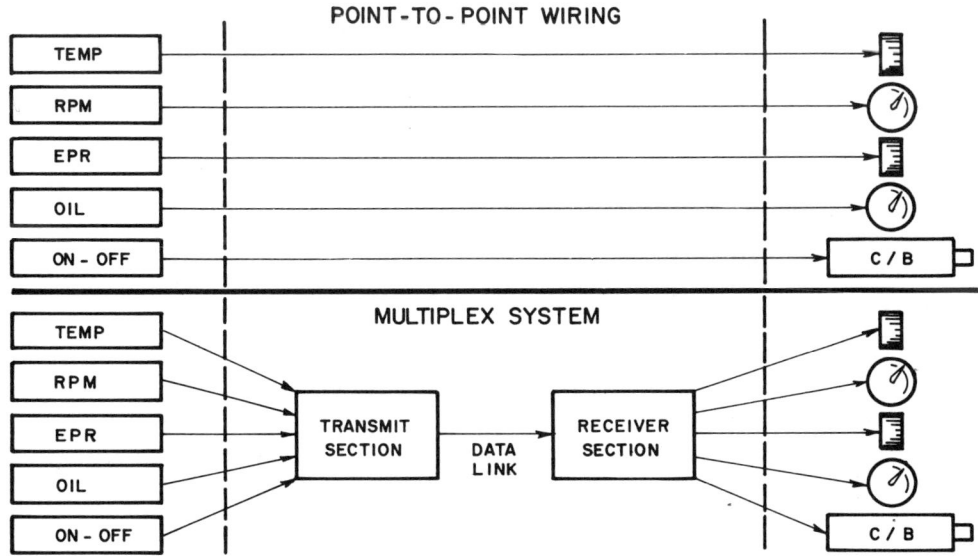

Figure 1-6. Point-to-point wiring (analog) for a multiplex system. *(Courtesy of American Multiplex Systems, Fullerton, California.)*

toring signals; therefore, the data transmission cables are simplified and less costly [11]. One tiny cable is all that is required.

In some applications, analog transmission may be satisfactory. In most cases, however, digital transmission has benefits that greatly tip the scales in its favor. The technique of remote analog to digital (A to D) and digital to analog (D to A) conversion provides the lowest cost initially, while facilitating the future addition of digital minicomputers, cathode ray tubes (CRT's), and other computer peripheral devices. When large quantities of signals are involved, time division multiplexing (TDM) is the most efficient approach. In this technique, all commands and data are transmitted serially down the cable. Digital transmission provides superior data accuracy even compared to dedicated point-to-point wiring. We don't have to detect how large or small the signal is; we merely need to know whether it is a 1 or a 0. Analog transmission techniques, in which the information is in the amplitude of the signal, the duration of the signal, and the pulse position or frequency, are much more susceptible to noise effects than pure digital techniques of the type illustrated in Figure 1-7.

Each TDM digital word consists of address bits, data or command bits, and a check bit such as a parity bit. The parity bit permits each word to be checked for the insertion or dropout of bits. When errors are detected, the system rejects the data or command to ensure that no bad data are used.

The noise environment of the data transmission system is one of the principal considerations affecting the design of the system. Noise within the transmission system consists of undesirable electrical signals induced from sources that may be either external or internal to the system. The noise may be random (caused by a large number of elementary disturbances occurring randomly in time), or it may be impulsive (occurring at regular intervals). In addition, there may be cross-talk, running from other parallel electrical circuits. The use of Manchester biphase encoding for the data reduces the susceptibility of the transmission system to DC and low frequency noise. This result is achieved by concentrating the information in a frequency band above the high noise environment band.

A good digital transmission cable is a two-conductor twisted and balanced transmission line with an overall shield around both conductors. Twisting the two balanced signal-carry-

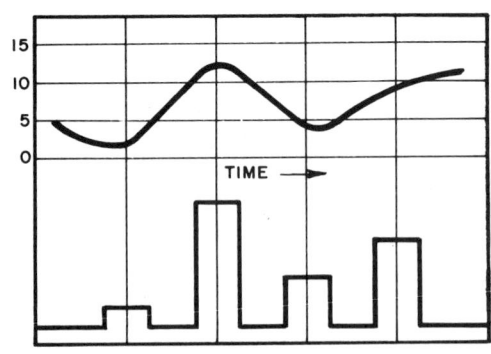

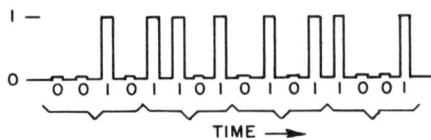

Figure 1-7. Time division multiplexing (TDM) transmission. *(Courtesy of American Multiplex Systems, Fullerton, California.)*

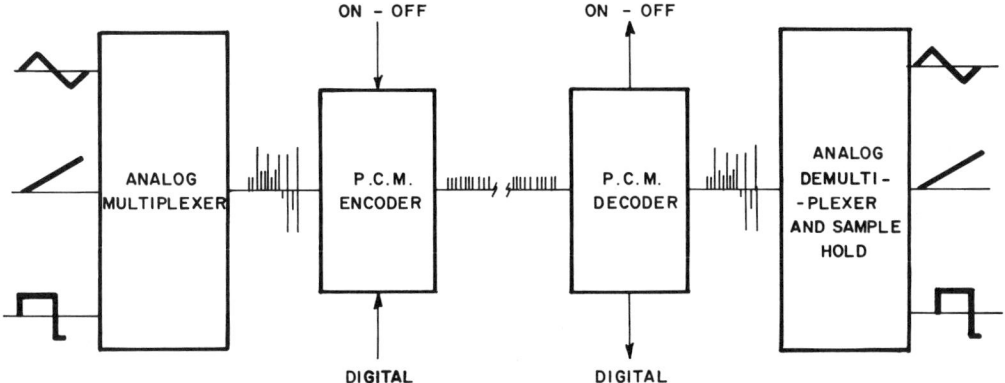

Figure 1-8. Pulse code modulation (PCM) technique for a multiplex system. *(Courtesy of American Multiplex Systems, Fullerton, California.)*

ing wires provides high common-mode noise rejection, thereby protecting against a low frequency magnetic noise field. When used properly, it also provides protection against ground loops and electrostatic noise fields.

The digital transmission cable should be a transformer coupled for DC isolation and common mode voltage rejection. High signal accuracy and noise rejection can also be provided through use of time division multiplex PCM transmission techniques. The serial bit stream that results is easily interfaced with a digital processor (or minicomputer), eliminating expensive centralized signal conditioning, conversion, and multiplexing equipment. Refer to Figure 1-7 for a schematic representation of a pulse code modulation and pulse amplitude techniques. Figure 1-8 illustrates how the pulse code modification approach works conceptually.

Automated computer monitoring and environmental control systems for utility and manpower economization and optimization are an inevitable necessity in medium to large buildings and building complexes. Wise use of the computer will require accurate performance of the data transmission system. True digital techniques are essential as the building blocks of future requirements. The computer programs developed and the acquisition of valuable data will determine how effective automatic optimized control will be. The computer, however, will be used in many types of systems prior to its full implementation because it is capable of doing many of the functions currently being accomplished with more expensive techniques of hardwired logic.

Each individual application must be studied to determine if the addition of a computer is economically justified. Consequently, building system designers and consultants must familiarize themselves with multiplexing technology so they can advise their clients on energy management and monitoring systems.

XII. NETWORKS AND DISTRIBUTED DATA PROCESSING

It may be helpful to understand how a minicomputer network distributes computer functions among its elements in the most cost-effective arrangement for a specific application [12]. Such communications software must go one step beyond the so-called intelligent terminal in that it must be able to arrange priorities in such a manner as to interrupt certain activities of low importance when necessary. Among other major functions, such software must:

1. Control the hardware interface that links the computer to the transmission line.
2. Assure that incoming/outgoing messages conform to prearranged data communications protocol.
3. Create logical data links among the various network processes.

4. Route various messages over the network.
5. Acknowledge and diagram message flow through the network.

In short, network software often makes the difference between a multiplicity of independently acting computers and the same number of computers integrated into a smooth-working, distributed data processing network [13]. One of two general methods of data transmission is selected in such systems:

1. Dedicated computers can store all of the data generated and employ high-speed synchronous communications to transmit the information to a central computer.
2. Data can be transmitted piecemeal as determined by the program, using synchronous data transmission techniques.

No matter which approach is employed, all of the various computers must be compatible. Compatibility is normally achieved by means of communications software designed to follow a so-called protocol. (By protocol we mean that all processors linked in a network must understand *all* transmitted and received messages.) Basically, there are five principal network types from which even the most sophisticated or hybrid forms are made up:

1. Point-to-point type
2. Hierarchy or tree type
3. Star type
4. Ring or loop type
5. Fully connected or hybrid type

The point-to-point type is characterized by Figure 1-9. It employs a communications line to link two computers together. Generally, in this arrangement, one minicomputer arranges the data into a format used to make a forecast or some decision, perhaps, while the other minicomputer performs communications control message switching, and data concentration.

In the hierarchical or tree arrangement illustrated in Figure 1-10, several minicomputers, arranged to perform a dedicated function, are linked to another generally larger-capacity computer that both monitors their activity and acts as a backup. The latter computer may, in turn, be monitored by yet another larger computer arranged to carry out higher-order strategies.

The star type arrangement illustrated in Figure 1-11 is similar organizationally to a multi-tiered hierarchy in which the remote computers all report to a central computer and tap its data file. Basically, the star type arrangement enables the smaller computers to function autonomously while still being able to update the system periodically. This, in effect, imparts centralized control to the overall arrangement.

In the ring or loop type arrangement illustrated in Figure 1-12, several processors are linked together to form the equivalent of a ring in which data lines connect each processor only to adjacent units. Communications between more distant units must be routed through intermediate processors. Since physical proximity is required to make a ring arrangement cost-effective, it is seldom employed on a

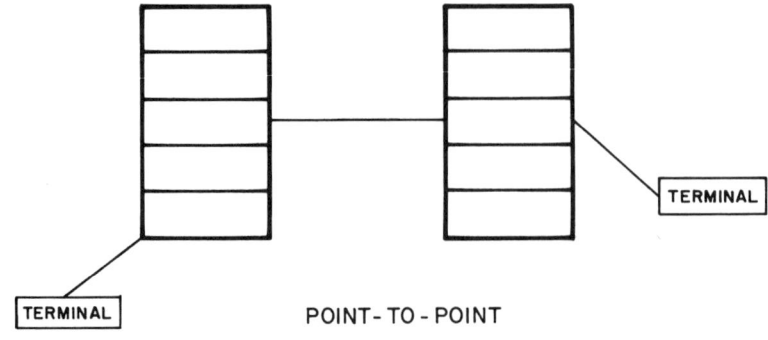

Figure 1-9. Representative point-to-point schematic network diagram. (From Ref. 12.)

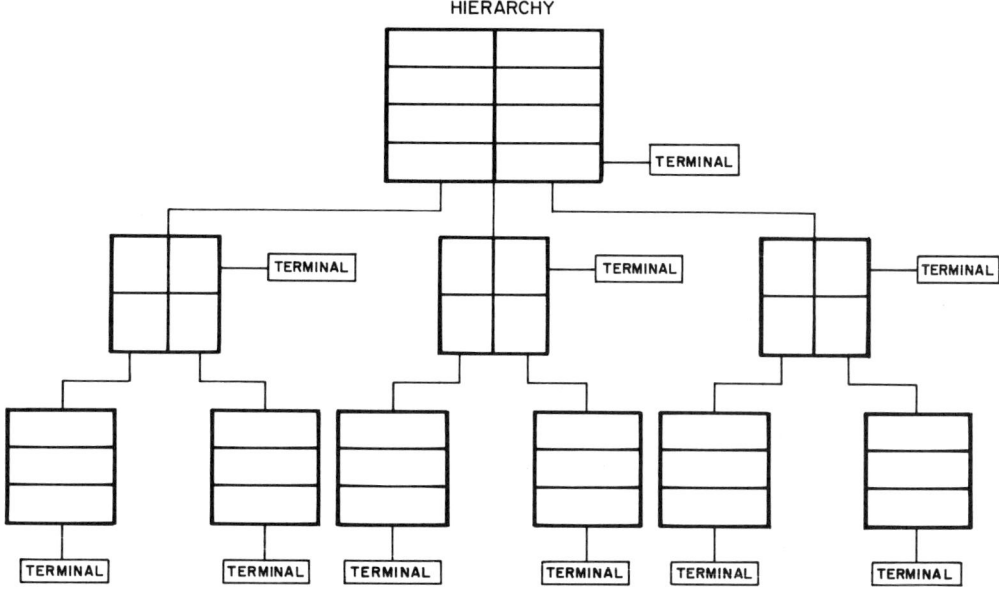

Figure 1-10. Representative hierarchy schematic network diagram. (From Ref. 12.)

stand-alone basis. It is, however, generally a useful concept in building up hybrid or fully connected types, as illustrated in Figure 1-13. In such networks, two or more central computers, each surrounded by a separate star network of satellite processors, can also be connected in a star of their own. Such a multi-star type arrangement permits main computers to be tied together so that only one of these units becomes the host to the other main computers in addition to its own satellite processors.

XIII. COMPARING AND SELECTING PROCESSORS

The importance of software language selection cannot be overstated. It is a measure of the "value added" provided by a manufacturer,

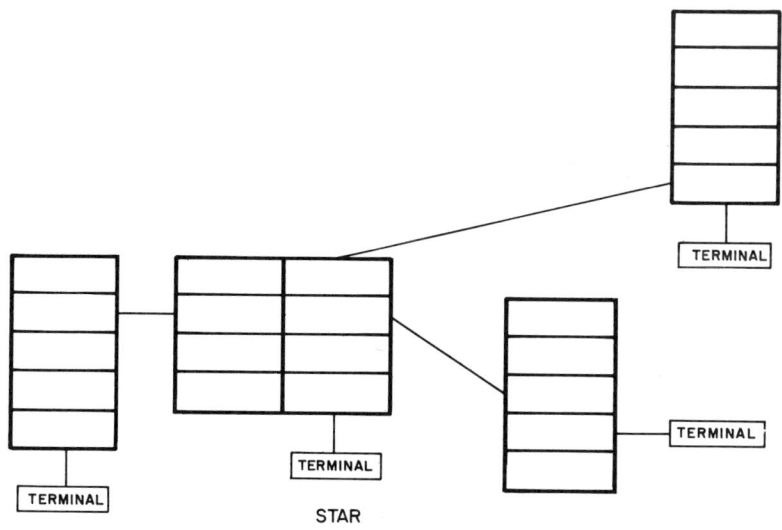

Figure 1-11. Representative star schematic network diagram. (From Ref. 12.)

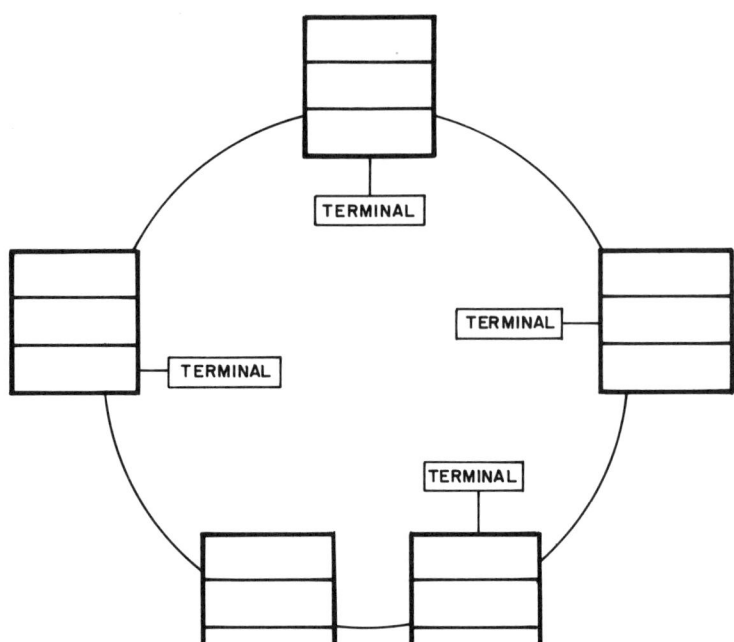

Figure 1-12. Representative ring schematic network diagram. (From Ref. 12.)

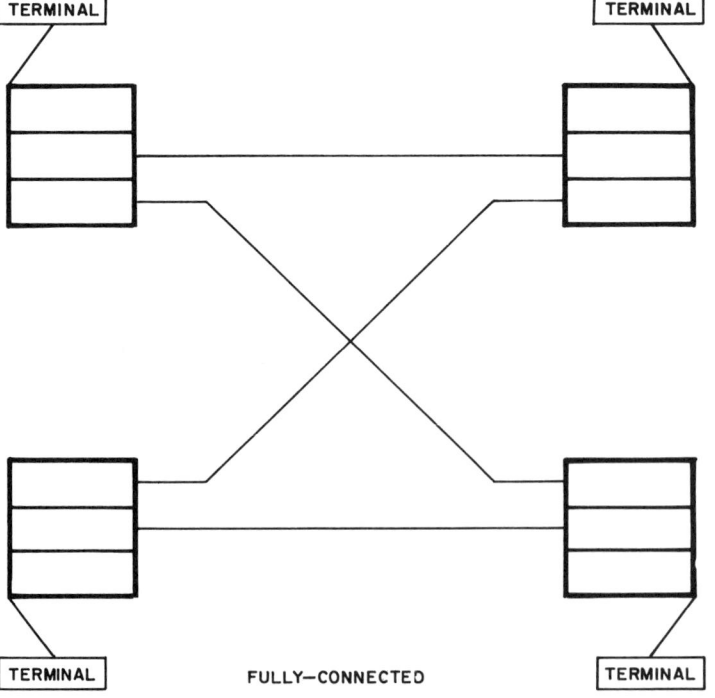

Figure 1-13. Representative fully-connected schematic network diagram. (From Ref. 12.)

Computer Manufacturer: _____ Model: _____

Rack Requirements: Computer — 19WX _____ HX _____ D _____

Other _____

                          Number of Model _____ in service

Hardware

32K 16 Bit Words   Type of Memory _____
Auto-power Fail/Restart   ROM — Auto Program Load
Hardware Integer Multiply/Divide   16 Digital Inputs and 16 outputs
CRT Terminal   Line Printer Speed: _____
Real Time Programmable Clock   32 Channels A/D High Level,
                                              12 Bit

Mass Storage Device _____   Cost _____

Notes: _____

	Single Qty.	2 Quantity	5 Quantity
Discount Percentage	_____	_____	_____
System Hardware Cost	_____	_____	_____
System Software Cost	_____	_____	_____

Optional Hardware

Memory Protect _____ Increments _____ Cost _____
Hardware Floating Point Arithmetic _____ Cost _____
Sealed Disk Drive _____ K Words Cost _____
Cartridge Disk Drive _____ M Words Cost _____
   Disk System Software Cost _____
System Cost w/Cartridge Disk _____ (1) _____ (2) _____ (5) (Qty.)
System Cost w/Sealed Disk _____ (1) _____ (2) _____ (5) (Qty.)

Machine Instructions

Number System _____ Direct Addressing Capability _____ K Words
Number of Vectored Interrupts _____ Macros _____
Registers _____
Instructions   Total Number _____
   I/O _____ Load/Store _____ Branch _____
   Integer Arithmetic _____ Floating Point Arithmetic _____
   Byte _____ Bit _____ Status _____ Memory Protect _____
   Immediate _____ Logical _____ Shift _____ Compare ___ Move Data _____
   Words/Instruction _____ 1 Word _____ 2 Word _____ 3 Word

Fortran

Level _____ Assembly Language in Code _____ Subr. _____
Double Precision _____ Mixed Mode Arithmetic _____
Run-Time Library _____ K Words _____ Memory Resident _____

Notes: _____

Basic: _____

Operating System

Name _____ Size (Words) _____
Number of Concurrent Active Tasks _____
Mass Storage Media Supported for Overlays _____

Figure 1-14. Minicomputer selection summary sheet. (From Ref. 15.)

perhaps a more important measure than any other of a computer-based product. Software maintenance is by far the largest single cost factor in software development.

With skillful software, a vendor can add great functional capability. Through the use of standard languages, moreover, the user simplifies a great deal of the future software maintenance that will be required in all applications.

Before specifying the type of energy management and monitoring automatic systems for a given project, we must determine minimum computer requirements in programming, hardware, physical size, service and software [14]. In comparing different vendor submissions, it is often helpful to develop summary sheets for major equipment items. These sheets provide a checklist of key features to ensure that system operation and architecture are thoroughly reviewed and understood prior to purchase. An example of such a selection summary sheet developed [15] for a minicomputer is given in Figure 1-14.

Often vendor-supplied marketing and sales information does not accurately reflect equipment performance regarding computers, I/O devices, peripherals, and so on, for a specific application. The following steps may prove helpful in comparing various major elements of the proposed energy management automation systems:

1. Check quoted system cost, software and program development costs, hardware and data transmission costs, etc.
2. Establish system compatibility with proposed computer architecture, i.e., hardware and software structure.
3. Interview current users of proposed equipment and systems to discuss reliability and performance.
4. Obtain system service listings, including number of units of the type proposed in service and how long.
5. Develop estimates of the required time to put the overall system into operation.
6. Evaluate the extent and capability of proposed software.
7. Confirm thoroughness and readability of available documentation.

Energy conservation controls can be subdivided into two principal product categories corresponding to the dual nature of electric utility billing: (1) power demand controllers concerned with reducing or minimizing demand charges (load-duty cycling features are increasingly being integrated into these designs for reducing overall energy consumption); and (2) energy management systems concerned with both demand limiting control and the reduction of overall energy consumption.

Electrical energy controls currently being marketed employ four types of logic and control technology: special-purpose (solid state), programmable logic controllers, microcomputer, and minicomputer. Suppliers of electrical energy controllers now include minicomputer manufacturers, specialized manufacturers, electrical equipment companies, and companies selling industrial controls. It should be clear that microcomputer use will increase significantly in the years ahead.

REFERENCES

1. Air Conditioning Research Institute, Arlington, VA. (Statistical Division).
2. William J. Coad, "Energy Management," *Heating/Piping/Air Conditioning,* January, 1977.
3. "Evaluation of Building Characteristics Relative to Energy Consumption in Office Buildings," National Technical Information Service, PB-254 774, September, 1975.
4. National Electrical Manufacturers Association, "Total Energy Management: A Practical Handbook on Energy Conservation and Management," U.S. Department of Commerce, National Technical Information Service, PB-254 683, prepared for Federal Energy Administration, March, 1976.
5. M. Meckler, "Speed Retrofit Decisions," *Buildings,* June, 1977.
6. "Presidential Report: Energy Management," *Air Conditioning & Refrigeration Business,* June, 1976.
7. M. Meckler, "Energy Conservation Experience on Large Scale HVAC Systems," CH-77-12, delivered at ASHRAE Semi-Annual Meeting, February, 1977.
8. "Energy Conservation Training Program: Pilot Program For Six New England States," Bolton Insti-

tute, Incorporated, PB-248 609, prepared for Federal Energy Administration, February 1, 1975.
9. L. Monrad-Krohn, "The Micro vs. The Minicomputer," *Mini-Micro Systems,* February, 1977.
10. Frost and Sullivan Inc., NYC, *Minicomputer and Microcomputer Market Study for United States,* December, 1974.
11. James Martin, *Telecommunications and The Computer,* 2nd ed., Prentice-Hall, Inc., Englewood Cliffs, N.J., 1976.
12. Stephen A. Kallis, Jr., "Networks and Distributed Processing," *Mini-Micro Systems,* March, 1977.
13. William G. Moore, Jr., "Going Distributed," *Mini-Micro Systems,* March, 1977.
14. J. J. Fling, 1/C Engineering Corp., "Specifying Multiplex Data Systems," *Power Engineering,* February, 1976.
15. D. L. Jameson, "An Engineering Approach to Minicomputer Selection," ASME Paper #76-Pet-24.

2. Economic Analysis of Energy Retrofit Projects*

Milton Meckler, P. E.
President, The Meckler Group
Encino, California

	Page
I. Introduction	24
II. Rethinking Construction Economics	24
III. Payback Analysis Methods	25
IV. Discounted Cash Flow Analysis Methods	25
V. Benefit/Cost Analysis	26
VI. Employing LCCA in Energy Decisions	27
VII. Applying the Algorithm	27
Illustrative Example	28
VIII. Examining Solar Energy Investment Opportunities	28
A. Solar Analysis	29
B. Economic Analysis	31
C. Sample Energy Cost Analysis	31
IX. Fuel Availability and the Future	32

I. INTRODUCTION

In planning new buildings, there are many ways to conserve energy that are not capital-intensive, but existing buildings often require unique measures that call for additional investment. Since every modernization job is different, a careful study of a mix of system modifications as well as operating or occupancy schedule changes is required to achieve meaningful energy savings.

Determination of the most productive use of such investments, particularly cost effectiveness in generating energy savings, is facilitated by using one or more of the life cycle costing methods proposed here. The uncertainties we face in these changing times clearly expose the weakness of employing traditional construction economics.

II. RETHINKING CONSTRUCTION ECONOMICS

We are entering an age of limits and uncertainties as to growth, the use and probable future costs of energy, and, above all, our expectations. Our past expectations in the building sector have generally been based on available energy and financial resources, provided a project was well planned and located, and a favorable demand in the near term could be demonstrated. Operating costs were predictable and could be expected to follow historical trends, with normal departures for inflation. Future availability of fuels and electricity was taken for granted, following receipt of a commitment letter from the serving utilities.

Economic analysis for project viability also followed one or more of the traditional depre-

---

*Portions of this chapter are reprinted from "After the Energy Audit, What?," *Buildings: The Construction and Building Management Journal,* November, 1976.

ciation cost-to-benefit schemes, adjusted for tax impacts and with primary emphasis on initial cost to maximize owner/developer leverage.

Against this background, a new scenario emerges. It is especially important that building owners look to life cycle costing (LCC) for better answers when evaluating building design and operations decisions, especially competing energy conservation and solar energy investment opportunities.

Life cycle costing can be thought of as a procurement or decision-making technique that considers operating, maintenance, and other ownership costs over the full life cycle, as well as acquisition or initial costs. An LCC approach provides a more systematic treatment of the full set of relevant costs when considering the development, production, and lifetime ownership of a building. It represents a valid method for evaluating alternative investment opportunities, often in terms of an uncertain or soft approximation of future energy cost, system efficiencies, annualized maintenance or operating cost, and the actual or opportunity cost of capital.

The life cycle cost analysis (LCCA) is based on relating all costs to present value—simply another way of saying that a dollar in hand today is more valuable than a dollar that may be in hand in the future. Some of the more difficult tasks for the architect and design engineer include selection of realistic criteria, organization of input cost data, and estimating performance of proposed but untried systems.

III. PAYBACK ANALYSIS METHODS

Various investment proposals can be ranked in one of several ways. Use of the simple payback, for example, involves computation of the number of years required to recover the initial investment. Payback analysis can also be improved slightly by modifying the approach to include the effects of debt service. For example, if we find that an initial investment of $30,000 results in an annual operating and maintenance savings of $6,000, to conclude that such a proposal has a simple payback of five years is to ignore the fact that either interest must be paid on a loan or that there is some opportunity cost for this money because it could be more productively utilized on alternate investments. To solve this deficiency, one may employ the following formula:

$$n = \frac{\log \frac{S/rC}{S/rC - 1}}{\log (1 + r)}$$

where

C = capital cost
S = annual operating and maintenance savings
r = interest rate
n = number of years to achieve payback.

IV. DISCOUNTED CASH FLOW ANALYSIS METHODS

Payback analysis corrected for the cost of monies borrowed still suffers from one serious drawback. Since this method does not account for the time value of money, for example, the effects of discounted cash flow, it is subject to major errors when applied to long-term investment decisions, subject to crossover.

One of the ways to overcome this problem is to employ the internal rate of return method (IRR), which is based on defining an interest rate (r) that sets the present value of the anticipated future receipts equal to the cost of the initial investment (C) as follows:

$$C = \sum_{t=1}^{N} \frac{R_t}{(1 + r)^t}$$

where

N = project's expected life (years)
R_t = net cash flow in the year t
k = cost of capital
C = initial investment cost
r = internal rate of return.

The sum of the discounted receipts is thus made equal to the initial cost of the proposed investment (C) at some interest rate, and this value of r is by definition also the value of IRR.

One can also employ the net present value method (NPV), which is used to establish the present value of the expected cash flow, for example, annualized energy savings (Rt) from a proposed capital outlay (C), discounted at some estimated cost of capital, for instance, cost of borrowing. This value is computed by substituting k for r in the equation for computing C, given above. To establish NPV, we merely subtract this value from C, the initial investment, required by the proposed system modifications. It must be understood that for it to be considered valid, one cannot obtain a negative value for NPV. When comparing a number of competing alternate investments of equal initial investment (C), the one yielding the highest value of NPV is selected, and the others can then be ranked in descending order. The same basic relationships are employed in computing IRR and NPV. The principal difference is that the IRR method allows one to evaluate the proposed initial investment in terms of the cost of capital available for such investment. If the value of IRR exceeds the interest cost of the indebtedness, the proposal would be considered profitable. In the event the value of IRR is less than the cost of the indebtedness, the investment would result in losses.

V. BENEFIT COST ANALYSIS

Another approach involves the use of benefit/cost analysis, which can be utilized in two ways: to determine whether it is worthwhile to replace an existing system with a new system, and to determine which of two or more alternative new systems provides the highest cost benefit ratio. For example, to determine whether it would be more advisable to retain existing equipment or obtain new, more efficient equipment, the following factors must be considered:

1. Maintenance savings to be obtained by installing the new equipment.
2. Operating and energy cost savings to be obtained by installing the new equipment.
3. Salvage value of old equipment.
4. Capital cost of new equipment, including legal fees, design and professional fees, installation charges, cost of equipment.
5. Costs of financing.
6. Changes in property taxes as a result of installing new equipment.
7. Income tax factors.
8. Changes in rental income as a result of installing new equipment.

Assume it can be shown that, on a strict monetary basis, installation of new equipment will result in some savings. Obviously, such measures should be strongly considered. However, one must also take into account the ease of maintenance and possibly the ability to eliminate necessary manpower, lowered number of tenant complaints, or greater reliability.

The so-called benefit/cost analysis (B/C) ratio or profitability index is computed by dividing the present value of future receipts (also discounted at the cost of capital) by the initial investment cost:

$$\text{B/C} = \sum_{t=1}^{N} \frac{Rt}{(1+k)^t} \div C$$

Since the numerator is directly equal to the present value of the expected cash flow, all proposals yielding a B/C greater than 1.0 are acceptable. In evaluating several candidate proposals, the one yielding the largest B/C ratio should be selected and the others ranked in descending order. While it is probable that most situations employing either IRR, NPV, or B/C methodology will yield identical rankings, there are some notable exceptions:

1. When investment requirements vary significantly among the various alternate investments compared.
2. When the direction of cash flow generation varies among the several alternate investments compared.
3. When there is wide divergence in the time frame—for example, variations in

project life associated with the various alternate investments compared.

VI. EMPLOYING LCCA IN ENERGY DECISIONS

We can now employ the above criteria in escalating energy conservation modifications in buildings, such as energy management or automation programs and systems controlling one or more buildings, and to evaluate economic performance of a proposed solar energy system.

Where energy cost savings are all positive with each of the earlier-described discounted flow methods, the difference is that cash flow yields should not now be considered important. Where possible, project lifetimes for various competing alternate energy conservation investment opportunities should be held the same.

It may be helpful to review briefly some of the strengths and weaknesses of these various discounted cash flow methods. Recall that the IRR method involves an iterative trial and error process, whereas use of NPV or B/C methods is often more straightforward. However, where the cost of capital is believed to be a major factor, the IRR method is often more useful than either the B/C or NPV methods. B/C methods are recommended for analysis in situations where capital is constrained and one is faced with a number of cost-effective energy conservation measures over the same project life. Under these circumstances, it would be possible to rank each of the proposed energy conservation alternatives, starting with the highest-ranked and cumulatively adding them down the list until all available capital had been committed. The NPV method is probably the simplest and most flexible of the recommended methods discussed, although most viable energy conservation proposals will show a positive net present value under the assumed life of the building, and thus make long-term financial sense.

Often, timing may prove poor for the building owner. Under such circumstances, it is usually desirable to establish an acceptable payback period and then compute the NPV for a given set of proposed energy conservation measures for that time period. If the NPV is found to be positive, the building owner's criterion is satisfied, and the proposed energy conservation measures should be dropped and a new NPV computed. This process is usually repeated until a positive NPV is found.

VII. APPLYING THE ALGORITHM

Another possible approach is to develop a life cycle cost algorithm. In a recent project, it became necessary to evaluate the net present value benefit resulting from installation of an automated energy management monitoring and control system (AEMS) for a major university campus in California. It was determined that this system could be characterized by three main elements of cash flow:

1. Initial investment
2. Annual maintenance cost attributable to the AEMS
3. Annual energy savings resulting from operation of the AEMS

If we assume AEMS salvage value at the end of useful life to be negligible, there will be no appreciable savings realized in the first year's operation due to normal start-up problems and familiarization needs. Since maintenance costs for the first year's operation are included as part of the initial AEMS investment, one may write the following cost algorithm:

$$\text{NPV} = S\left[\left(\frac{1+f}{1+i}\right)^2 + \cdots + \left(\frac{1+f}{1+i}\right)^n\right] - C$$

$$\text{minus } m\left[\left(\frac{1+h}{1+i}\right)^2 + \cdots + \left(\frac{1+h}{1+i}\right)^n\right]$$

where

C = AEMS initial investment ($)
S = annual fuel savings at present cost basis ($/year)
f = annual fuel cost escalation rate (%)
n = expected serviceable lifetime of AEMS (years)

m = estimated annual add-on maintenance cost, such as outside contract for AEMS ($/year)
h = annual inflation rate for AEMS maintenance cost (%)
i = investment discount rate (%)
NPV = net present value benefit for AEMS.

Illustrative Example

Employing the NPV algorithm developed above, it was determined that for an AEMS capital cost of approximately $950,000, an estimated annual fuel cost escalation, annual inflation rate for add-on maintenance, discount rates respectively of 15, 8, and 4%, and an estimated AEMS useful life of ten years requiring an initial annual maintenance add-on cost of $60,000, that the AEMS system had an effective payback of just under seven years and an NPV of approximately $925,250. This is calculated as follows:

n	$\dfrac{1 + f^n}{1 + i}$	$\dfrac{1 + h^n}{1 + i}$
1	0*	0*
2	1.23	1.08
3	1.37	1.13
4	1.52	1.17
5	1.69	1.22
6	1.87	1.27
7	2.08	1.32
8	2.31	1.37
9	256	1.42
10	2.84	1.48
Sum	17.47	11.46

The effective payback period of seven years was also computed by trial-and-error iteration using the above tabulation and searching for a value of n that makes the NPV equal to zero.

VIII. EXAMINING SOLAR ENERGY INVESTMENT OPPORTUNITIES

Present value evaluation algorithms can also be developed to evaluate profitability of proposed solar energy systems. However, there are some important distinctions to keep in mind when developing such models. Discretion must be used in avoiding overly optimistic annual energy savings in computing the applicable IRR, NPV, or B/C ratio—particularly for prototype solar systems for which few or no operating data exist.

Also, one must recognize that current tax provisions tend to mask advantages of employing solar energy as a supplemental source in buildings by allowing building owners to deduct conventional energy expenses from taxable income. Net effect is to substantially reduce the value of fuel savings from solar energy.

For example, if we assume a building owner's tax ratio to be nominally 50%, the practical impact of such savings is reduced by half. Furthermore, current capital depreciation allowance and tax credit provisions generally fail to credit building owners for loss of the tax deduction on energy saved by the proposed solar system.

Fortunately, state legislators and members of Congress are aware of the need to promote use of solar energy in both new and existing buildings. Legislation is currently under consideration to make solar energy systems more cost-effective and therefore to exhibit a more favorable impact in life cycle costing analysis.

In performing life cycle cost analysis for a solar energy system, key factors influencing cost effectiveness include:

1. Current fuel prices, future availability, and rate of escalation.
2. Selection of an appropriate interest rate to discount future costs to present value.
3. Useful life of the system.
4. Initial acquisition (investment) cost, and resale (salvage) value.
5. Determination of tax or incentive impacts on net cash flow.
6. Determination of useful energy that the solar system can supply over the useful system life. This value can vary widely as a function of the microclimate, mechanical system operating features, and building design, size, utilization, and occupancy.

If we look at reasonable variations in each of the above—within realistic limits—net benefits for any proposed solar system can vary in substantial amounts from positive to negative. Application of life cycle costing must follow a careful operating analysis of each proposed solar energy system, based on specific data. Also, one must take into account the sensitivity of solar system cost effectiveness and the impact of regular taxes and/or available and foreseeable incentive programs.

A. Solar Analysis

As a part of a feasibility study for a school, a preliminary schematic design for the proposed solar-augmented domestic hot water (DMW) and space heating system was prepared. This was accomplished by modeling the facility on a computer using a solar analysis simulation program called TRNSYS.

TRNSYS (a Transient Simulation Program) was developed by the Solar Energy Laboratory, University of Wisconsin-Madison, under grants from the RANN program of the National Science Foundation and from the Energy Research and Development Administration.

In applying the TRNSYS program, basically the solar system is modeled to represent a set of components, interconnected in such a way as to accomplish a specified task. For example, a typical solar water-heating system may consist of a solar collector, an energy storage unit, an auxiliary energy heater, a pump, and several temperature-sensing controllers. One obvious characteristic of such a system is its modularity. Because the system consists of components, it is possible to simulate the performance of the system by collectively simulating the performance of the interconnected components.

The performance of a system component will normally depend upon characteristic fixed parameters, the performance (or outputs) of other components, and time-dependent forcing functions. It is important to realize that time-dependent forcing functions can be thought of as outputs of specialized system components, and can thus be treated in the same manner as any other component.

The modular simulation technique greatly reduces the complexity of system simulation because it essentially reduces a large problem to a number of smaller problems, each of which can be solved independently with relative ease. In addition, many components are common to different systems with little or no modification. This feature makes modular simulation most attractive.

With the program such as TRNSYS, which has the capability of interconnecting system components in any desired manner, solving differential equations, and facilitating information output, the entire problem of system simulation reduces to a problem of identifying all of the components and formulating a general mathematical description of each.

For this project, the building loads and occupancy profiles used were those previously developed for fuel analysis computer simulations. The primary components for the solar model were:

1. Solar collector panels
2. Space heating storage tank
3. Space heating storage heat exchanger
4. Auxiliary space heating heater
5. Domestic hot water storage tank
6. Double wall DHW storage heat exchanger
7. Auxiliary DHW heater
8. Recirculating pumps

Linking these components together, using TRNSYS, four basic configuration schemes were investigated. These were:

Scheme A:

Collector area = 2,350 sq ft (218.31 m^2)
Space heating stores = 3,500 gallons (556.5 m^3)
Heating storage tank height to diameter ratio = 3.8
Relationship of solar heating and auxiliary heating systems—in parallel
DHW Storage = 700 gallons (111.3 m^3)

Maximum DHW demand = 330 GPH (346.5 ml/s)

Scheme B:

Collector area = 2,350 sq ft (218.31 m²)
Space heating storage = 3,500 gallons (556.5 m³)
Heating storage tank height to diameter ratio = 3.8
Relationship of solar heating and auxiliary heating systems—in series
DHW storage = 700 gallons (111.3 m³)
Maximum DHW demand = 330 GPH given (346.5 ml/s)

Scheme C:

Collector area = 2,350 sq ft (218.31 m²)
Space heating storage = 2,500 gallons (397.5 m³)
Heating storage tank height to diameter ratio = 1.0
Relationship of solar heating and auxiliary heating systems—in series
Maximum DHW demand = 330 GPH given (346.5 ml/s)
DHW storage = 700 gallons (111.3 m³)

Scheme D:

Collector area = 1,744 sq ft (162.02 m²)
Space heating storage = 2,500 gallons (397.5 m³)
Heating storage tank height to diameter ratio = 1.0
Relationship of solar heating and auxiliary heating systems—in series
DHW storage = 600 gallons (95.4 m³)
Maximum DHW demand = 330 GPH given (346.5 ml/s)

The orders of magnitude used as initializing parameter values for the above schemes were determined using accepted industry guidelines. The solar panels were assumed to be at the latitude plus 20° or at 52° tilt above the horizon facing due south to maximize winter solar gain. It was implicitly assumed here that any solar panels on the site would not be shaded by any other buildings, trees, or mountains. This should be positively ascertained in detail design of this system.

Each of the above schemes was analyzed using the TRNSYS simulations to determine just which alternative was the best for the particular location and occupancy.

It should be noted that this solar analysis was accomplished under the assumption that heating oil temperatures were to be from 180°F to 140°F (82.22°C to 60°C). These temperatures required the use of double-pane solar collectors instead of single-pane, which does increase initial investment in the system substantially. Four different schemes were simulated to determine the optimum configuration.

Within these basic configurations, four principal parameters were varied in the computer simulations: size of the collector surface, space heating storage tank size, relationship of the solar space heating system and the auxiliary heating system, and the height to diameter ratio of the storage tank.

The critical parameter in the resulting analysis was the utilization ratio of solar energy collected compared to the total energy used for each scheme used. The following table summarized the computer findings for the four schemes:

	Collector surface area	Tank size	Auxiliary relationship	H/D ratio	Space heating utilization	D.H.W. utilization
Scheme A	2,350 sq ft (218.31 m²)	3,500 gal (556.5 m³)	Parallel	3.8	92%	95%
Scheme B	2,350 sq ft (218.31 m²)	3,500 gal (556.5 m³)	Series	3.8	87%	97%
Scheme C	2,350 sq ft (218.31 m²)	2,500 gal (556.5 m³)	Series	1.00	86%	98%
Scheme D	1,744 sq ft (162.02 m²)	2,500 gal (397.5 m³)	Series	1.00	86%	95%

In calculating the utilization of each of the schemes, it was assumed that the school and therefore the solar system did not operate on weekends and holidays and during the summer recess.

B. Economic Analysis

Based on unit costs obtained from MEANS Building Construction Cost Data and the M-B-M Building Cost File or from actual vendor quotes, price estimates were prepared for Schemes C and D. These estimates were:

Scheme C = $148,000
Scheme D = $108,000

Estimates were not prepared for Schemes A and B because these schemes could be seen by inspection to be non-cost-effective based on the relatively small increase in utilization ratio compared to the substantially increased capital costs.

A life cycle cost estimate was developed for the solar augmentation systems under study. The objective of this analysis was to determine the amortization period of this system considering all of the anticipated major cash flows over the life of this project. Formulation of this analysis relationship and subsequent analysis employed the following generalized equation:

$$\text{Net present value} = S\left[\left(\frac{1+f}{1+i}\right)^1 + \cdots + \left(\frac{1+f}{1+i}\right)^n\right] - I$$

where

I = estimated initial investment in the system
S = estimated annual fuel savings at present costs based on the actual solar Btu gain as estimated by the TRNSYS simulations and priced at prevailing rates
f = estimated average annual fuel cost escalation rate over the probable life of the system
n = estimated serviceable life of the system—expected to be 15 years for the quality of equipment proposed to be furnished
i = estimated opportunity cost of investment capital, assumed to be zero.

In this calculation, the estimated average annual fuel cost escalation rate was estimated to be 8%. Although there were numerous differing estimates of the future fuel costs, both above and below this figure, this was seen to be a reasonable conservative approximation of what the future might hold for energy rates within the anticipated life of the proposed solar system.

Our analysis did not consider additional maintenance charges for the proposed solar system. Such costs would be minimal compared with the overall maintenance program for the class of mechanical system already planned for this school. Therefore, the following values were used for this analysis:

	Scheme C	Scheme D
I =	$148,000	$108,000
S =	$ 2,887	$ 2,742
f =	8%	8%
n =	15	15

Net present value for Scheme C = $63,500
Net present value for Scheme D = $28,000

Using a simple payout calculation of

$$N = \frac{I}{S_1 + S_2 + \cdots + S_n}$$

with a fuel inflation rate of 8% per year, the number of years for system payback are:

Scheme C = 21 years
Scheme D = 18 years

It should be noted that these cost estimates were conservative.

C. Sample Energy Cost Analysis

In the section below, the required number of remote sensing points was developed for several central control/monitoring schemes. In this section, information will be combined to determine the economic payback that is developed by the application of these remote sensing points to receive the benefit of the energy savings.

The initial step in this analysis was to determine the utility cost of energy. There were three types of energy used at this project, a military base: electricity, natural gas, and fuel oil. Using information provided by the facility operators, the following energy cost determination was made:

Electricity Cost

Assume: Highest heat loads are June through September.
Average 88.43% of total kWh is provided from the Bureau of Reclamations.
Average Bureau of Reclamations unit cost = $0.0061/kWh.
Average community unit cost = $0.252/kWh.

Natural Gas Cost

$$\text{Natural gas cost} = \underline{\$1.79/\text{MBtu}} \left(\frac{\$1.70}{1{,}000 \text{ kJ}}\right) = 1{,}050 \text{ Btu } (1{,}107.75 \text{ kJ}).$$

Fuel Oil Cost

$$\text{Fuel oil cost} = \underline{\$2.40/\text{MBtu}} \left(\frac{2.27}{1{,}000 \text{ kJ}}\right)$$

based on 140,000 Btu/gal (39,060 kJ/l).

IX. FUEL AVAILABILITY AND THE FUTURE

Analyses of fuel oil costs will be meaningless in the future if current problems of availability and delivery are not remedied. And the situation appears fairly bleak. Supplies are dwindling, and political realities are increasingly grim.

In view of all this, it would seem prudent for the farsighted engineer or manager, interested in retrofitting, to study other potentially significant sources of energy that may become important in the future. Retrofitting, in essence, is a type of strategy that emphasizes flexibility and imagination in its approaches and solutions. Keeping abreast of important developments in all areas of the energy field makes this kind of flexibility possible.

3. Utility Rates and Retrofit Economics

J. Bob Roberson
Southern California Edison Company
Rosemead, California

	Page
I. Introduction	33
A. Historical Background	34
B. Today's Reality	34
II. Building Operations	35
A. Impact of the Utility System	35
B. Impact of Utility Rate Changes	35
III. Types of Utility Rates	35
A. Commodity Rates	36
B. Socially Motivated Rates	36
C. Demand Rates	36
D. Time-of-Use Rates	37
E. Interruptible Rates	37
IV. Technology–Rate Interface	37
A. Demand Control	38
B. Storage	38
V. The Decision-Making Process	39
References	39

Owners and operators of facilities have adopted the term "retrofit" from the jargon of space technology, where it is used to refer to the solution of unforeseen high-technology problems through the use of new, often innovative technical "fixes," or retrofits. This borrowing of a jargon expression is most appropriate, since more conventional language would have been inadequate under the circumstances.

Most of today's existing inventory of buildings—homes, stores, offices, factories—were built prior to the recognition of what we now call the energy crisis. They were designed and built during a time when energy utilization was seldom even on a list of design constraints. Energy was plentiful, and, more significantly, it was very inexpensive relative to the other costs associated with designing, building, or operating a facility. It is well proved that those halcyon days are past, yet their legacy of an inherently inefficient building inventory remains.

Thus, buildings and processes that were originally designed with very little thought to making them even reasonably efficient users of energy, today in operation yield energy operating costs beyond anyone's definition of reasonable. To be competitive, perhaps to survive, the owner/operator of such facilities must correct the deficiencies of the various energy-using systems—he must retrofit. The reward for energy efficiency, discounting patriotic or moral motivations, is lower operating costs resulting from lowered energy consumption. Consumption and any savings in consumption are measured in therms of gas, or kilowatt-hours of electricity, or tons of coal, or some other measure whose meaning is equally obscure to the owner/operator. The common denominator, of course, is money—cash. And the translator of such difficult-to-comprehend measures of energy into the universally understood term, money, is the utility rate structure. "Ah, there's the rub," said the bard [1], and right he was! Utility rates are neither simple nor straightforward, nor do they yield to simple analysis. In spite of this, it is incumbent on the owner/operator to come to grips with these mysterious utility rates, to do his homework, and to become comfortably familiar with exercising those rates to his benefit. Failure to take this step will likely result in retrofit decisions that may be technical successes, but do not translate through utility rates to the desired common denominator, money saved.

It is the purpose of this chapter, then, to familiarize the reader with the form and types of utility rates along with their rationale, not to dwell on rate specifics. Because, as you will see, utility rates are very sensitive to local climate, geography, utility system configuration, and local/regional politics, and thus are utility-specific.

A. Historical Background

In order to understand and thus deal with utility rate evolution as it is likely to impact the retrofitter of the 1980s and 1990s, it is helpful to understand the generic concepts that underlie utility rates. Fundamental concepts such as rate level and rate structure, and such nuances as rate discrimination, have been dealt with by Wilcox [2] and many others, while the economics of public utilities, as opposed to our subject here, the economics of retrofitters as impacted by the utilities, has been treated often and well, as for example by Clemens [3]. The underlying fundamentals of demand, supply, and pricing of products and services were covered well by Leftwich. [4].

We need not re-enroll in economic studies in order to deal with retrofit economics. However, the better our understanding of historical and fundamental concepts is, the more easily can we assimilate the implications of innovative, special-purpose rates. And make no mistake, federal impetus [5] to such developments makes them a near certainty for every part of the United States, and less certain but still likely elsewhere. One of the five pieces of legislation making up the National Energy Act, the Public Utility Regulatory Policies Act of 1978 (PURPA), had as one of its principal purposes the conservation of energy by consumers of electricity. To that end, both utility rate structures and load control schemes must be studied in depth.

B. Today's Reality

It is a reality in today's world that a large segment of the population is distrustful of and unsatisfied with its established institutions, "the establishment." This phenomenon extends beyond church and state to "big business," which is exemplified by the utility industry. This unrest, which seems the norm today, is a near perfect backdrop for giving widespread support to critics of utility rates. The common thread of these criticisms is that conventional, historical, cost-based rates are "unfair," or "encourage waste," or "give the wrong price signals." The solution proposed to the "problem" thus identified takes on a wide range of forms, some with wide support, some mere curiosities.

Because traditional cost-based rates are the only approach to utility rate-making for which there is any significant body of experience, all of the many new, innovative, nontraditional, non-cost-based approaches to utility ratemaking are theories—theories begging to be tested. Several such theories have found jurisdictions willing to experiment, and more will follow. Those meeting with success will find quick acceptance in less daring jurisdictions.

Along with nontraditional rate structures, there has come an alternate philosophy to conventional cost spreading. Conventionally and traditionally average costs have been used as the underlying data base. This means that all customers have shared equitably in the average cost of utility plants from the least expensive, turn-of-the-century hydroelectric facility to the most expensive, inflation/energy-crisis-priced nuclear facility. The new alternative that is receiving wide support is marginal costing [6-8]. This concept has been articulated in such a wide variety of dissimilar specifics that it remains unclear which, if any, of the specific approaches will prevail.

II. BUILDING OPERATIONS

It has often been said (by designers) that inept building operators can defeat the best, most energy-efficient building design. It has also long been recognized that conscientious, intelligent building operators are frustrated by buildings that are incapable of being operated efficiently because of inherently inefficient designs. To solve this latter problem and to ensure that buildings capable of being operated efficiently would become the design norm, a standard to establish minimums of building energy efficiency was established [9] and has been adopted either in the original form or with locally appropriate modifications in many states. A set of standards to similarly address retrofit is also being developed [10].

It is less well understood, however, that building operations, efficient or otherwise, have a measurable effect on the utility system, and that this impact translates back to the operator through rates. How, when, and how much building operations impact the utility system is to a great extent dependent on the particular utility system. An even more obscure but nonetheless real relationship has to do with the impact of changes in utility rates on the operators' freedom of action.

A. Impact of the Utility System

Although utility systems are generically quite similar, several utility-specific characteristics are very sensitive to the utility–building interface. Perhaps the most significant and most sensitive of these relationships revolves around the utility generation mix, which is the mix of resources used in the generation of electricity. Attempts to document this important characteristic have met with only limited success [9]. This is so because the scale has been much too large; regional data were used instead of more appropriate and useful utility-specific data.

The capacity of the utility system, its ability to meet the maximum coincident load demanded of it, and the ability of its transmission and distribution circuits to deliver power when and where required, are major constraints that guide the structuring of the rates. It should be fairly obvious, upon reflection, that the above-named constraints are almost always time-sensitive. They are also sometimes seasonal, sometimes influenced by the day of the week, and most frequently affected by the time of day. In order to treat this coincidence time factor, demand rates including the time-of-use and interruptible varieties have been developed. In some utility systems these variables are complicated even further by changes in generation mix on a scheduled or at least on a time-predictable basis.

B. Impact of Utility Rate Changes

For the retrofitter, change in utility rate is perhaps the most ominous of the utility actions that are certain to occur. The change may be as simple as an adjustment to match the inflation-driven rise of fuel costs, or it may be as complex as a basic structure change, such as from an energy-only rate to a demand rate, or from a simple demand rate to a time-of-use rate. The point here is that change is inevitable, and that retrofit decision analysis must include an assessment of likely future rate structure changes in the area, their most likely timing, and their most likely magnitude. Adjustments in building operations may mitigate against such changes if done in a timely fashion, and may be reasonable provided the retrofitter took due note of likely forthcoming rate changes at the time of retrofit, and made appropriate provisions for rate change eventualities in the retrofit design. Such provisions, unfortunately, are often easier to describe in the abstract than to evaluate and specify meaningfully in a functional design.

III. TYPES OF UTILITY RATES

Utility rates, which are a useful vehicle for accomplishing a number of unrelated ends, are administered by a wide variety of regulatory jurisdictions with but one thing in common: they are politically controlled. That control may be direct and overt in some jurisdictions, or it may be indirect, subtle, almost invisible in others. But it is always there. Utility rates are too convenient an economic/political tool to be long ignored by most political action groups.

The following discussions will address only a select few of the many types of utility rates that exist today, in the belief that others, and those yet to be proposed, are but variations on, or combinations of, these fundamental types.

A. Commodity Rates

Commodity rates are the simplest and easiest to understand. This is the rate form most familiar to most people. The market advertises melons at 50¢ each, three for $1.00. This is a pure commodity rate with a volume discount. Utilities use such rates more consistently than any other form, and usually with the convention of what are called "descending blocks." For example, a simple two-step gas rate might be $2X$ per 100 cu ft for the first 300 cu ft, and X per 100 cu ft for everything over the first 300 cu ft. Similarly, because the first block includes the fixed charges associated with such constants as meter reading, bill processing, and so forth, such charges are often spelled out separately as a "customer" charge. In the gas example above, if X were assumed to be $1.00, such a rate might be expressed as $1.50 customer charge, plus $1.50 per 100 cu ft for the first 300 cu ft, plus $1.00 per 100 cu ft for everything over the first 300 cu ft. This is a more generally preferred form because the fixed charges are captured independently of the variable charges. A customer utilizing 300 cu ft or more in a billing period would receive the same billing under either scheme. On the other hand, a customer utilizing only 50 cu ft during a billing period would receive a bill for only $1.00 from the scheme where customer charges are "buried" in commodity charges, but would receive a bill for $2.25 when called upon to pay a separate customer charge. In electric rates, the same scheme prevails, but the "commodity" is a kilowatt-hour of electricity instead of a cubic foot of gas. This sort of rate is generally applicable only to residential and small commercial applications.

B. Socially Motivated Rates

Occasionally, social objectives are directly addressed through the utility rate structure. Because of the irrationality of the relationship between such goals and energy use, forecasting the impact of such rates, or, even more difficult, forecasting the probability of the future creation of such new rates, is particularly hazardous. The example that most clearly demonstrates this phenomenon is the residential "lifeline" rate. In order to come to grips with the Orwellian [11] rhetoric surrounding such a rate, it is necessary first to accept the idea that both motivations and expected results are most often clouded or totally obscured. In the lifeline example, the announced but phantom target for "rate relief" was the poor, the elderly, the unemployed, and the disadvantaged of all sorts who were being adversely impacted by rapidly escalating electricity costs. A worthy cause, a documentable problem, a clear need—but the "solution" turned out to be nothing more than a transfer tax. The purpose of the rate, as originally espoused and defended, was almost completely lost in the actual rate.

For the retrofitter, such anomalies are probably beyond prediction and seem exempt from rational preventive planning.

C. Demand Rates

Demand rates are perhaps the most misunderstood, or least understood, of the common rate forms. In their most usual form, simple demand rates segregate their charges into three categories: a customer or minimum charge, a commodity charge, and a demand charge. The last charge is the source of the regularly occurring misunderstanding, although it represents a relatively straightforward concept. Demand may be established in a variety of ways. A typical approach [13] establishes the maximum demand in any month as the measured maximum average kilowatt input during any 15-minute metered interval in the month. Other intervals may be used, of course, such as a 5-minute interval where the demand is intermittent or subject to violent fluctuations. In this approach to the demand rate, the customer charge (minimum charge) is the monthly demand charge, which is $860 for the first 200 kW or less of billing demand, and $4.30/kW for all excess kilowatts of billing demand. That is, if, for example, the monthly billing demand were 300 kW, the demand charge would then be $860, plus (300 − 200) × $4.30 = $430, for a total of $1,290. To extend this example, assume the energy consumption is 135,000 kWh. This particular rate calls for 0.730¢/kWh for the first 150 kWh/kW of billing demand, 0.530¢/kWh for the next 150 kWh/kW of billing demand, and 0.330¢/kWh for the kilowatt-hours over 300 kWh/kW of billing demand. Thus, (150 kWh/kW × 300 kW × 0.730¢) + (150 kWh/kW × 300 kW × 0.530¢) + (150 kWh/kW × 300 kW × 0.330¢) = $328.50 + 238.50 + 148.50 = $715.50. Therefore, for the full example, the total charge would be the sum of the demand charge and the energy charge, $1,290 + $715.50 = $2,005.50.

The application of such a rate in an actual case would be complicated by a number of additions or deletions, such as voltage discount, power factor adjustment, energy cost adjustment, tax change adjustment, and conservation load management adjustment.

D. Time-of-Use Rates

The time-of-use rate is a concept whose time has come. The notion is reasonable enough. The utility's system peak is built upon the coincidence of many customers' individual demands. Should not the use of energy at a time when that usage contributes to the utility system's peak be at a higher rate than the use of energy at other times? Time-of-use rates address the equity of this idea. One such rate [14] sets a monthly customer charge of $1,075, a demand charge to be added to the customer charge of $5.05/kW for all kilowatts of on-peak billing demand, plus $0.65/kW for all kilowatts of mid-peak billing demand and no charge for all kilowatts of off-peak billing demand, and an energy charge to be added to the demand charge of 0.530¢/kWh for all on-peak kilowatt-hours, plus 0.380¢/kWh for all mid-peak kilowatt-hours, plus 0.230¢/kWh for all off-peak kilowatt-hours. For this example rate, the daily time periods are defined as follows:

On-peak: 12:00 noon to 6:00 P.M. summer weekdays except holidays.
5:00 P.M. to 10:00 P.M. winter weekdays except holidays.

Mid-peak: 8:00 A.M. to 12:00 noon and 6:00 P.M. to 10:00 P.M. summer weekdays except holidays.
8:00 A.M. to 5:00 P.M. winter weekdays except holidays.

Off-peak: All other hours. Off-peak holidays are New Year's Day, Washington's Birthday, Memorial Day, Independence Day, Labor Day, Veteran's Day, Thanksgiving Day, and Christmas.

While the definition of on-peak, mid-peak, and off-peak will be dictated by the characteristics of the local utility system load curves, and the appropriate holidays will vary according to local custom, it can be seen that considerable incentive is provided by such rates to shift the use of energy away from the utility's peak and toward off-peak periods.

E. Interruptible Rates

Some utilities offer a contract rate [15] to the large industrial and/or commercial customer in exchange for the customer's allowing service interruptions during on-peak and mid-peak periods with appropriate notice to the customer prior to the beginning of such an interruption. In the example cited in the reference, not less than 10 minutes' notice would be given for the company-controlled interruptible load, and not less than 30 minutes' notice would be given for the customer-controlled interruptible load. This rate takes the form of reductions from the charges established under a standard time-of-use rate [14].

IV. TECHNOLOGY–RATE INTERFACE

Retrofit technology covers the full spectrum of the construction disciplines. All of the energy-related retrofits can be equated to utility rates and by that means can express energy savings in the common denominator of money. Some of the techniques are so closely tied to utility rate structures that they deserve special notice here.

A. Demand Control

An important industry has grown up around the technology of controlling systems in a fashion intended to optimize energy utilization. The evolution of miniature, solid state electronics and the digital computer have been the driving forces in this control-industry growth. Such controls may be segregated into user-controlled and utility-controlled categories.

1. User-Controlled Loads. The most common, the most widely and actively promoted innovation in control systems has occurred in the area of user-controlled systems to load and unload in patterns intended to minimize the charges resulting from utility demand rates. This same technology has also been employed to address time-of-use rates. These products are now firmly established in the marketplace in competition with other products and subject to normal market forces. This development is referred to as the "sustained commercial activity" step [16] in the diffusion of new innovations.

The trick in these control schemes is *always* to set up an objective and then design systems and select hardware to meet just that objective. Failure to follow this sequence foredooms the control system to almost certain "overkill," which tends to remove, or at least reduce, the economic benefits to be derived through use of the system. Unfortunately, many applications where the unstated objective could have been realized by a simple time-clock installation were rendered ultimately uneconomic by the purchase of elaborate and expensive hardware capable of control far beyond the needs of the basic objective. However, this is not to imply that there are not many economically justified applications for advanced technology control systems. To the contrary, the optimization of control strategies using computer technology is receiving increasing attention [17].

Also, potential retrofitters should be aware that some electronic systems devote their innovative efforts to "tricking" the utility rate, or more precisely, the utility meter, rather than attempting real load control. Such efforts are ultimately self-defeating as they are met with utility electronics and with reduced and/or floating demand intervals.

2. Utility-Controlled Loads. This technology, which to some has called forth images of "Big Brother" [12], has been marked by the development and widespread testing of systems for utility control of customer-owned energy utilization hardware. These systems have generally been employed to interrupt or cycle the operation of customer devices such as air conditioners, water heaters, pool circulating pumps, and space heaters on the command of a utility-initiated signal. Such signals may be delivered via carrier, radio, or some hybrid variation and are scheduled on the basis of the needs of the utility system. In theory, and in practice on the basis of as yet limited testing, significant savings in generation capacity requirement may be realized through these systems without any undue inconvenience or discomfort to the controlled customers.

B. Storage

An old, long neglected technology is being recycled in response to the introduction of time-of-use rates. Heat storage utilizing off-peak electricity has had widespread application in some winter-peaking American utility systems, and is a way of life in a number of areas outside the United States, notably Australia and parts of Europe. Yet little has been done with cool storage since the ice bank application for theaters and churches went out of fashion 25 or more years ago. Today, in time-of-use rate areas, ice, chilled water, or brine storage systems are making a strong comeback, without a great deal of fanfare. This is proven technology utilizing proven off-the-shelf hardware; only the right price signal was needed to get it dusted off and in use again. It seems clear that as word of these new, economically successful applications spreads [18], more and more system designers will join the parade. To further that end, the electric utility industry has sponsored research [19] to quantify the interface effectiveness between utility systems and customer cool storage systems.

V. THE DECISION-MAKING PROCESS

According to Drucker [20], "The first managerial skill is the making of effective decisions." He also warns that "managers who make effective decisions know that one does not start with facts. One starts with opinions." In the case of retrofit decisions, the original opinions are often clouded by overly enthusiastic salesmanship, wishful thinking, inaccurate technical assessment, or, perhaps worst of all, a preconceived decision in search of justification. The scope of this chapter does not extend to a discussion of cost-benefit analysis, life-cycle costing, or engineering economics, but must include some guidance for the use of utility rate information as input into those analysis techniques. Therefore, the following checklist is offered in the hope that its use will assist the potential retrofitter in translating energy savings into dollars:

1. Contact the local utility serving the project to become familiar with its rate schedules and with its forecasts of probable rate changes over the expected life cycle of the retrofit system.
2. Either exercise your retrofit alternatives through all reasonable combinations of present rates and probable future rates, or have a professional consultant do so for you. Be sure to assign probability corrections to the various future rate alternatives using the best available opinion.
3. Do as much analysis as you can, playing the "what if" game in your analysis. Continue this exercise until the decision resultant of the analysis is self-evident, until you feel *comfortable* in the decision. "In everything a prudent man acts with knowledge, but a fool flaunts his folly" [21].
4. As much as is possible, keep your options open. Allow enough flexibility in your selected systems and hardware to accommodate future rate changes that are not presently thought to be most likely, but which, if they did come to pass, would materially alter the economic consequences of your retrofit system. As Drucker has said, "the irretrievably wrong decision is the most dangerous course."
5. And finally, to reiterate an earlier caution, before alternative systems are considered, before utility rates are studied, before economic analyses are begun, identify clearly and unequivocally the *objective* of the proposed retrofit. You must know where you are headed before you have any chance of selecting the proper road to get there.

In summary, the retrofitter is urged to become thoroughly familiar with the utility rates that will apply to the project in question. There is no substitute for knowing the rate impact of potential decisions, and that certain knowledge can come only from doing one's homework.

REFERENCES

1. W. Shakespeare, *Hamlet,* Act III, Scene I.
2. C. Wilcox, *Public Policies Toward Business,* Richard D. Irwin, Inc., Homewood, Ill., 1966.
3. E. W. Clemens, *Economic and Public Utilities,* Appleton-Century-Crofts, Inc., New York, 1950.
4. R. H. Leftwich, *The Price System and Resource Allocation,* Holt, Rinehart and Winston, New York, 1966.
5. H. F. Perry, "Electric Rates: A Federal Perspective," in *Energy Users & Government Regulations, Proceedings of the Third Annual Energy Users Law Seminar, January 1978, Washington, D.C.* (L. E. Buck, ed.), Government Institutes, Inc., Washington, D.C., 1978.
6. C. J. Cicchetti, W. J. Gillen, and P. Smolensky, *The Marginal Cost and Pricing of Electricity; An Applied Approach,* Ballinger Publishing Company, Cambridge, Mass., 1977.
7. D. Frishberg, "The FERC Rules on Marginal Costs," *Public Utilities Fortnightly,* September 13, 1979.
8. D. Frishberg and R. G. Uhler, "The Neo-Classical Costing Controversy," *Public Utilities Fortnightly,* August 3, 1978.
9. Standard 90-75, American Society of Heating, Refrigerating and Air Conditioning Engineers, Inc., New York, 1975.
10. Proposed series of standards, nominated Standard 100, American Society of Heating, Refrigerating,

and Air Conditioning Engineers, Inc., New York, as yet unpublished.
11. George Orwell, author of *Animal Farm, 1984,* and other social comment fiction.
12. G. Orwell, *1984,* Harcourt Brace Jovanovich, Inc., New York, 1949.
13. Schedule No. A-7, *General Service,* California P.U.C. Sheet No. 5123-E, Southern California Edison Company, effective January 1, 1979.
14. Schedule No. TOU-8, *General Service—Large,* California P.U.C. Sheet No. 5225-E, Southern California Edison Company, effective July 22, 1979.
15. Schedule No. TOU-8-I, *General Service—Large—Interruptible,* California P.U.C. Sheet No. 5149-E, Southern California Edison Company, Effective January 1, 1979.
16. R. Schoen, A. Hirshberg, and J. Weingart, *New Energy Technologies for Buildings* (J. Stein, ed.), Ballinger Publishing Company, Cambridge, Mass., 1975.
17. S. M. Zvolner, in *Proceedings: Third International Symposium on the Use of Computers for Environmental Engineering Related to Buildings, 10–12 May 1978, Banff, Alberta, Canada,* National Research Council of Canada, Ottawa, Canada, 1978.
18. K. L. Heitner, "Energy Storage Systems for Improved Load Management," *Power Engineering,* September 1979.
19. Research Project 1089, Electric Power Research Institute, Palo Alto, Calif.
20. P. F. Drucker, *Management: Tasks, Responsibilities, Practices,* Harper & Row, New York, 1973.
21. Proverbs 13:16 (RSV).

Part 2

Energy Audits and Surveys

4. Energy Audits and Energy Management

Larry W. Bickle, Ph.D., P.E.

The Bickle Group
Houston, Texas

	Page
I. Introduction	43
II. Types of Energy Audits	43
III. Limitations of Energy Audits	43
IV. Energy Management Program	45

I. INTRODUCTION

Energy conservation is a popular subject—so popular, in fact, that the federal government appropriated millions of dollars in 1978 to conduct energy conservation programs in public schools and hospitals. Energy audits play a major role in these and other federal programs; but so much attention has been given to energy audits that they have become almost synonymous with energy management. This is a dangerous trend. Energy audits do not in themselves produce energy savings. They are simply one of several essential steps in an energy management program.

Energy audits do have a useful role and should be conducted. This chapter, however, takes a critical look at the subject. The goal is to raise important issues and limitations for both owners and auditors to consider. It is hoped that this will allow more effective use of energy audits within the framework of a meaningful energy management program.

II. TYPES OF ENERGY AUDITS

A wide range of activities are loosely described as "energy audits." At one end of the scale, simply collecting utility bills, calculating energy consumption per square foot, and comparing this consumption to "normal" or "average" values might be termed an audit. At the other end of the scale, the audit of a professional engineer and/or architect might include spending several man-days inspecting a building, testing HVAC equipment, measuring lighting levels, computing theoretical performance, determining life-cycle costs, and preparing retrofit construction documents.

The exact level or type of energy audit is not pertinent to the issues raised in this chapter. In general, however, most of the points relate to audits in which a professional makes site visits and performs technical calculations. In the terms used in various federal programs, this would be a Class A type audit and would include some aspects of the Technical Assistance Program (TAP) type of technical and economic calculations.

III. LIMITATIONS OF ENERGY AUDITS

With this general background, consider the following:

A. *Energy audits do not save money by themselves.* It costs money to conduct any kind of energy audit. Unless the findings of the audit *are used,* the audit itself will produce no savings. Conducting an audit before there is top-level management commitment to implement the results can be a serious waste of money.

B. *It is possible to do an energy audit too early in the program.* If a detailed energy audit is conducted too early in the overall energy management process, there is possibility for misdirection. Many of the most important first steps in the energy management program, such as changes in administrative policies, cannot be easily quantified. Unfortunately, neither the true cost nor the mathematical relationship can be easily determined. However, there is evidence suggesting that these actions may have benefit/cost ratios 50 to 100 times greater than capital improvements.

While most energy audits do identify low- and no-cost actions, the energy audit with its calculations and "precise" numbers for capital improvements can divert attention from these more important early action areas. The net result may save energy but not be the most cost-effective program.

C. *There are significant limitations in the engineering techniques used to compute energy consumption in buildings:*

1. Engineering analysis techniques to analyze the long-term average impact of small actions are not readily available. For example, it is not possible accurately to estimate the yearly savings that would result from replacing a specific piece of weatherstripping or from installation of edge seals on outside air dampers.

But many of these so-called minor capital improvements can, when taken collectively, produce substantial energy savings. These "minor" actions also involve considerable cost, and thus it is important not to apply them indiscriminately in every case. In technical terms, the result of each action is smaller than the uncertainties in the calculation techniques themselves.

2. Even if there were not uncertainties in the engineering techniques, there would be uncertainties in the input data. The input data that are difficult to obtain precisely include a description of the building, environmental variables (such as temperature, wind, and solar radiation), and internal loads (such as occupants and lights).

In an older building there are usually uncertainties about wall insulation, control system set points, *in situ* efficiencies, and other "details." Even when this is not the case, reducing a real physical building to a set of idealized nodes, conductances, and terms in a calculational model introduces simplifications and loss of precision.

Shade trees, obstructions at ground level, local ground reflectants, small lakes, etc., change the specific microclimate for an individual building. In most cases, engineering calculations will use macroclimate information from nearby weather stations, with resultant uncertainty and imprecision regarding the weather data specifically applicable to the building being studied.

How many people are in the building at the same time? During what hours is it operating? Are exhaust fans switched off at certain times? Do occupants turn off lights when they leave the room? How are drapes used? These and other important questions about the interaction between occupants and the building affect the precision of the energy consumption calculations. These occupant use patterns are extremely difficult to determine with any precision because they change from hour to hour, day to day, month to month, and year to year.

Assumptions are added to assumptions, and so on and on. Even the most complex, comprehensive computerized methodologies such as DOE-2, BLAST, TRACE, and AXCESS can rarely predict actual energy consumption in a building within 10 to 15%. Discrepancies of 25 to 50% between theoretical calculations and actual consumption are not uncommon. An occasional difference of 100% or greater is not unknown. It is unclear how much these differences are due to calculational techniques, and how much they are due to errors and lack of precision in input data. What is clear is that the overall precision of the process is not much better than the total combined savings of possible modifications.

3. Another engineering problem is that few

simplified methods exist for predicting the interaction between energy conservation methodologies, and generally modifications are not cumulative; two modifications that each would save 10% probably would not save a total of 20% if both were implemented. In the simplest case, the first modification saves 10%, and the second saves 10% of the remaining 90% or 9%.

But the situation can be worse: both modifications can compete to save the same energy. Consider, for example, the combination of double-glazing and a night setback thermostat. Both modifications reduce heat conduction through the windows, but the interaction is highly complex. If the night setback temperature is 55°F (13°C) and the nighttime outside air temperature is 55°F, there is no temperature difference to cause a heat flow. Thus, the double-glazing has zero added benefits at that particular instant. Clearly the savings obtained by making both of these modifications is not the sum of the savings that would be calculated for each modification individually.

These complex interactions can be modeled to some extent using sophisticated computer simulation programs such as DOE-2. However, there need to be simpler ways to evaluate the interactions between potential modifications.

D. *Another major limitation of energy audits is cost estimating.* Most of the standard cost estimating methods and data files are designed for use with either new construction or major remodeling. Much of the energy conservation retrofit work is really "odd jobs" that are handled by small independent contractors. These costs tend to be highly localized and difficult to predict. While no single retrofit project is large, there can be a large percentage error in cost estimating. These cost estimating errors can accumulate to produce large errors in the total project cost.

IV. ENERGY MANAGEMENT PROGRAM

In spite of the limitations raised in this chapter, energy audits do have an important role to play in an overall energy management program. However, an energy management program needs to focus on more basic issues.

Exactly what is basic varies from one client to another. Based on our own past experience, and a review of available literature, we would propose the following as building blocks for a successful energy management program. Whether you agree or not, an internal discussion of these fundamental issues will help focus energy management efforts for maximum results.

A. *Energy conservation should be viewed as an upper-level management responsibility.* Energy conservation involves improved operations and maintenance and investments in hardware. These are but pieces of the broader management problem of controlling energy costs. (See Figure 4-1.) A well-balanced program will cut across internal divisions and require policy changes, integrated administrative practices, improved operations and maintenance practices, public relations programs, and, finally, capital investments.

B. *An energy conservation program should be financially sound.* Energy conservation actions cost money. Weigh these costs carefully against potential savings so only cost-effective actions are taken. The definitions of cost-effec-

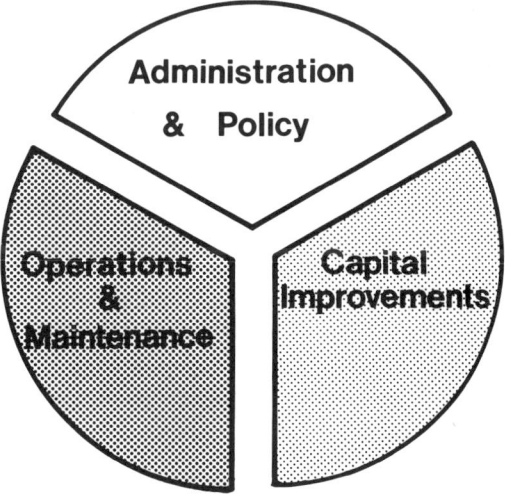

Figure 4-1. Components of controlling energy costs.

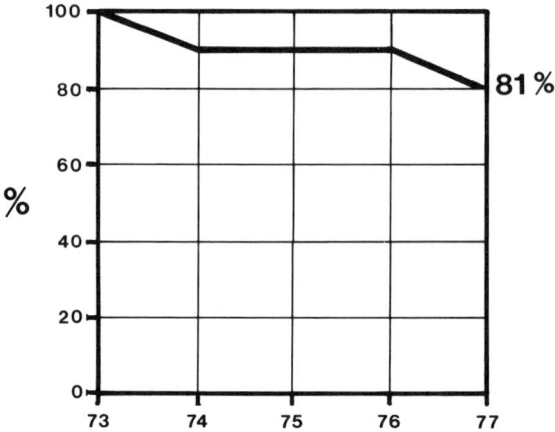

Figure 4-2. Savings achieved through administrative, policy, and operations changes.

tive must be formulated at the policy level using opinions from as many interested parties as possible.

Administrative and policy changes tend to be the least costly changes to implement, are the most cost-effective, and should be undertaken first. Improvements in operations should be undertaken next. And capital improvements should be deferred until last. (See Figure 4-2.)

By progressing from less expensive changes to funding the more expensive improvements out of proven future savings, a client could have a self-financing program.

C. *Establish goals and priorities.* State precisely the goals of the energy conservation program. Is the object to reduce costs? Improve the public image? What are the relative priorities of the various goals? Insofar as possible, these goals should be quantitative.

D. *Energy conservation is a team activity.* Whatever program evolves must be a cooperative program among all participants. Recognize that energy management is a sensitive activity. The program defined for a particular client must be unique and responsive to local needs, conditions, personalities, requirements, and restraints and the broader goals of the client's organization.

E. *Focused accountability and responsibility are critically important.* While energy conservation is a team activity, one person must be in charge. This person must provide clear, strong leadership. The energy management program must be well defined, and specific responsibilities and authority agreed upon by all participants.

F. *Motivation and evaluation of progress are vitally important.* Energy conservation is really a collection of many small actions, like turning out the light when leaving an unoccupied room. Motivation and constant feedback are essential to a cost-effective energy conservation program.

One of the best ways both to motivate and to provide feedback is to implement an effective, easily understood, and highly visible "scoring system" for measuring the success of the energy management program.

An added benefit of this scoring system is that energy conservation actions—policy changes or capital investments—can be evaluated, and quantitative estimates of actual savings can be made.

G. *Energy conservation programs are site-specific.* They must be tailored to specific climatological conditions and building types.

H. *A Master Plan for Energy Management (MPEM) should be developed.* This plan should identify specific responsibilities, establish evaluation criteria, and systematically rank actions into logical priorities for implementation.

Clearly these basics are not absolute, but they do illustrate how the common-sense management techniques of commitment, motivation, education, and evaluation can be integrated into a meaningful program to reduce energy costs. Perhaps more important, implementation of an effective energy management program provides the proper perspective for interpretation and use of energy audit results.

5. Managing the Energy Use Survey and Audit

Kenneth M. Clark, P.E.

Burns & McDonnell
Kansas City, Missouri

	Page
I. Presurvey Organization	48
II. Use of Thermography during the Energy Survey	53
III. Other Important Considerations in the Energy Audit	53
IV. Obtaining Data for Computer Use	54
V. Waste Energy Measurement	54
VI. Economic Aspects of Energy Audits	55
VII. Survey Process (by Facility)	55
VIII. State and Federal Requirements of 1978 National Energy Act	57
IX. Who Should Perform Energy Audits	57

I. PRESURVEY ORGANIZATION

A systematic approach to the survey or audit is essential. Before the survey or audit is performed, a complete checklist should be made. The alternative is more time-consuming data gathering and the likely omission of some information. It is difficult to obtain all the data needed without overlooking something.

Organization is the key. A checklist of all items to be investigated must be made. Here is a list of items that should be addressed in making an energy audit:

1. Data Gathering.

(a) *Utility Bills.* Utility bills provide a history of the energy consumption in a building. This historical record can be quite informative in evaluating the use of the facility, the growth of the facility, usage of energy, and the cost of energy. It is important to get as many years' past records as possible. At least three years are necessary to obtain a trend in energy consumption. This avoids the use of data from an atypical year. Utility bills should include electricity, gas consumption, fuel oil purchases, propane purchases, and water bills.

(b) *Lighting Systems.* Lighting has a major impact on energy in any plant or facility. A thorough survey should be made to determine what types of lighting are used (i.e., fluorescent, incandescent, HID, etc.). It is also important to determine the efficiency of the lighting system. This requires answering the following questions: Is lighting in the right location for tasks? Were the ballasts removed when the fluorescent tubes were? Is the lighting intensity reasonable? It is also important to obtain the lighting schedule (when the lights are on and off).

(c) *Heating and Air Conditioning Systems.* Considerable data must be gathered regarding the heating, ventilating, and air conditioning systems. These data should include the capacity of the system, the condition of the equipment, the operating schedule of the equipment, the operation effectiveness of the outside air dampers, and temperature control diagrams and information.

(d) *Ventilation and Exhaust Systems.* Much

heat can escape up the stack, consuming great quantities of energy. It is important to determine the exhaust quantities of energy. In many cases, the exhaust requirements are necessary and essential. However, in some cases the exhaust is not required all of the time or not in the quantities that are being used. When gathering the exhaust data, get the fan size, the motor horsepower, and the rpm if possible. If these data are not available, the manufacturer's nameplate data, model number, etc., should be obtained so that the design capacity of the fan system can be determined.

Data on ventilation systems are also important. Again, the size of the system, the horsepower of the fan, and the manufacturer's nameplate data are important information. From these data, it is possible to ascertain just how much ventilation is being used. Also, it is important to determine the condition of the ventilation and exhaust systems. A system in poor condition may or may not be operating at the design air flow quantities. A check of all ventilation system filters is important to determine their condition. Clogged filters can significantly affect the static pressure of the system and, therefore, the horsepower required. The condition of the coils in an air-handling unit should also be investigated. If the filters are clogged, there is a good chance that the coils are also in poor condition. An overall system evaluation is important to determine the efficiency and level of maintenance of the system.

(e) *Temperature Controls.* Heating and air conditioning control systems can significantly affect the energy consumption in a building. The temperature controls should be investigated from two standpoints: (1) the function of the controls and (2) their condition. Temperature control diagrams should be obtained to determine if the temperature controls were installed according to the plans and specs. If this is the case, then proceed with the analysis to determine if the function of the controls is in the owner's best interest. As far as the condition of the controls is concerned, it is important to determine whether the controls are operating as designed. In many cases, building operators adjust their temperature controls to satisfy some particular building needs. Therefore, the outside air quantity, for example, may be substantially different from the design. Other items to look for include dampers not closing tightly or control functions jumpered out of the circuit for the convenience of the operator.

(f) *Plant Process Loads.* Process loads can play an important function in the building energy picture. Process loads consume energy that must be accounted for and may be a potential heat source for energy recovery. Exhaust quantities from any plant processes should be determined in order to evaluate the potential of recovering this waste heat.

(g) *Potential Heat Recovery.* Heat recovery can come from two sources: (1) the plant process loads mentioned above and (2) building exhaust. If a high outside air ventilation rate is required in a plant, there is a good opportunity for recovering heat from this exhaust by preheating incoming outside air. The potential for heat recovery also exists any time there is a simultaneous heating and cooling requirement. That is, one plant process may require cooling while the other process requires some form of heating. This situation could occur over the full range of the temperature spectrum anywhere from a refrigeration process up through steam heating requirements. Obtain all information possible on waste heat and its potential uses.

(h) *The Building Shell.* The wall and roof construction must be determined, and the size, quantity, and condition of all windows must be tabulated. Also, size, quantity, and types of all overhead doors should be observed (including the frequency of opening and length of time that they are open). An overall observation of the condition of the building shell should be made to determine if there are any large cracks or leaks in the building. Thermal leaks from the building can be costly and are less apparent than water leaks.

(i) *Building Schedules.* The building schedule greatly affects the total amount of energy consumed in a building.

First, the number of shifts of operation should be determined. The payback period of energy conservation modification may be significantly faster in a plant operating on three shifts than in a plant operating on only one shift. Also, the energy consumed may be considerably less in one shift of operation.

Second, the occupancy of the building should be determined, and the schedule of the occupancy obtained; ascertain the number of personnel occupying the building during each of the 24 hours of the day.

Lighting Schedules. It is quite possible that the lighting schedule will not follow the occupancy schedule. Cleanup crews at night may require lighting even though the building is not occupied. Skylights in the building may permit the lighting to be shut off during daylight hours.

HVAC Equipment Scheduling. This information needs to be determined not only on a daily basis but on a weekly and seasonal basis. HVAC equipment may be shut down on the weekends, nights, and holidays. Also, the cooling season may be locked out until a certain time of the year; this needs to be verified with the owner.

Process Equipment Schedules. The actual operating schedule for each piece of equipment in the plant needs to be determined. This schedule is important, but it is also important to determine the schedule required for each piece of equipment. Some pieces may be operating 24 hours a day, but they may be required for only a few hours of that day.

(j) *Plumbing Systems.* The number and types of plumbing fixtures need to be tabulated. The quantity of domestic hot water required must be determined. These data should be compared with utility bills for water. Great discrepancies between the water utility bill and the calculated consumption may indicate either a leak or wasteful consumption.

(k) *Fuel Purchase Records.* Usually, utility bills are readily available for electricity and natural gas. However, don't overlook the purchase of propane or fuel oil as a standby fuel. These fuels can have a significant impact on the winter energy consumption in the plant, especially if the plant is on an interruptible gas supply.

2. Preparing a Checklist. We have talked primarily about what should be included in the checklist. Some thought needs to be given to the way the checklist should be set up. The purpose of the checklist is for organization before the survey and to minimize omissions during the survey. Therefore, it is important that the checklist be efficient, organized, and an effective reminder of all the items that should be observed during the audit. Several lists may be required. A list of items to be covered when talking with management is helpful. A list of items to be observed while touring the plant and a list of items to be covered with others, such as utility companies, are desirable. Perhaps you would prefer to break down the list by mechanical items, electrical items, etc. The way checklists are prepared will depend upon the number of people making the audit and their areas of expertise.

3. Equipment Required. The sky is the limit on the amount of equipment that can be used in an energy audit. It is nice to have every tool available that you need. However, from the practical side, carrying a suitcase of equipment around can be exhausting. The equipment required will depend somewhat on the type of audit being made. As a minimum requirement, the following equipment is recommended:

(a) *Thermometers.* A thermometer is needed to measure space temperature as well as air temperature leaving air handler hot and cold decks. Also, thermometers for measuring hot and chilled water temperatures are needed. These instruments may range from a pocket mercury bulb thermometer to an electronic sensing device. The electronic sensing device is handy for measuring air temperatures, plus hot or chilled water temperatures. A sling psychrometer is also valuable to determine the wet bulb for humidity conditions in the air.

(b) *Light Meter.* Light level readings need to be taken in all occupied spaces. These data are among the easiest to gather, since most readings are taken at eye or desk level.

(c) *Tape Measure/Folding Rule.* Both these measuring devices are extremely handy during the audit procedure. The tape measure should be a minimum of 50 feet long, and preferably the 100-foot length. A folding rule is essential for measuring shorter distances; it can be handled by one person.

(d) *Flashlight*. Invariably, the data needed will be on a nameplate in a deep, dark, barely accessible corner. A flashlight is thus an essential tool. It is also important to carry a pocket knife. Invariably, the nameplate data illuminated by the flashlight will have been painted over by maintenance personnel, and the paint will have to be scratched off to get to the data.

(e) *Hand Tools*. A pair of pliers and a screwdriver can save valuable time. When you are the greatest distance from assistance or tools, the data needed may be behind an access panel mounted with screws or "stuck" wing nuts.

The above list is the minimum equipment required for an effective audit. Several other very helpful items are:

(f) *Camera*. Dozens of times after returning to the office to begin the analysis, you will kick yourself for not photographing certain items that you thought would be indelibly etched on your mind. Take as many pictures as plant personnel will allow. Be sure to clear all photography with the plant operations people first. Many people are very sensitive about the confidentiality of their processes, etc. Also, avoid taking pictures of plant personnel. This tends to be disruptive to plant operations and, therefore, tends to displease the client.

(g) *Walkie-Talkie or Two-Way Radio*. This is a handy item if there is more than one person making the survey in a large facility. A two-way radio can also be a valuable tool for observing a piece of equipment that requires someone at a remote location to start and stop it.

(h) *Air Flow Meters*. Some audited facilities will be in older buildings, where the air flow has probably changed substantially from its original design. Taking data from a set of building plans may give you a false idea of the real energy consumption or the real problems. Therefore, it is valuable to make spot checks of air flow for air measurement readings.

The easiest device to carry in an energy audit is the rotating vane anemometer. This is a small instrument approximately 6 inches in diameter that measures the air flow over a period of time. This instrument will allow you to take air flow readings across louvers, through doorways, and across filter banks, and also give you a rough idea of the air coming from diffusers. It is an all-around tool, lightweight, compact, and easy to carry.

(i) *Volt-Ammeter*. This instrument is almost an essential tool. It is easy to take nameplate data, and this may be sufficient for the level of the survey, but, in many cases, the nameplate horsepower on the motor may be significantly higher than the actual operating horsepower. I have reviewed energy audits in which it was assumed that all equipment was running at 100% of nameplate horsepower. A grossly inaccurate picture of the actual energy consumption can be presented in such cases. Volt-ammeters are handy tools in determining electrical loads.

4. Security Clearance. One of the first things to consider before auditing a facility is the security clearance. An initial meeting with the client should clarify whether special badges are required, or special keys for access to certain areas, or if plant personnel must accompany you at all times.

The plant security force should be alerted to the fact that audit personnel are on the premises; they should be told when the audit will take place. It may be necessary to do part of the audit at night, to have special access to the facilities.

On military bases, clearance to certain facilities may be more complicated; it may be necessary to carry a letter of identification at all times. If the energy audit takes place in a multi-building facility, it is a good practice to alert the appropriate people that audit personnel are coming, instead of walking in cold, expecting them to be at your service. Proper identification is essential while making the audit; this may consist of a plant identification badge, a letter of clearance from the client, or your own company identification card. It is essential that proper security arrangements be agreed upon with the client and then adhered to.

5. Meeting with the Client. Before the survey is started, meetings should be held with the client to gain as much knowledge about the facility and operation as possible. The following items should be discussed during this initial meeting. If the audit work is being performed entirely "in house," the same information needs to be clarified with management.

(a) *Owner's Future Plans.* Obtain the owner's projections for future plans at this facility (plans for expansion, production changes, etc.). This information needs to be obtained for near-term and long-term planning. Any contemplated change in plant schedules or number of shifts should be obtained. Records of the plant's increase in production should be obtained, including future projections. Also, gather data on any new equipment or processes the owner anticipates installing or deleting.

(b) *Specific Problem Areas.* Often the owner will have a particular problem area in his building or systems that should be addressed in the energy review. These problem areas may also suggest some potential energy-saving areas.

(c) *Scope of Study.* It is important that the client and auditor both understand exactly what is to be accomplished, what the energy audit survey is to cover, and what it is not to cover. Also clarify what the report will include or exclude and what details the report will address (i.e., computerized analysis, life cycle costing, etc.)

(d) *Occupancy-Usage Profiles.* Obtain schedules from the client on building occupancy and number of personnel by areas. Get not only daily schedules but weekly, monthly, and annual ones as well. Determine if there are any unusual shutdowns, such as an annual vacation schedule, when the entire plant is shut down at once.

(e) *Historical Information.* Obtain all the information available about the building or facility: when it was constructed, when additions were added, when modifications were made, what problems have been experienced by the owner, etc.

6. Utility Information. During the energy survey, all the utilities serving the client should be contacted. Some of the information that should be obtained directly from each utility is: (1) the availability of future service to the client, (2) a projection of how long the utility will be able to serve the client, and (3) how much additional capacity can be provided, or, if a reduction in service is anticipated, the expected future curtailments. Obtain price projections in the future, both short- and long-term. Also investigate specific items relating to the client, such as additional transformer capacity, etc.

It is valuable to obtain fuel information for fuel sources not currently being used by the client. For example, standby fuels such as propane, heavy or light oil, etc., might be considered. Coal may be a viable alternate fuel in the future. Information should be obtained on each of these potential fuels, including delivered price, anticipated availability, and problems relating to storage.

7. Time of the Survey. When the survey is conducted can have an impact on the data gathered. Obviously, operating conditions are going to be different between the summer and winter months. Some things will be more apparent during winter, such as building air leaks. It is important to remember the seasonal considerations when making the survey. During a summer audit, try to anticipate the conditions that may be different during other seasons of the year. Consult with building operating personnel about variations and operations during other seasons. The audit should include surveys when the facilities are unoccupied as well as occupied. A nighttime survey can provide additional information regarding building system operation, lighting levels, etc.

8. Weather Data. Design weather data can be obtained in the ASHRAE handbooks; however, more detailed information may be required, depending on the scope of the study. Check with the client; at the location of his facility, weather conditions may vary from the normal or official weather data for that area.

9. Maintenance Operation. An important consideration in the evaluation of building energy systems is the maintenance factor. Information should be obtained on the type and frequency of maintenance performed. It is important to make a general observation of the condition of equipment and the apparent level of maintenance throughout the facility. Maintenance procedures followed can have a significant impact on energy usage. Some maintenance procedures will be obvious. However, other information will have to be requested, such as chiller tube cleaning, boiler tube cleaning, control systems checked, etc. Information regarding frequency of filter changes is useful.

II. USE OF THERMOGRAPHY DURING THE ENERGY SURVEY

1. General. The use of thermography, or infrared scanning, is becoming more and more popular in the evaluations of building energy systems. Infrared scanning equipment is available at prices ranging from several hundred dollars up to many thousands of dollars. There are many advantages to the use of infrared scanning equipment. One is expressed by the old saying that a picture is worth a thousand words. If the owner can see visually where great quantities of heat energy are lost, that may mean more to him than many pages of data and calculations. Also, infrared scanning can detect heat leaks and problems not visible to the naked eye, such as lack of insulation in walls and roofs.

2. Building Shell. An evaluation of the building shell is probably the most common use for infrared scanning equipment. Whenever a significant temperature difference exists between the ambient air and the building inside temperature, an infrared scan can detect hot spots through the surface. Scanning equipment is more accurate the colder the outside temperature. An area where the insulation has failed will show up as a significant heat difference compared with the adjacent surfaces. The same thing holds true for windows: if the windows are single pane or if they have many leaks around them, these leaks will show up on the infrared scanning equipment. The scanning equipment will also point out heat losses from louvers and other building openings, which may not have been apparent in the past.

3. Mechanical Systems. Infrared scanning equipment can be effectively used in the evaluation of mechanical systems. An infrared camera is a valuable tool in checking the operation of steam traps, especially when it would be difficult to determine if a steam trap were stuck in the open position without disassembling the trap. An infrared scanner can determine this instantly.

A scan of piping insulation is informative. Any failure of the piping insulation can be determined quickly with an infrared camera. A visual inspection of piping insulation systems may not reveal any problems with the insulation, but if it is wet or if the insulation has borken down under the vapor barrier, the thermal effectiveness of the insulation may be nil. The same applies to equipment insulation and boilers, etc. In many equipment and piping systems, it is not uncommon to have aluminum jacketing over the insulation. A failure in this insulation cannot be detected easily without the use of an infrared camera.

Inspection of bearings is another good use for the infrared scanning equipment. This is true not just from an energy perspective but from a maintenance standpoint as well. Overheated bearings will show up on the infrared scanner.

4. Electrical Systems. The use of infrared scanning is not limited to mechanical systems. Electrical system evaluation can also be made. The infrared scanning of electrical systems may be done primarily for maintenance reasons rather than for energy conservation. Infrared scanning is a valuable tool for finding "hot" spots in electrical gear and wiring. Such hot spots will appear in gear and wiring on the verge of failure or drawing excessive current. Motors should be checked. Hot spots in motor windings or bearings will reveal potential problems.

III. OTHER IMPORTANT CONSIDERATIONS IN THE ENERGY AUDIT

1. Overview. It is important to get an accurate picture of the energy usage in the building. To do this, you cannot rely solely on opinions of the owner or management personnel or, for that matter, an operating engineer. All their information is very valuable, but you must make your own observations to verify the actual operating conditions. Your observations may verify what the owner's personnel have told you. If that is the case, fine. Otherwise, further evaluation and "digging" into the problem might be required.

2. Rapport with the Client. Another important consideration in the energy audit is the way in which you relate to the client's employees, including both management and plant personnel. At the very outset, use of the word "audit" can have negative connotations to management personnel. This word typically relates to financial investigations where the company's books are evaluated and the financial management ability of the staff is scrutinized. For this reason, it is sometimes better to use the words "study" or "survey" rather than "audit." It is always good to reassure management personnel that you are not there to pass judgment on the building operation but merely to help them conserve energy and save money, and to lend your expertise to that end.

The same holds true for plant personnel. Plant personnel can become very defensive and cautious with a person walking around with a clipboard, making observations and taking notes. It is not unusual for them to feel that they are being evaluated personally. It is a good policy to let the plant people know what you are doing if management has not already informed them. Let them know that you are friendly, and not the enemy. Valuable data and insight can be obtained from plant personnel on the operation of the facility. However, it is important not to spend much time with them. This takes time away from their duties and can generate a negative reaction from management. It is best merely to state what you are doing and move on. If they have some energy-related information, they will be eager to tell you. You will not have to prod them for information.

I don't mean to belabor the matter of personnel relationships. However, I mention it because of an experience I had at a plant survey in Kentucky. After we briefly explained what we were doing in the plant to some employees, they seemed to understand, and we exchanged some ideas. But as we started to leave, one individual asked, "Well, did we pass or fail?" Obviously we didn't get our message across; this was not a pass-or-fail examination, but merely an evaluation to see how energy was consumed.

IV. OBTAINING DATA FOR COMPUTER USE

Many energy surveys and studies require the use of a computer to evaluate a building or facility, and computer modeling is becoming more and more important in building evaluation. If a computer model is being used, learn all the input data requirements for that computer program before the survey is made. In generating a computer model, some data are required for the input that would not normally be gathered during an energy audit. One of these factors is the building configuration, the physical shape of the building, rather than just the overall wall area. How the structure might shade other portions of the structure is important. Shading from adjacent buildings or trees must be determined on all sides of the building for proper input into a computer program's solar routine. Another important data input requirement is infiltration. Depending on the computer program, there may be several different ways to enter the infiltration data, through the crack method, the air change method, etc.

You will need a weather tape for the computer program; therefore, it might be necessary while you are in the area of the facility to obtain information for weather input.

V. WASTE ENERGY MEASUREMENT

1. Exhaust Systems. Substantial quantities of heat will escape through the building's exhaust system, and during an energy audit determining how much heat is escaping through the ventilation and exhaust systems can be a problem. Exhaust through gravity ventilators can be estimated by measuring the throat area, the building height, and the wind velocity and then referring to catalog data. A rotating vane anemometer can also be used in the throat of a gravity ventilator to determine the air flow.

Powered exhaust systems can present a greater problem. In most cases, fan nameplate data must be relied upon to determine air flow quantities. It is usually impractical to attempt air flow measurement on the fan discharge. However, it might be possible to determine air flow quantities if the exhaust fan is connected to a hood or paint spray booth, etc., where you can take air flow measurement readings across the face area of the hood opening or the area associated with the exhaust system intake. Measuring the exhaust gases can be a problem, because of high temperature and contamination. The exhaust may be too hot or dirty for delicate sensing equipment. Look for temperature gauges already installed on equipment or exhaust stacks that can provide the data needed. Usually plant operators will have a good idea of the exhaust temperatures.

In an effort to determine the energy utilization in a process system, it may be necessary to rely on plant operators, maintenance people, or engineering personnel for information. Shop drawings and engineering drawings can also be a useful data source.

2. Solid Waste. Solid waste is being increasingly recognized as a potential fuel source. It behooves the energy auditor to make an analysis of the potential use of solid waste. This requires a determination of the types of trash available and the quantities involved. Solid wastes include not only the trash from the plant but also by-products from plant production or processes. Plant personnel can probably provide a good estimate of the quantity of trash hauled away from the site, or the amount incinerated at the plant. If this information is not available, some visual observations can be made of the amount of trash or by-products of production available. Get the help of plant management; they can require that the trash be weighed before dumping, for a better estimate of the quantities involved.

VI. ECONOMIC ASPECTS OF ENERGY AUDITS

Preparing an estimate of the cost of an energy audit involves many factors. The key to minimizing the cost is organization. It is important that you be well organized before making an audit trip to a plant or facility. If the facility is out of town, make arrangements to ensure that all necessary personnel will be available during the audit. If meetings with representatives of the utilities are required, make sure they will be available also. Even if the facility is in town, it is most economical to try to make the survey in one visit. This is not always possible; it will depend upon the complexity of the survey and the size of the facility.

If several people are involved in the audit, organize the work into teams before the survey begins in order to make most efficient use of personnel. Before committing your firm to a fee for an energy audit, be aware of travel costs as well as the cost of special equipment that might be required.

Computer modeling will have an impact on job expenses. This can be a significant expense in a large, sophisticated program. If a new, unfamiliar program is used, be careful of the costs. As with any computer program, budget several reruns for errors and modifications.

Obviously, the cost of an audit and analysis can vary drastically, depending on the scope of the work. We have found a cost based upon annual energy consumption to be a reasonable gauge. As an average, a fee of approximately 6% of the client's annual energy bill is realistic. Travel expenses and other expenses peculiar to the particular project should be added to that fee. The percentage will vary, depending on the size of the audit; it is usually somewhat higher for a small study and lower for a very large study.

VII. SURVEY PROCESS (BY FACILITY)

1. General. The following is a general outline for an energy audit. Different types of facilities are addressed, after the general outline, and particular requirements for each type of facility are given.

    I. Building Envelope Data
       A. Physical survey of the facility
          1. Type of construction
          2. Dimensions

3. Orientations
4. Construction materials
5. Fenestration systems
6. Shading techniques
   B. Building area
   C. Inspection of building plans
II. Building Systems Data
   A. Type of mechanical systems
      1. Subsystems
      2. Components
      3. Age of equipment
      4. Primary conversion systems
      5. Energy distribution systems
      6. Control systems
   B. Review of shop drawings
      1. Manufacturer's machinery capacities
      2. Input energy requirements
   C. Lighting systems
      1. Lighting fixture locations
      2. Types
      3. Sizes
      4. Switching circuits
      5. Illumination levels
      6. Functions served
      7. Schedules
III. Occupancy–Usage Profile
   A. Population and occupancy schedules
   B. Building utilization data (hours/day, days/week, weeks/year)
   C. Special conditions of usage
   D. Facility use patterns
IV. Operating Practices–Procedures
   A. Identification of intermittent and sequential loads
   B. Equipment operation and schedules
      1. HVAC systems
      2. Setback control systems
      3. Winter heating mode
      4. Summer cooling mode
      5. Maintenance practices
      6. Process loads
V. Equipment
   A. HVAC–plumbing equipment
      1. Equipment inspection
      2. Controls
   B. Electrical equipment
      1. Electrical wiring
      2. Distribution systems
      3. Lighting conditions for each area
VI. Utility Data
   A. Historical bills or invoices
      1. Billing period
      2. Consumption
      3. Demands
      4. Cost
   B. Energy flow study
      1. Type
      2. Quantities
      3. Periods when energy is entering a facility
VII. Weather Data

2. Schools. Some important items to look for in schools are:

(a) Individual classroom controls
(b) Night setback
(c) Air handling systems for the auditorium, gymnasium, and cafeterias (controls and schedules)
(d) Fresh air shut off at night, etc.

3. Hospitals. Hospitals are notorious energy hogs. The energy consumption in a hospital may be several times that of other types of offices and industrial facilities on a per-square-foot basis. This is partly because of the high fresh air requirements. These air requirements, combined with high ventilation rates, restrictions on cross contamination, and humidity requirements, generally mean a high cost for HVAC systems. It is not unusual to find a reheat type system or dual duct system in a hospital. Investigate the following possibilities for energy savings:

(a) Reduction in 100% OA requirements for some areas
(b) Eliminating or applying energy conservation controls to reheat systems
(c) Heat recovery
   (1) Double bundle chillers
   (2) Exhaust systems
(d) Use of a computerized control system
(e) Building usage schedules (many areas of a hospital are not used at night)

Caution must be exercised in the evaluation of hospitals, because of hospital code requirements (both Hill-Burton and fire codes).

4. Industrial Plants. Generally, many opportunities for energy conservation exist in industrial plants. Some of the areas for investigation are:

(a) *Lighting systems*
   (1) Plants with fluorescent or mercury vapor lighting should be considered for conversion to high pressure sodium lighting.
(b) Energy management control systems
(c) Heat recovery systems (from building exhaust or process heat requirements)

5. Office Buildings. Important items to investigate in an office building are:

(a) Scheduling—of people, equipment, HVAC systems, lighting, outside air, etc.
(b) Mechanical equipment rooms
(c) Electrical rooms
(d) Occupied spaces (Determine such things as windows left open, lighting left on in unoccupied areas, or other equipment used in the space.)

6. Military Installations. Many branches of the service are in the process of making extensive energy studies of their facilities, requiring a substantial amount of work. However, there are unusual requirements peculiar only to military installations:

(a) Security (Advance clearance must be obtained before one tries to make an energy audit on a military base.)
(b) Access to buildings (Keys become a real hassle. It is best to prepare a schedule with the facilities engineer before beginning an audit.)
(c) Central plants and utility distribution systems (Type, condition, utilization, age, automation of controls, etc., are assessed.)

7. Residential Buildings. A residence is much smaller, and will require much less time for a survey, compared with commercial or industrial buildings. However, there are still many factors involved in a residential audit:

(a) Insulation (in both the walls and the roof)
(b) Windows (general condition, tightness, and type)
(c) Whether the building is caulked and sealed tightly (The exterior shell is probably the biggest factor in residential energy consumption.)
(d) The furnace and air conditioning units
(e) An observation of types and usage of lights
(f) Overall usage of the residence

VIII. STATE AND FEDERAL REQUIREMENTS OF 1978 NATIONAL ENERGY ACT

Currently, federal monies are available to provide energy audits and construction projects in hospitals, schools, local government buildings, and public care facilities. This money is being provided by the federal government to the states. Each state administers its own program individually. All phases of the grants program are funded on a 50–50 matching basis, with half of the funds coming from the federal government and the other half from the participating institutions and local government.

There are three levels of audits: the Preliminary Energy Audit (PEA), the Energy Audit (EA), and an energy engineering feasibility study called the Technical Assistance Program (TAP). The individual school or hospital facility applies to the state for an audit. Energy auditors should contact their own states to determine the requirements for their particular area. The individual states will certify the energy auditors and administer the act.

IX. WHO SHOULD PERFORM ENERGY AUDITS

It is important that a person performing an energy audit be well versed in all phases and components of building systems. This requires someone with an engineering background and several years of experience. The individual need not have a professional engineering registration, but should have a great deal of experience and be certified if he or she is not a P.E. An energy auditor should understand wall and roof systems, mechanical systems, electrical systems, and control systems, and be able to communicate effectively with clients and the client's employees.

6. Industrial Energy Management

Paul W. O'Callaghan

*School of Mechanical Engineering
Cranfield Institute of Technology
Cranfield, Bedford, U.K.*

	Page
I. Energy Auditing	59
II. How Is the Energy Utilized?	60
III. How May Energy Most Profitably Be Saved?	61
IV. Asymmetries and Periodic Behavior	63
V. Conclusions	63

I. ENERGY AUDITING

Before any worthwhile attempt can be made to identify cost-effective energy-conserving options within a manufacturing organization, a fully comprehensive audit of energy inputs, throughputs, and outputs must be constructed. Figure 6-1 demonstrates such an audit for the heating period of a typical large factory in the United Kingdom. The unit $\overline{kW}$, adopted throughout, has been obtained by normalizing the data using the total number of heating hours per annum. The fuel energy delivered to the boilers, expressed in terms of gross calorific value, is modified by the aggregated annual mean conversion efficiency of the boiler plant, and discounted by the distribution losses through pipelines and heat transfer equipment. Sundry heat generators within the plant are identified, and the proportion of sundry heat gains (i.e., that fraction that is rejected into the internal environment) is added to the space heating heat delivered. All the electrical or process energy that is used to produce mechanical work or to provide lighting or high temperature process heating (including welding and flame cutting) within the factory ends up as a sundry heat gain, unless local cooling systems are provided. The net distribution losses shown in the audit have also been reduced by the amount of heat energy "lost" by high temperature pipework systems but accepted by the internal environment. The remaining sundry energy is rejected directly to the environment (e.g., from exhaust air extract fans, waste hot water, battery charging, the condensers of air conditioning and refrigerating systems, or the cooling banks of air compressors) and so does not add to the space heat supply.

It must be emphasized that, during the heating season, sundry gain energy performs two useful functions in tandem: it fulfills that purpose for which the energy is primarily released (e.g., process work or heating, machining, lighting, etc.) and also provides bonus heating. Where, however, the predominant space conditioning requirement is for cooling, the sundry heat gains must be offset by additional capacity in the refrigerating and air conditioning system, and are, therefore, a disbenefit. When this is the case, efforts should be made to remove excess heat at the source, using water or air at the prevailing outside environmental temperature. This course of action prevents the transfer of a sundry gain to the internal environment, which must then be cooled using higher energy grade chilled water or refrigerant.

60 2/ENERGY AUDITS AND SURVEYS

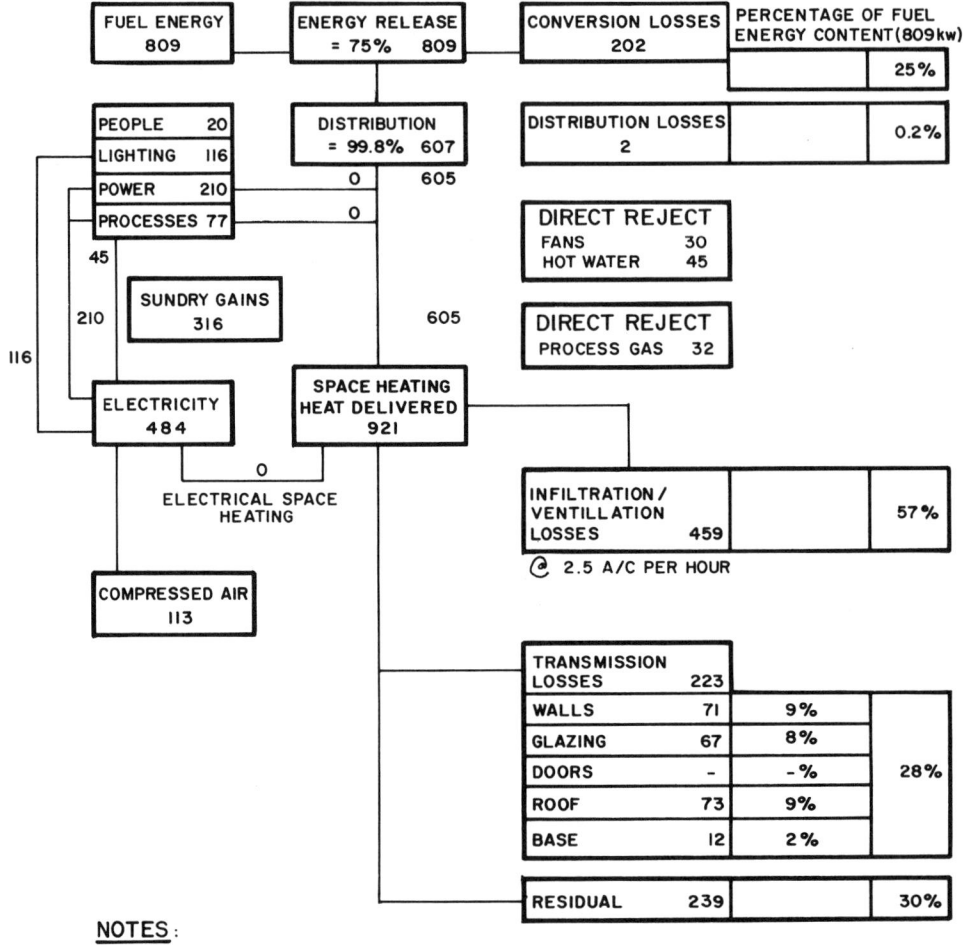

Figure 6-1. Summary of heat flows through building.

After the total space heat delivered is estimated, infiltration/ventilation and fabric transmission losses are then computed from a survey of the building, its usage, and occupancy characteristics.

In the ideal case, the total amount of heat delivered to the conditional space exactly matches the sum of the heat rejections via these two mechanisms.

II. HOW IS THE ENERGY UTILIZED?

Table 6-1 summarizes output data from five energy audits recently constructed for industrial manufacturing systems in the United Kingdom. It is seen that although the overall energy consumptions vary from 1,000 $\overline{kW}$ to 90,000 $\overline{kW}$, the usage patterns are startlingly similar when normalized using the base area or, in particular, the enclosed volume (Table 6-1a).

This similarity is also evident in considering the G-value (the total energy rejection by fabric transmission and ventilation divided by the product obtained by multiplying the enclosed volume times the mean annual temperature difference between inside and outside environmental conditions), but is not indicated by the

Table 6.1. Examples of Energy Audit Summary Data.

(a) *General Data*

FACTORY	OVERALL ENERGY CONSUMPTION	MEAN ENERGY CONSUMPTION, $\overline{kW}$ PER		
		M^2 (BASE AREA)	CAPITA	UNIT VOLUME
A	1,300	0.195	2.72	0.0268
B	11,000	0.209	4.21	0.0278
C	22,000	0.198	11.00	0.0274
D	90,000	0.377	6.6	0.0312
E	3,600	0.180	6.87	0.015

(b) *Energy Indices*

FACTORY	OVERALL FABRIC U-VALUE, $Wm^{-2} K^{-1}$	OVERALL G-VALUE, $Wm^{-3} K^{-1}$	TRANSMISSION/VENTILATION LOSS
A	2.02	1.28	0.49
B	2.07	1.20	0.44
C	1.33	1.44	0.19
D	1.50	1.44	0.15
E	3.24	1.00	0.54

(c) *Input/Output Data*

			% OF TOTAL ENERGY USAGE					
FACTORY	FUEL ENERGY, kW	SUNDRY GAINS, kW	CONVERSION AND DISTRIBUTION LOSSES	INFILTRATION/VENTILATION LOSSES	FABRIC TRANSMISSION LOSSES	COMPRESSED AIR	RESIDUAL FROM BALANCE	ALL OTHER USES
A	800	300	15	35	17	9	18	6
B	6,000	2,400	12	44	18	1	3	22
C	11,000	4,000	11	35	9	12	18	15
D	56,000	14,000	13	44	7	11	16	9
E	3,000	800	20	45	31	0	3.6	0.4

more commonly cited U-value. The ratio of transmission to ventilation losses indicates the relative advantages that may be gained by insulating or recovering "waste" heat from the rejected ventilating air.

By far the greatest proportion of energy is utilized in every case in offsetting ventilation losses. Typical losses are as follows:

Infiltration/ventilation losses	35–45%
Transmission losses	7–31%
Conversion and distribution losses	11–20%
Compressed air generation	1–12%
All other energy uses	0.4–22%

It is notable that all other uses of energy (i.e., those that do not add to the sundry heat gain) account for no more than 22% of the total energy consumed.

III. HOW MAY ENERGY MOST PROFITABLY BE SAVED?

1. Eliminating the Residual. A large discrepancy between the net space heat delivered and the total of the ventilation and fabric transmission losses indicates that the system operation does not correspond to the design specification. A negative residual implies that less than the design minimum air change rate is being achieved. A positive residual can result from:

(a) *The Maintenance of Excess Air Temperatures.* The temperatures are above those stipulated in the design. Temperature measurements should be made for substantiation. Closed-loop local thermostatic controls should be introduced. When heating and cooling services are provided, the "deadspace" bandwidth should be selected to be as wide as is compatible with thermal comfort considerations.

(b) *Overventing.* This problem is often difficult to rectify. Random infiltration must be minimized by weather stripping and attending to doors and windows. A balanced inlet/exhaust fan system should be provided and air flows continuously monitored and controlled to adhere to minimum ventilation requirements during the heating season. Excess capacity can be installed for summer use.

(c) *Internal Vertical Air Stratification.* This common occurrence can result in vertical temperature gradients of up to 20°C per 30 meters of height while the set design level is maintained at floor level. The existence of stratification causes far greater ventilation and transmission losses than the assumption of a uniform internal design temperature implies. The gradients may be destroyed by mixing the air, using fans, blowers, etc.

2. Insulation. The options for insulating should be identified and ranked in order of cost-effectiveness for sequential implementation.

3. Reclaiming Heat from Exhausts. The exhausts include all those items listed as direct reject in the audit, the heat from air conditioning or refrigerating condensers, the heat dissipated from air compression, and the ventilation exhaust. The last is by far the greatest single possibility for heat reclaim. Suitable heat recovery equipment (e.g., recuperators, runaround coils, regenerators, heat pipes, or heat pumps) is available to reclaim up to 70% of the heat in contaminated extract air. The major problem in installing a waste heat recovery system arises from the need to collect and redirect exhaust fluids to common ducts containing the fresh air heaters. Straight payback periods commonly quoted for waste heat recovery systems are often less than three years.

4. Improved Boiler Efficiency. Most boiler systems are correctly designed and well maintained. Thus it is often difficult to obtain substantially increased efficiency without introducing recuperative devices. Care must be taken that individual boilers do not overmodulate as a result of unbalanced system load factors. Close control of boiler groups is necessary, and short-term thermal accumulation might be beneficial. Distribution losses are commonly less than 2% of the total energy throughout. It should be appreciated that any energy savings accruing from the application of conservation techniques in the factory result in greater savings of fuel at the boiler house. The energy accountant should, however, also consider the effects upon boiler efficiency of the resulting reduced demand loads.

5. Compressed Air Services. Compressed air is the most expensive form of energy—3.5 times the cost of electricity, kilowatt for kilowatt at 100 psi (689 kPa)—and should thus be used with the utmost frugality. The operating pressures maintained should be questioned, and the possibility of heat recovery from coolers should be investigated.

6. All Other Uses. Only after the major single energy loss sectors have been dealt with should close attention be rendered to the various dispersed and different individual processes. Energy savings via improved monitoring and controls, insulating, and recovering waste heat are always possible. The optimal grade of energy (i.e., type, temperature, and pressure) should be adopted for each station in a process. Attempts should be made to cascade energy through the various levels of process systems, finally rejecting heat to become a sundry gain, if this is beneficial. Process savings that reduce bonus sundry gains are financially worthwhile only if the cost of primary process energy per useful kilowatt-hour exceeds that for space heating fuel. Descriptions and analyses of energy-conserving modifications to individual items of process equipment tend to be esoteric and thus defy generalization.

IV. ASYMMETRIES AND PERIODIC BEHAVIOR

The auditing technique described considers the system in an imaginary quasi-steady state where inputs and outgoings are balanced over a representative annual heating cycle. Further energy savings are possible by considering the periodicity of energy demand. The energy should be provided where and when it is required. An efficient distribution system and the elimination of maldistributed energy (i.e., such as results from stratification, heating unoccupied areas, employing blanket design parameters to include warehouse areas as well as workshops) is vital. Many control systems exist to regulate local air temperature and air change rates, to modulate groups of boilers to maximize overall conversion efficiency, to peak lop the demand curve via load scheduling, and to optimize pull-down periods. Energy storage devices are being researched to further reduce energy wastage as a result of load factor mismatches.

V. CONCLUSIONS

The importance of considering not merely the boiler plant or the load distribution system, or the building characteristics in isolation, but an entire energy-consuming system as an integrated whole has been stressed. A method of energy-auditing the overall system has been described that leads to the identification of conservation options in order of cost-effectiveness. Survey data from recent energy audits have been included and analyzed to highlight the major areas for further attention.

Part 3

Energy Analysis Techniques and Strategies

7. Estimating Energy Consumption in New and Existing Buildings

L. G. Spielvogel, P.E.

Lawrence G. Spielvogel, Inc.
Wyncote, Pennsylvania

		Page
I.	Introduction	67
II.	New Buildings	67
III.	Predictions of Energy Consumption	68
	A. Degree-Day Method	68
	B. Bin Method	69
	C. Hour-by-Hour Calculations	71
IV.	Existing Buildings	72
	A. Energy Use Evaluation	73
	B. Examples	74
V.	Utility Rate Analysis	83
VI.	Methods of Economic Analysis	84
	A. Simple Payback	85
	B. Life Cycle Costing	85
VII.	Conclusion	87
	References	87
	Reading List	88

I. INTRODUCTION

Engineers all too often get so involved in the minute details of calculations that they forget how accurate the overall answer is going to be. They use technology to its limit without thinking about how practical that technology is.

In looking at energy use in buildings, we find that only a fraction of that energy use is due to the building itself, that is, the heating and cooling and lighting energy within the building. To an equal or greater extent the quantity of energy used in a building depends upon how intensively the building is utilized and how long various areas of the building are used, as well as how well the building is operated and maintained. In each of these categories the engineering precision is poor at best, resulting in predictions of energy use that are only partially based upon the detailed calculations for energy consumption for heating, cooling, and lighting.

II. NEW BUILDINGS

It is necessary to differentiate between the types of calculations done in conjunction with the design of new buildings and the types of calculations done to estimate the energy performance of existing buildings. In the case of existing buildings, the answers are already known, since metered energy consumption data are usually available. Quite often it is difficult to obtain agreement between the energy calculations on an existing building and the metered data because it is necessary to make trial-and-error assumptions on intensity of use and hours of use in order to get the answers to agree.

In the case of new buildings, the engineer is not constrained by having the answer. He can make the answer turn out to be anything he wants because he is free to make assumptions about how the building is going to be utilized. Therefore, in new building design and analysis, the burden falls most heavily on the process of estimating intensity and hours of use, rather than on the precision of the calculations for heating and cooling loads, etc.

Taking heating and cooling energy consumption under consideration and recognizing that the heating and cooling applications are typically not the most significant influence on the energy use of a building, the methods of calculation to be utilized can be addressed.

All designers of buildings are quite familiar with making "load calculations" for the purpose of sizing heating and cooling equipment. An underlying philosophy behind making load calculations is that one must make the necessary assumptions regarding those factors that are going to establish the peak or maximum cooling and heating requirements in any given space in order to ensure adequate capacity. Quite often those maxima never occur, for a wide variety of reasons. Rare is the case in an engineering design when the capacity of the heating and cooling system is inadequate.

For example, almost any designer who was asked for the time of the day and day of the year at which the maximum cooling load would occur in a building would say at 4 P.M. on August 21, since this information is the basis for design calculations in the various handbooks used. In contrast, by talking to operators of buildings, it is found that, in fact, the maximum load on air conditioning systems occurs at just about any other time and day of the year, most often being 7 A.M. on a Monday morning or on a cloudy or rainy day when the wet bulb temperature is very high.

This tells us that the actual heating and cooling loads on a building can and do differ quite significantly from what the designer thought the peak or design load would be. This difference is due to the hours of use and intensity of use of the building. Thus, in order to make reasonable energy calculations it is necessary to take these factors into account to a more significant extent than the details of the peak load calculation. Thus it can be concluded that energy calculations are not the same as load calculations.

It can be further concluded that the degree of precision required in load calculations for the purpose of calculating energy consumption is not very high and need only be on the same order as the precision used in estimating the hours and intensity of use of the building.

III. PREDICTIONS OF ENERGY CONSUMPTION

The following discussion will explore the techniques for calculating energy consumption and the relative accuracy and precision to be obtained by their use. The standard reference on this subject is Chapter 28 of the 1981 ASHRAE handbook [1]. It describes three basic techniques to be utilized for energy calculations, the degree-day or single measure method, the bin or multiple measure method, and hourly calculation techniques.

A. Degree-Day Method

A reading of Chapter 28 of the ASHRAE handbook will indicate quite clearly that the degree-day method is applicable only to single-family homes. In further examining this method, it can be seen that there is an underlying assumption that the energy requirements for heating are directly proportional to the design heating load. Recognize that this assumption applies principally to single-family homes and to a very few other types of buildings, namely, those buildings whose heating requirements are almost totally sensitive to outdoor temperature. The degree-day method does not apply, nor was it ever intended to apply, to nonresidential buildings, since their heating energy requirements are influenced by so many factors other than the weather, the most significant of them being the internal heat gain which offsets the need for heating.

The degree-day method has been "verified" over the years in numerous studies as being representative of the average heating energy consumption when taken over a large number of houses. Therefore, the degree-day method can be used for making relative evaluations of the energy performance of single-family houses with some degree of confidence. Recognize that what is really being compared is not energy but load. Therefore, the degree-day method inherently says that anything that can be done to reduce the load will reduce the energy consumption in the same proportion. While it is known that this is fundamentally untrue, this method is still used extensively.

One example would be that of a house that has a 50,000 Btuh (15 kW) heat loss. Using the degree-day method, one could easily calculate the energy consumption for any type of fuel in any location. If the areas of windows facing south are substantially increased, the design heat loss would also increase substantially, say to 75,000 Btuh (22 kW). The degree-day method would show that the annual energy consumption would increase by 50%. This fails, however, to take into account the passive solar heat gain, which could well result in a substantially lower consumption than the house with the 50,000 Btuh (15 kW) heat loss. Thus one of the many weaknesses of the degree-day method is exposed.

There are numerous other weaknesses to the degree-day method. For example, this method does not allow comparisons or predictions of energy consumption based upon room-by-room thermostatic control compared with single thermostatic control. Tests over the years have indicated that this alone can make a difference of as much as 20 to 30% in the annual heating energy requirements of a house.

The degree-day method does not permit duct losses to be taken into account. Here it matters where the duct work is located, whether it is inside of the insulation envelope of the house or outside. Various tests have shown that duct losses can amount to 20 to 35% of the annual heating energy requirements, independent of the house, its heating system, and the form of energy used.

One unpublished test took place at a suburban housing development in Pennsylvania. Halfway through the construction of the houses, the original heating contractor went bankrupt. The builder then hired a second heating contractor to do the second half of the houses. In the first half of the houses, the duct work serving the second floor was run between the studs in the exterior wall. In the second half of the houses, the ducts were run vertically in the inside walls. The difference in heating energy consumption between the two groups of houses was 8%. Any of the energy calculation methods available would have predicted equal energy consumption in those two groups of houses. Yet, when the reasons why the energy consumption differed so much are examined, it becomes quite evident that the houses that had vertical duct work in the outside wall did not have any insulation in those stud spaces, and the hottest part of the heating system was operating against the coldest part of the house.

The degree-day method does not permit night setback to be taken into account, nor does it permit adequate evaluation of infiltration, since the amount of infiltration depends upon wind direction and velocity which are not even considered in the computation of degree-days. In Europe there is a move toward looking at a concept called the wind-temp factor as an alternative to the degree day method. The wind-temp factor is a statistical correlation of both dry bulb temperature and wind direction and velocity for the purpose of establishing a proportionality between design heat loss and energy consumption.

B. Bin Method

Recognizing the limitations on the degree-day method even for single-family homes and recognizing a need for energy calculations on a manual basis for other types of buildings, the bin or multiple measure method has evolved over the years. The bin method provides a technique for estimating the energy consumption of a building to the degree of engineering precision deemed necessary by the user. It makes use of weather data available from various sources in bins, typically of 5°F (3°C) or 10°F (6°C). Each bin consists of the number

Table 7-1. Example of Bin Weather Data, Newark, New Jersey. Mean Frequency of Occurrence of Dry Bulb Temperature with Mean Coincident Wet Bulb Temperature for Each Dry Bulb Temperature Range.

TEMPERATURE RANGE		NOVEMBER					DECEMBER					JANUARY					FEBRUARY				
		OBSN HOUR GROUP			TOTAL OBSN	MCWB °F	OBSN HOUR GROUP			TOTAL OBSN	MCWB °F	OBSN HOUR GROUP			TOTAL OBSN	MCWB °F	OBSN HOUR GROUP			TOTAL OBSN	MCWB °F
°C	°F	01 to 08	09 to 16	17 to 24			01 to 08	09 to 16	17 to 24			01 to 08	09 to 16	17 to 24			01 to 08	09 to 16	17 to 24		
41/43	105/109																				
38/40	100/104																				
35/37	95/99																				
32/34	90/94																				
29/32	85/89	0		0		67															
27/29	80/84	1			1	66															
24/26	75/79	1	0		1	66															
21/23	70/74	4	0		4	60						0		0	64		0				
18/21	65/69	1	12	5	18	59		1	0	1	54	0	0	0	0	60	1	0			
16/18	60/64	6	22	11	39	55	0	3	1	4	55	1	1	1	3	57	2	0			
13/15	55/59	18	40	31	89	52	5	10	6	21	53	2	3	1	6	53	1	5	2		
10/12	50/54	32	47	45	124	47	6	15	8	29	48	3	10	5	18	48	2	11	7		
7/9	45/49	38	44	44	126	42	12	25	21	58	43	8	19	14	41	43	6	25	14		
4/7	40/44	46	38	46	130	37	25	47	39	111	38	18	42	33	93	38	20	42	37		
2/4	35/39	51	21	39	111	33	49	56	55	160	34	43	57	55	155	33	49	55	58		
−1/1	30/34	36	8	14	58	29	53	40	50	143	29	53	50	58	161	29	50	38	52		
−4/−2	25/29	8	1	4	13	24	39	28	32	99	24	48	31	39	118	24	42	20	24		
−7/−4	20/24	3	1	1	5	19	27	15	22	64	19	32	20	23	75	19	24	13	16		
−9/−7	15/19	1	0		1	15	21	7	10	38	15	21	9	14	44	15	14	7	8		
−12/−10	10/14						8	2	3	13	10	12	2	5	19	11	10	3	4		
−15/−13	5/9						2	0	1	3	6	5	1	1	7	6	4	1	2		
−18/−16	0/4						0		0	0	1	1	0		1	2	1	0	0		
−21/−18	−5/−1																0	0			

of hours of weather experience in a month or year in a given temperature range. For example, in one city there may be an average of 223 hours per year in the 5°F (3°C) bin from 40°F to 44°F (4°C to 7°C).

These bins can be broken down further into months of the year and even into hours of the day if desired. Bin weather data are also available for wet bulb temperature and cloud cover if that much detail is desired. Condensed bin weather data are included in Chapter 28 of the 1981 ASHRAE Handbook. The most detailed bin weather data are published in U.S. Air Force Manual 88-29, *Engineering Weather Data* [2]. A sample is shown in Table 7-1.

The basic premise behind the bin method is to determine by engineering calculations the amount of energy that a building will require at any given outdoor temperature, typically the dry bulb temperature. Then the energy consumption can be determined by simply multiplying the energy requirement at any given temperature by the number of hours at that temperature, and summing. This can be done in as little or as much detail as necessary, on an annual or monthly basis.

The key element then becomes how the energy consumption at any given outdoor temperature is determined. This is where the user of this technique must apply judgment and determine how much precision is required. Typically the heating and cooling loads of a building are plotted against outdoor temperature, and then the energy requirements of the systems and equipment that meet those loads are determined based upon the part load performance data available for the equipment and the unique characteristics of the control sys-

		MARCH					APRIL					ANNUAL TOTAL					
			OBSN HOUR GROUP					OBSN HOUR GROUP					OBSN HOUR GROUP				M C W B	
TOTAL OBSN	MCWB °F	01 to 08	09 to 16	17 to 24	TOTAL OBSN	MCWB °F	01 to 08	09 to 16	17 to 24	TOTAL OBSN	MCWB °F	01 to 08	09 to 16	17 to 24	TOTAL OBSN	°F	°C
													0		0	75	24
												0	0		0	76	24
													15	4	19	74	23
						68	0		0	0		0	60	17	77	74	23
						66	2	0		2		2	143	51	196	71	22
						63		6	2	8		17	237	124	378	69	21
		1	0	1		58	7	3		10	60	102	279	227	608	67	19
0	58	2	1		3	55	1	17	7	25	58	261	256	295	812	65	18
1	57	0	3	1	4	53	4	18	13	35	56	304	233	272	809	60	16
2	53	0	10	4	14	51	12	32	22	66	52	286	226	247	759	56	13
8	50	3	16	10	29	48	21	44	33	98	49	261	208	236	705	51	11
20	46	8	30	21	59	44	43	45	56	144	46	247	205	237	689	47	8
45	41	17	47	34	98	40	60	44	53	157	42	239	225	223	687	42	6
99	37	43	54	60	157	37	56	20	38	114	38	253	247	266	766	38	3
162	33	77	47	63	187	33	31	4	10	45	33	313	241	284	838	33	1
140	29	53	25	33	111	28	10	2	2	14	29	256	163	209	628	29	−2
86	24	27	9	12	48	24	1	0	0	1	23	165	89	111	365	24	−4
53	19	16	3	6	25	19	0	0	0	0	19	102	52	68	222	19	−7
29	14	3	1	1	5	14						60	24	33	117	15	−9
17	10	2	0	2	9							32	7	12	51	10	−12
7	6											11	2	4	17	6	−14
1	1											2	0	0	2	1	−17
0	−2											0	0	0	0	−2	−19

tem, as well as the estimated hours of use of the building. It is in this area that the detail and sophistication of these calculations can grow exponentially, depending upon the complexity of the systems and their use.

In order to utilize the bin method properly, to satisfy the judgment of the user, it becomes a substantially complex and time-consuming process. However, it does provide a manual calculation technique that is capable of handling the most unique situations in a building, provided that the user is willing to put forth the necessary engineering effort to do so.

One of the big drawbacks of the bin method is that once the energy consumption has been established for a particular set of circumstances, it becomes a repetitive technique when alternatives are to be evaluated. Since energy calculations are used most often as the basis for comparisons of many alternatives, the repetitive calculations can be extremely time-consuming and provide an opportunity for miscalculation and error.

C. Hour-by-Hour Calculations

The tedious nature of bin calculations quickly leads the user toward more time-efficient methods of calculation—computer programs. A host of energy analysis computer programs are on the market today. These programs have been developed by a wide variety of organizations for a wide variety of purposes. Table 7-2 lists some of the more popular energy analysis computer programs now available.

Most of these programs make calculations each hour of the year, since weather data are typically available from the U.S. government

Table 7-2. Energy Programs.

Ross Meriwether Associates, Inc.
ECUBE—American Gas Association
AXCESS—Electric Energy Association
TRACE—Trane Company
HACE—WTA Computer Services
MEDSI—McClure Associates
NBSLD—National Bureau of Standards
MACE—McDonnell Douglas
BEEP—American Electric Power Corp.
Westinghouse Corporation
POST OFFICE—U.S. Postal Service
NECAP—NASA
SCOUT—GARD/GATX
FLEET—State of Florida
CAL-CON—State of California
DOE-2—Department of Energy
BLAST—Corps of Engineers
ESP1—APEC
SEE—Singer Company

[3] on an hour-by-hour basis. Each and every program differs substantially from the others in its ability to handle the many and varied situations found in buildings today. One can easily conclude that no two programs are alike, or will produce identical results, or even results that are within a few percent of each other.

Evaluating computer programs is a time-consuming process, but is necessary in order to establish that the computer program in question is capable of analyzing and evaluating the particular building being studied as well as the various alternatives being considered for that building. For relatively simple buildings with relatively simple systems, almost any of these programs provides adequate results. With more complex buildings and more complex systems, the number of programs that can adequately handle these situations becomes very small very quickly. Thus it is incumbent upon the user of these programs to determine whether or not each program is capable of evaluating the particular questions at hand.

In evaluating and comparing programs the following conclusions can be drawn:

1. The results to be obtained by using several computer programs on the same building will range from very good agreement to no agreement at all. The degree of agreement is dependent upon the interpretations made by the computer program user and by the ability of the computer programs to handle the building in question.
2. Several people using several programs on the same building will probably not get good agreement on the results of an energy analysis.
3. The same person using several programs on the same building may or may not get good agreement, depending upon the complexity of the building and its systems and the ability of the computer programs to handle the specific conditions in that building.
4. Several people using the same program on the same building will probably not get good agreement on the results of an energy analysis.

Various studies and research have shown that the fundamental engineering techniques used in these computer programs are as accurate as necessary in order to predict the energy consumption of a building. The problem comes in making the assumptions on intensity of use and hours of use of a particular building. Only when this information is known in at least as much detail as the heating and cooling loads will the results of a computer analysis agree with measured energy consumption.

It is very costly and impractical to know what these assumptions are in order to estimate energy use accurately. Therefore, in utilizing computer programs the agreement of results is only as good as the judgments made by the user on how the building is going to be utilized. It is becoming generally accepted that the results of a computer program can be considered "good" if they are within 15 to 20% of the actual measured energy consumption in a building.

IV. EXISTING BUILDINGS

The romance with computer programs is increasingly extending to the energy analysis of existing buildings. People who have used computer programs for evaluating the energy performance of existing buildings have found great difficulty in obtaining agreement between the results of the computer program and the utility billing records. While a part of this problem may be due to their not using weather data for the particular year in question, that aspect depends upon how weather-sensitive the building and its energy-consuming systems are. Where the building has a lot of glass and/or uses a lot of outdoor air, then the weather data that are used in the computer analysis can have a significant impact on the results. Where these factors are not so important, the weather data utilized will make much less difference and indeed can be even less significant than the assumptions made about how intensively and how long the building and its systems are in operation.

The most significant advantage to analyzing energy use in existing buildings is that the answers are already available. Since it is difficult to get good agreement between any type of calculation, manual or computer, and the measured energy consumption, the question then becomes which type of energy calculation is most appropriate for existing buildings? Unfortunately, the answer to that question is not very simple.

Generally, it is found that computer analysis provides the most convenient technique when one is evaluating very complex systems and buildings and when a large number of alternatives are being considered, especially those that have a substantial degree of energy interaction. On the other hand, for relatively simple buildings with relatively few alternatives under consideration, the same types of manual calculations used for new-building analyses most often are sufficiently accurate and adequate.

A. Energy Use Evaluation

Since at least some energy consumption data are available for most existing buildings, a great deal of information can be obtained from an analysis of whatever data exist. The quantity and type of data available depend upon the type of metering that is installed in the building. Therefore, before doing any analysis of the building, one should obtain as much information as possible from the metered data for that building. Where no such records are available, it may be beneficial to install some temporary recording meters in order to obtain that information.

Plotting and analyzing these data permits rather precise evaluation of the quantities of energy used for various purposes in a building. This provides a convenient method of knowing what the actual energy consumption is for each major purpose, thus obviating the need for any energy calculations to establish those quantities. Since the measurements are available, any manual or computer analysis could only hope to come close to the facts. Also, the available factual information frequently indicates that the energy consumption of a building is influenced by many more things than are traditionally taken into consideration in making energy calculations.

A wealth of information can be obtained from energy consumption data. The data are frequently available from fuel and energy suppliers for many different periods of time—annually, monthly, weekly, daily, hourly, and for even shorter periods. Information is usually available for each type of fuel or energy used in a building, and on occasion some buildings are found with numerous meters that permit precise determination of how much energy is used for specific purposes. Even in the residential category, many electric companies have special rates for water heating and therefore have separate meters. In order to determine how much energy is being used for water heating, it is a simple matter to look at data from the meter. Having this type of information eliminates much of the guessing or estimating of energy use.

Many things can be done with an energy use evaluation before one even sets foot in the building. By gathering together utility company billing data and the supporting information, one can construct various types of curves, plots, and breakdowns. In examining measured energy consumption data, several categories of that data are worth looking at.

The first category is the "base load," which is defined as the non-weather-related uses of energy, typically for lighting and equipment. The difference between the total consumption of energy and the base load is the energy used for heating and/or cooling. This permits a rather precise determination of how much energy is being used for heating and cooling, and how much energy is used for all other purposes.

Another category is "occupancy-related" energy consumption, which is defined as the quantity of energy that is consumed in a building during those hours when it is occupied for its ordinary purpose or function. In the case of a commercial office building, it would be from 8 A.M. until 6 P.M., five or five and one-half days per week. Similarly, the "non-occupancy-related" energy consumption is the energy consumed during all other times. The reason for breaking down consumption like this is that in many buildings more energy is used during the non-occupancy period of time than when the building is occupied. Office buildings and schools frequently fall in this category.

Another thing to look at is the "no use" energy use, which is defined as the minimum hourly rate of use in a year or month. Designers have traditionally been concerned with installing sufficient equipment capacity for the peak load. Utility companies most often bill for peak demand for electricity, steam, and even natural gas. Obviously, the peak rate of energy use influences how much the energy costs, but the minimum rate of energy use can be and usually is a very significant influence on energy consumption. If you wish to conserve dollars, look at the peak rate of use or the demand. If you wish to conserve energy, take very careful note of what happens during the hour of the year in which the minimum use occurs.

At a community college, it was determined that 20% of the total annual electric consumption was used to keep the transformers warm. The college got a bargain rate from the electric company because it bought at high voltage. The peak annual demand was 1,000 kW, yet there were over 10,000 kW of transformers installed. There was one bank of 5,000 kW from 33,000 to 13,200 volts, distributed to four locations on campus, each of which had a 1,000 kW transformer bank from 13,200 to 480 volts, plus a number of smaller dry transformers from 480 to 120 volts throughout the campus. There was over ten times as much installed transformer capacity as the peak annual demand. While some of this capacity is necessary with multiple levels of transformation, oversizing of transformers is also necessitated by the requirements of the National Electric Code, by the design conservatism of the designer, and by the potential for future expansion or future load growth. The fact still remains that the energy penalty for transformation can be substantial. By assuming 1% no load transformer losses, it is possible to show that a substantial portion of the total annual electricity consumption being used to keep transformers warm. This is just one element of the no use energy in a building.

Most utilities have much more detailed information on their customers' energy consumption than appears on the bills. Some utilities use meters with magnetic tapes or circular demand charts that provide 32 days of hour-by-hour electric consumption detail. Hourly energy consumption data come in a wide variety of forms. In most parts of the country, in most nonresidential buildings, a great deal of this type of information is available from the utility either free of charge or for a few dollars a month. Use of this type of information enables examination of the energy consumption characteristics and patterns of a building. Most important, it frequently provides the ability to determine what is worth looking at in a building and what is not worth looking at.

So the purpose of an energy use evaluation is not only to determine how much energy is being consumed for each purpose, but when, and also how much energy is being consumed that probably shouldn't be.

B. Examples

Since the subject of energy use evaluation in existing buildings is a relatively new one, a formal methodology is yet to be developed for doing these analyses and for evaluating the results. Some examples will show some of the benefits of this type of analysis.

1. Apartment House. The first example is a 300-unit high rise apartment house, almost 20 years old, in Philadelphia. The construction is typical of that of 20 years ago, a brick building with no insulation and aluminum sash single-glass windows. It has oil-fired boilers providing space heating, humidification, domestic hot water, and some reheat. Cooking and incineration are accomplished by natural gas. Electricity is used for air conditioning and general light and power. The air conditioning system is a two-pipe induction system with a single primary air unit located in the basement.

Figure 7-1 shows the month-by-month fuel oil consumption for a two-year period. These data are not very precise, since they are based upon fuel truck delivery records and not upon metered or measured fuel consumption to the boilers. This is the best information available. It can reasonably be presumed that there is at least some relationship between the quantity of fuel delivered and the quantity of fuel used. Also plotted in Figure 7-1 are monthly degree-days for the two years in question. Since this building is predominantly residential, the degree-days will have some bearing on heating energy consumption.

It should be noted in Figure 7-1 that the general character of the fuel oil consumption data matches that of the weather data except in the summer months when fuel oil is used for domestic hot water and reheat. By subtracting the non-weather-related uses of fuel oil, it is possible to obtain the net consumption of fuel used for heating. It is possible to obtain not only much closer agreement on fuel consumption per degree-day, but, most important, it is now known with great precision exactly how much fuel is used for the purpose of heating this building. Thus, any alternatives that are being considered for saving fuel can be compared to the actual fuel used, rather than to some hypothetical calculation.

Figure 7-2 shows the month-by-month electric demand and consumption for a two-year period. One of the first things to be noticed is the consistency from year to year, which in this case is quite close. The consistency of energy consumption from year to year will determine the confidence that one can have in projections of energy savings into future years. In this particular example, any projection of energy sav-

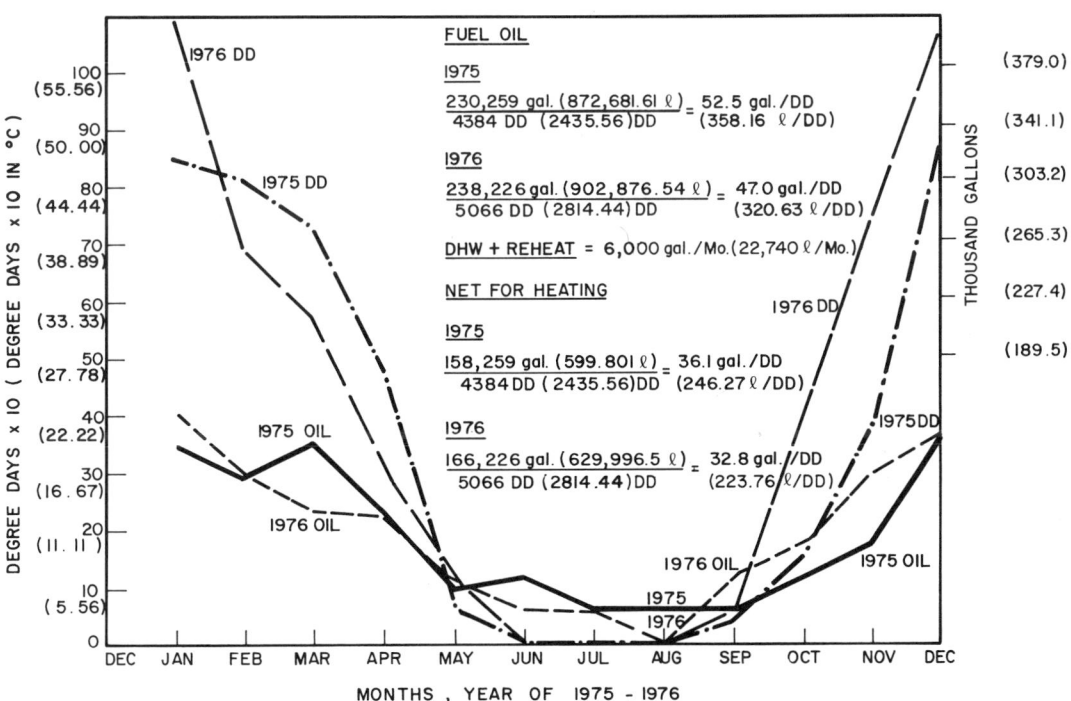

Figure 7-1. Month-by-month fuel oil consumption for a two-year period.

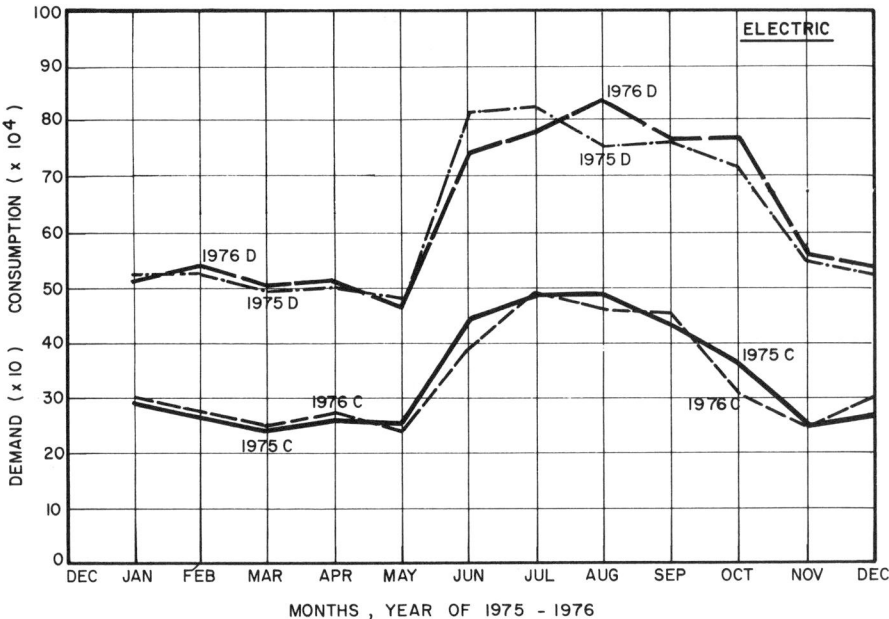

Figure 7-2. Month-by-month electric demand and consumption for a two-year period.

ings can be done with great confidence, since the year-to-year energy consumption data are almost identical. Were the year-to-year data not identical, some discount of future savings would have to be made based upon the degree of variation in year-to-year data. Thus, for buildings that are not used consistently, one must reasonably assume that a building that is a high-energy consumer one year will not necessarily be a high-energy consumer in successive years, necessitating an evaluation of energy saving based upon some of the lower data rather than the higher.

Referring again to Figure 7-2, it can be seen that there is a summer hump in both the demand and the consumption of electricity. This is obviously due to air conditioning, which in this building is based not only upon the weather but also on the calendar. The reason is that the leases in this building provide for air conditioning from May 1 to October 1 of each year. More specifically, that means that the air conditioning system is turned on one hour after the electric meter is read in May and is turned off one hour before the meter is read in October. Thus, it is reasonable to presume that the demand and the energy consumption in the summer months, over and above what is used in the winter months, are solely attributable to air conditioning. Based upon Figure 7-2, one can very easily determine the quantity of electricity used for air conditioning, so that it is not necessary to do energy calculations to establish what it might be.

Moving on to another level of energy use analysis, Figure 7-3 shows an hour-by-hour measurement of the electricity required on a typical winter day. It can be seen that the electricity consumption is highest from 6 P.M. until 9 P.M., which is typically what is expected in residential buildings, since this is the time during which the building and its energy-consuming systems are most intensively used. It is also interesting to note that the midday energy consumption is not significantly lower than the highest energy consumption. The reason for this is that a substantial number of occupants in this building are retired, so that they spend a great deal of time at home and in so doing continue to use energy at a fairly high rate.

It is also interesting to note that on a winter day the energy consumption all night long is about half the peak rate of use. There are three reasons for this. First, the induction unit fan runs 24 hours a day. Second, the parking lot lights are on all night long. Third, the lights in the lobbies and corridors are on 24 hours per day. Knowing this then indicates that a substantial fraction of the overall electricity consumption, at least in the winter, is due to things that are hardly related to the number of people or the number of apartments rented. Thus, when trying to save energy in a building such as this, one should concentrate his efforts on those components that represent the largest fraction of energy consumption, namely, those things that are outside of the apartments and those that are typically in operation 24 hours a day, or at least for long hours each day, the no use energy use.

Figure 7-4 shows the hourly electricity consumption for a summer day, and, in particular, the summer day in which the peak annual electric demand occurred. Also plotted in Figure 7-4 are the dry bulb and wet bulb temperatures obtained from local airport weather data. Note that the hour-by-hour consumption of electricity is similar in character to that of Figure 7-3, with the exception that this is much higher, obviously because of the air conditioning.

Also note that the peak demand occurred from 7 until 7:30 P.M., not at the 4 P.M. that calculations would have indicated. Also note the temperatures at the time the peak demand occurred. While the design conditions in Philadelphia are 95°F (35°C) dry bulb and 78°F (26°C) wet bulb, the annual peak electric demand occurred when the dry bulb temperature was 78°F (26°C), and the wet bulb temperature was 74°F (23°C), hardly a "design day." Note that the wet bulb temperature was much closer to "design." Since the induction system in this building utilizes 100% outdoor air, there is a hint that quite possibly the wet bulb temperature has a significant influence on the cooling energy requirements for this building.

Figure 7-5 shows the summer–winter difference in electricity consumption on an hour-by-hour basis that is obtained by subtracting the hourly electricity requirements on a typical winter day from the hourly electric requirements on July 29, 1976. It is reasonable to presume that this figure represents the hour-by-hour energy requirements for air conditioning. Note that the electricity consumption for air

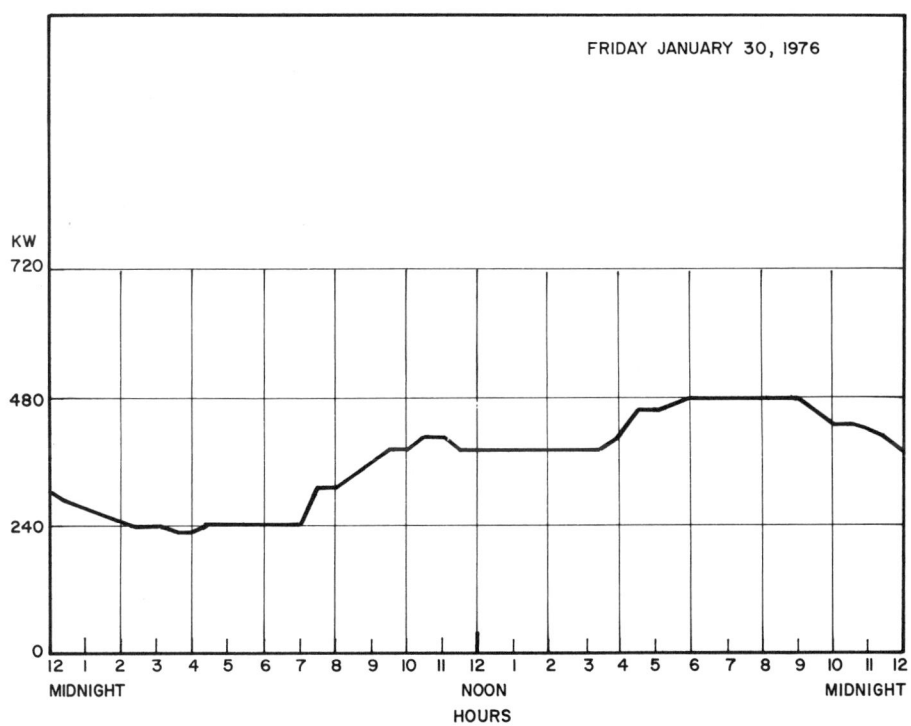

Figure 7-3. Hour-by-hour measurement of the electricity required on a typical winter day.

78 3/ENERGY ANALYSIS TECHNIQUES AND STRATEGIES

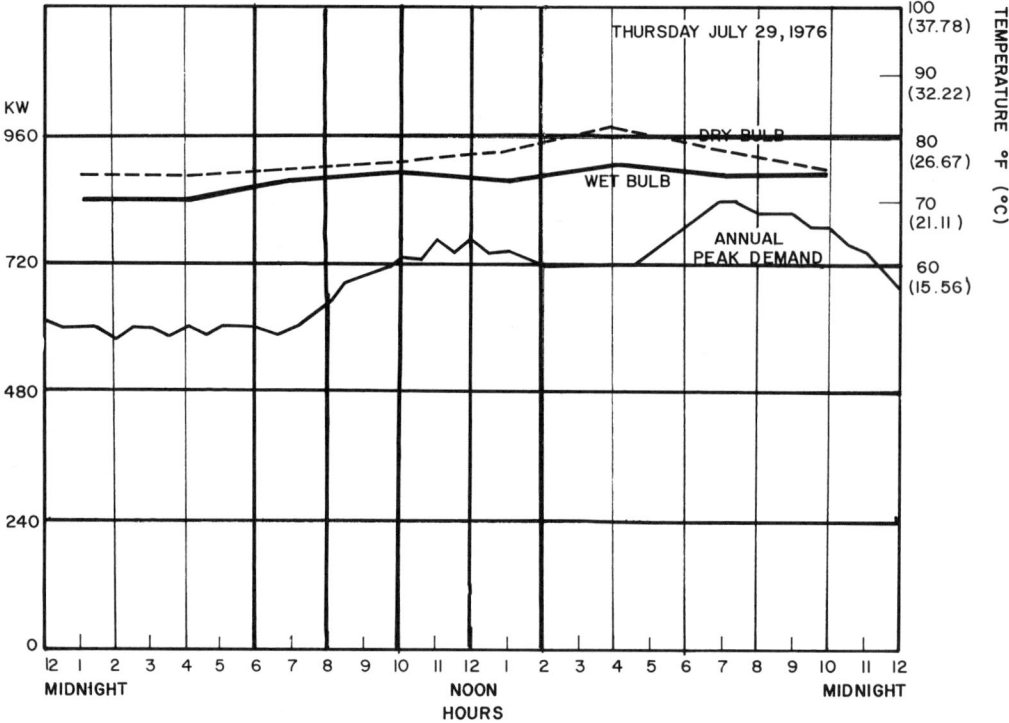

Figure 7-4. Hourly electricity consumption for a summer day.

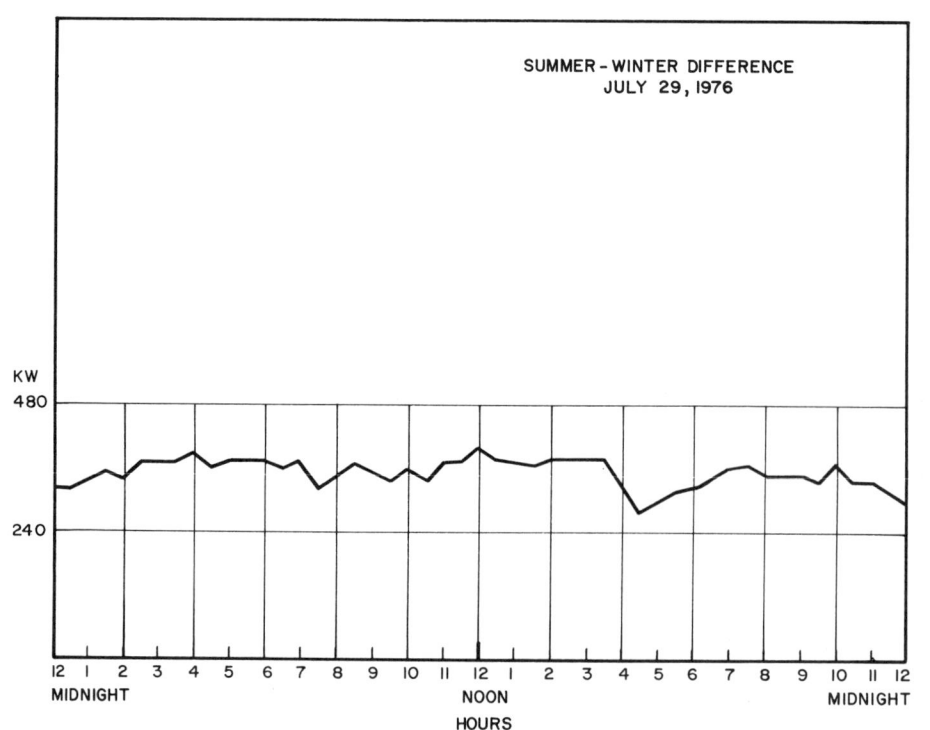

Figure 7-5. Summer–winter difference in electricity consumption on an hour-by-hour basis.

conditioning is almost constant around the clock. This is contrary to the theory that says that the air conditioning requirements peak in mid- to late afternoon and are minimum in the middle of the night. Also note that the minimum requirement for air conditioning occurred at 4:30 P.M. on that day.

What these data indicate is that the energy consumption for air conditioning is almost constant around the clock. Referring to Figure 7-4, it can be seen that the wet bulb temperature is relatively constant around the clock, and that is typically more so the case for wet bulb temperature than for dry bulb temperature. Since the induction system utilizes 100% outdoor air, its energy consumption is thus very heavily dependent upon the wet bulb temperature of that air.

The interesting part of this analysis is what it doesn't say but implies, namely, that the traditional techniques that are applied for conserving energy will have virtually no effect on the energy consumption of this building for air conditioning. Even though the building is uninsulated and has single glass, all of the insulation and double and triple glazing that could be added to this building would not change the cooling energy consumption to any great extent. Even with insulation added and double or triple glazing, the heating energy consumption would not change very much because a significant portion of the heating energy consumption is for the heating of the primary air in the induction system.

This example shows one of the many, many exceptions to the rule. It leads toward the conclusion that there is no rule. If anything, the world of buildings is made up of exceptions, rather than rules, thus requiring individual analyses and evaluations, rather than the generalized or simplified evaluations that are the rage today.

2. Branch Bank. Figures 7-6 and 7-7 show data for a small branch bank in Philadelphia. We determine the energy consumption for cooling by subtracting the base load from the total; in this case it is 18,000 kWh per year. At 4¢/kWh, the cost is $720 per year. In this particular building there is no economy cycle, and the standard approach to conserve energy is to install an economy cycle. This might save as much as 25% of the cooling energy, which in this building would amount to $180 per year,

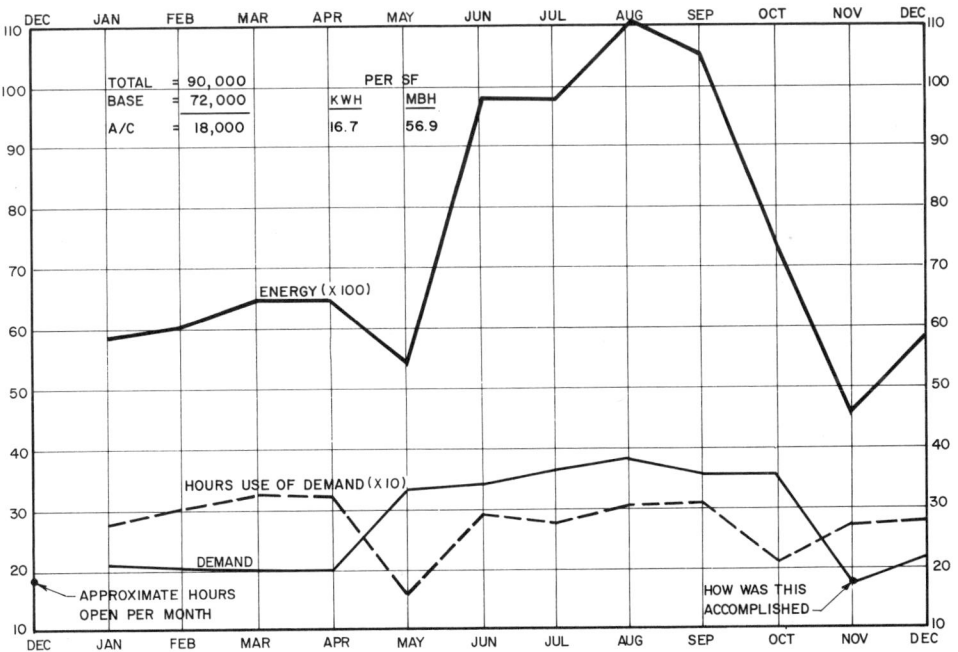

Figure 7-6. Electricity consumption by a small bank in Philadelphia.

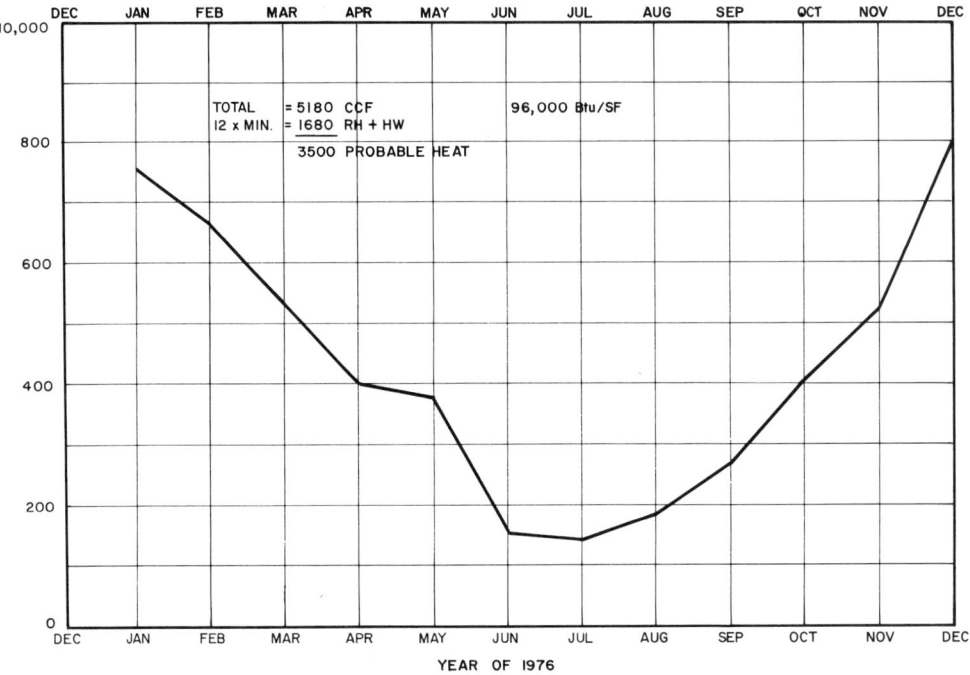

Figure 7-7. Gas consumption by a small bank in Philadelphia.

except for the fact that this building has a basement mechanical room.

Just outside and above the mechanical room is the drive-in teller window. It would probably cost $10,000 to install an economy cycle in order to save, at most, $180 a year. By examining the energy consumption data, it is possible to determine how much energy is consumed for each purpose and how much that energy costs, before setting foot in the building. One can simply look at a lot of possible opportunities and reject them out of hand as not being worth spending even the time to calculate their energy savings, because of the high capital cost or the impracticality of implementing them.

Note that the summer hump is broader, since this is an owner-occupied building with no concern about a lease. When it gets hot, the owner turns on the air conditioning. Plotted in Figure 7-6, with the dashed line, is the month-by-month hours use of demand, which was obtained by simply dividing the energy consumption by the demand. One of the biggest advantages to being in the banking business is bankers' hours. Banks are open for business less than 200 hours per month. But the dashed line shows that in most months of the year, the bank is using 300 hours' worth of electricity. This raises the question, "If the bank is only open for business 200 hours per month, why is it using 300 hours' worth of electricity?" The answer comes from the bank's security department: "If there is to be anything left in the vault the next morning, the lights had better be kept on all night."

Note also that in the month of November the bank's demand was lower than it was in any other month of the year. What were the bankers able to do in November that they were not able to do in any other month? This may be worth looking at.

In order to determine the relative energy health of this building, one can look at the energy budget in terms of either kilowatt-hours per square foot or Btu's per square foot for electricity. This building uses 16.7 kWh/sq ft (180 kWh/m^2) or 57,000 Btu/sq ft per year (650 MJ/(m^2·yr)). Looking at figures for comparable branch bank buildings, we find that this particular branch bank is about in the middle. It's not a real dog, but it's not the most efficient bank either. This means that there may be some reduction possible, but it will not turn out to be a gold mine.

Figure 7-7 shows that this building has an energy budget for gas of 96,000 Btu/sq ft per year (1,100 MJ/(m²·yr)). For branch banks, that's poor. That indicates there should be great potential for savings.

A gas-fired boiler provides space heating, domestic water heating, and some reheat. Doing an analysis, it turns out that the cost of heating this bank is about $1,400 a year. This building has 14-ft (4.3 m) high ceilings and a floor-to-ceiling, wall-to-wall single glass facade. Use of double glass could probably cut the heating bill by 25%, which would mean saving $350 a year. However, the cost of replacing the single glass with double would be tens of thousands of dollars in order to save $350 a year. So this technique provides the ability to know what not to look at, and what not to waste time on. That alone can be very important.

3. Hotels. Figure 7-8 shows some data for a hotel in Philadelphia. The hour-by-hour electricity consumption in the top curve shows what happened on the day of the year when the maximum demand occurred: Monday August 29, 1977. The maximum temperature on that day was 90°F (32°C), but note when that demand occurred: between 9 and 10 o'clock at night, which is characteristic of some hotels, but not all. Why did it occur at that hour? Most business people start their traveling on Monday. They are typically in the room and using energy at 8, 9, and 10 o'clock at night. At the same time, most of the function rooms and public facilities are in use, with various groups having dinner meetings and educational sessions. It is the combination of people in the rooms and people in the public spaces that causes the peak to occur at that time. That kind of knowledge gives a valuable insight into how the building is used and when.

Also plotted is another typical Monday, which peaked at about 9 o'clock at night. Note

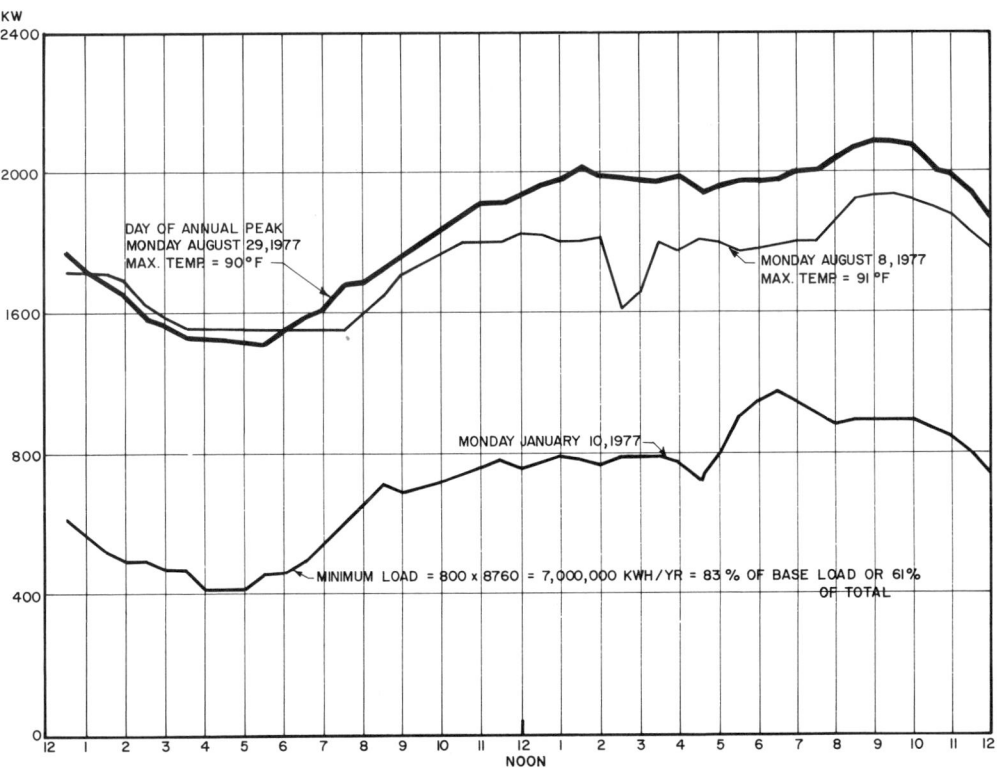

Figure 7-8. Electricity consumption for a large hotel on days of maximum, typical, and minimum demand.

that in the late afternoon there was a dip in energy consumption. The reason was determined by examining the engineers' logs—one of the chillers went off on high head. Note also that this day had a higher ambient temperature. The electric demand wasn't as high as on August 29 because the building wasn't being used as much as on August 29.

The lowest curve shows the no use energy use. This is the day of the year that had the hour in which the electric consumption was lowest, or lower than any other hour in the year, at about 3 or 4 o'clock in the morning. In this building over two-thirds of the energy consumed was attributed to what went on during those hours. This includes the common and public space energy use, heating and ventilating of the public toilet rooms, corridors, lobbies, stairwells, and meeting rooms, which in most hotels today operate 24 hours a day, independent of occupancy and use.

In the month-by-month demands and consumption shown in Figure 7-9, the summer hump is not nearly as pronounced for the hotel as it was in the apartment house or bank. In a hotel there are day-to-day leases, not year-to-year leases, so that there is much more liberal use of air conditioning in hotels. There is a high degree of consistency from year to year, which is expected in a hotel whose level of occupancy is consistently high.

By subtracting the base load from the total electricity consumption, it is determined that 3,000,000 kWh per year are used for air conditioning this hotel. There is no guesswork or calculations to be made—that's what it is, precisely.

Table 7-3 shows some data from another

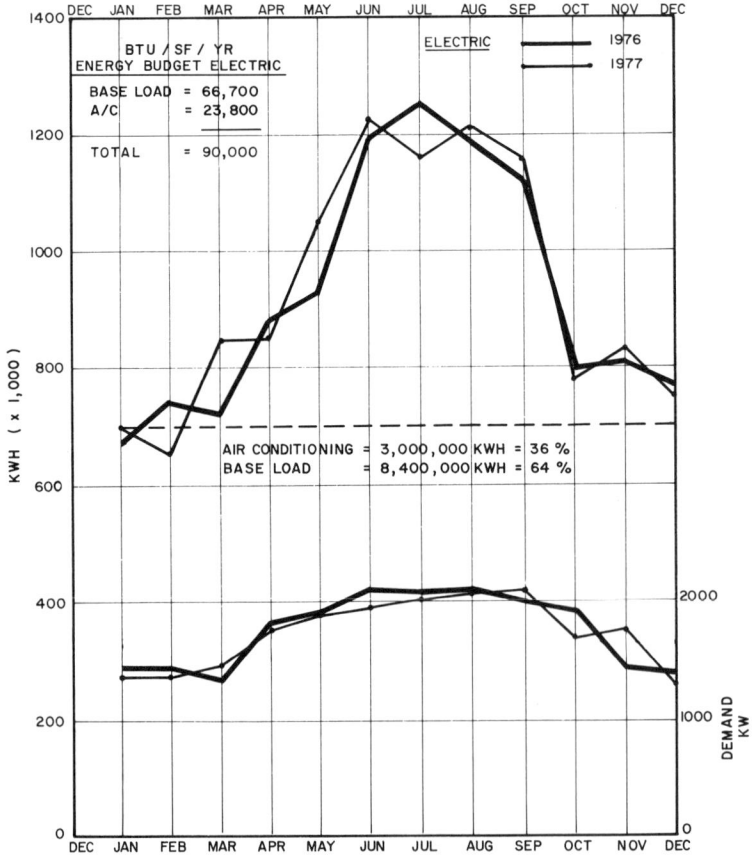

Figure 7-9. Month-by-month electricity consumption by a large hotel.

ESTIMATING ENERGY CONSUMPTION IN NEW AND EXISTING BUILDINGS 83

Table 7-3. Hotel Energy Budget Breakdown by Area and Function.

HOTEL ROOM FLOORS—69% OF AREA		
FUNCTION	MBtu/sf/yr	MJ/(m²·yr)
Cooling	7.6	87
Domestic hot water	11.9	136
Heating		
Corridor air	32.5	371
Rooms	11.5	131
Light and power	29.6	337
Totals	93.1	1,062

KITCHEN AND LAUNDRY—6% OF AREA		
FUNCTION	MBtu/sf/yr	MJ/(m²·yr)
Cooling	51	581
Process	1,093	12,460
Heating	228	2,599
Equipment	626	7,136
Totals	1,998	22,776

PUBLIC AND PRIVATE AREAS—25% OF AREA		
FUNCTION	MBtu/sf/yr	MJ/(m²·yr)
Cooling	29.5	336
Heating		
Lobby level	64.2	732
Service floors	12.8	146
Fans	66.7	760
Garage ventilation	5.1	58
Elevators	15.4	176
Storage	25.7	293
Misc. unaccounted	7.7	88
Totals	227.1	2,589

BY AREA APPROXIMATE—OVERALL				
	MBtu/sf/yr	MJ/(m²·yr)	% ENERGY	% $
Rooms	64	730	26	32
Kitchen and laundry	120	1,268	49	37
Public and private areas	60	684	25	31
Totals	244	2,682	100	100

hotel in Chicago which was equipped with 30 energy meters, including Btu meters for chilled water and hot water, and gas and electric meters. This is a precise breakdown of how much energy was used for each purpose. Note that the kitchen and the laundry, which occupy less than 6% of the floor area, used 49% of all the energy in the building. The hotel rooms, which occupied over two-thirds of the floor area, use less than a quarter of the energy. Yet engineers tend to shy away from the kitchen because that's the chef's domain, and put their effort into the walls and the windows and the efficiency of the heating and air conditioning. That's not where the action is in a hotel. Breaking energy consumption down like this tells where engineering efforts are going to be most worthwhile. Obviously, it is in those areas where the most energy is used.

These examples show the great value of doing energy use evaluations based upon factual metered data. Besides the capability of precisely determining the amount of energy used for various purposes in a building, it also provides a means for determining what items should and should not be addressed, in which order, and the relative energy significance of each.

V. UTILITY RATE ANALYSIS

Since the goal in most energy conservation programs is to save money, not necessarily energy, it is worthwhile to evaluate the rate structures under which energy is purchased. The preferred method of evaluating rate structures is shown in Figure 7-10, by plotting average cost per unit of energy, in this case per kilowatt-hour, against load factor. Rate GS is the general service or secondary voltage rate for small buildings, typically less than 100 kW in demand, and there is a summer–winter differential in cost because it costs more to serve in summer than it does in winter. Rate HT is the high tension or primary voltage rate, the lowest cost rate for large customers, typically in the range of 50 to 5,000 kW, or more.

There are some interesting things to be determined from a rate analysis such as this. The utility company's rate attitude or rate philosophy depends primarily upon two things. The first determinant is how it generates the energy that it sells, whether it be nuclear, hydro, oil, or some other type. The second thing that determines the rate philosophy is the nature of the customer mix, whether customers are predominantly industrial, rural, urban, or agricultural. Those two major elements make up the philosophy that determines what the shapes of these curves are.

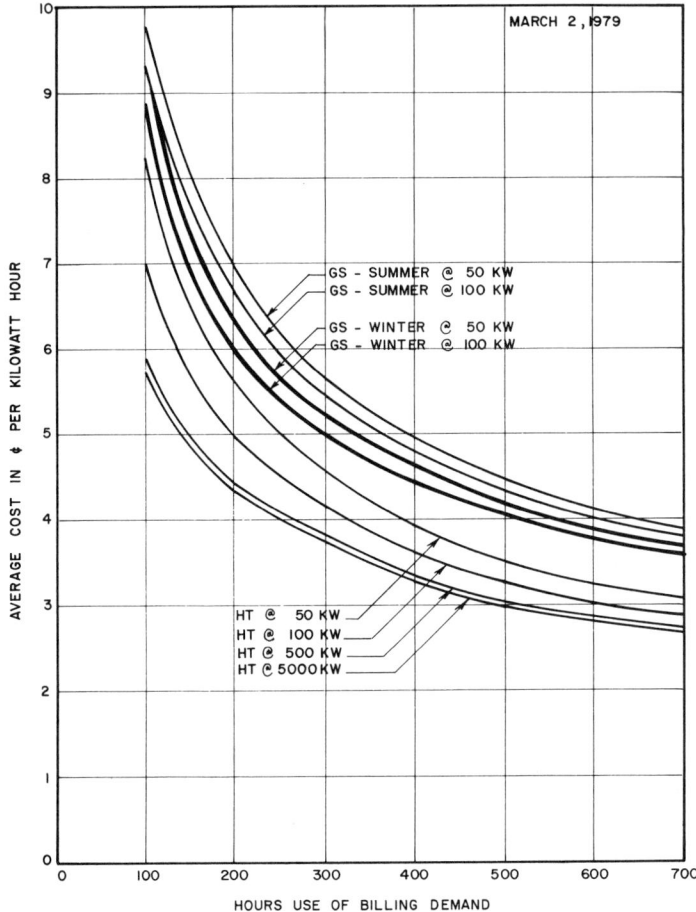

Figure 7-10. Preferred method of evaluating rate structures. Philadelphia Electric Company rates, with no fuel adjustment or taxes.

The purpose of doing this type of analysis is principally twofold. First, it determines the cost benefits to improving load factor; that is, depending upon how steep or how flat these curves are, one can make a determination as to how cost-effective it is to improve load factor, most often by controlling demand. By and large, with most utility company rate structures, demand control is most cost-effective for those buildings that have the lowest hours of use, like schools and office buildings. On the other hand, demand control is least cost-effective for those buildings that have the highest load factor or the highest hours of use, such as apartment houses and hospitals.

The second significant thing that this analysis indicates is the economic disincentive for conserving energy. As energy is conserved without changing demand, the load factor of a building drops. Therefore, the average cost of the remaining energy is increased. Again, depending upon how flat or how steep these curves are, it is generally found that conserving energy will typically result in a savings in cost only half as fast as the saving in energy. That will change from utility to utility, and from rate to rate.

VI. METHODS OF ECONOMIC ANALYSIS

When the results of the energy calculations are ranked, it is necessary to do some type of economic analysis. Table 7-4 presents a summary of the various methods of analysis.

Table 7-4. Economic Analysis Methods.

SIMPLE	SOPHISTICATED
Visual inspection	Discounted payback
Return on investment	Net present value
Simple payback	Percent present value
	Discounted cost/benefit
	Internal rate of return
	Life cycle costing

Since the very nature of the energy calculations used in establishing energy cost savings is not very precise to begin with, it doesn't pay to be much more precise than the simple methods in Table 7-4 allow. Too often there is a tendency to apply very precise and very sophisticated economic analyses to very imprecise and unsophisticated energy analyses. Thus, the question arises as to which method of economic analysis is most appropriate for the purpose of evaluating energy alternatives in buildings.

A. Simple Payback

The most commonly used economic analysis technique is the simple payback method, which is defined as the estimated capital cost of implementation divided by the annual energy cost savings at today's cost. It is well recognized that the simple payback technique does not take into consideration the effects of inflation, cost of money, escalation, taxes, and all of the other financial concerns that are emphasized in economics textbooks. The big advantage to the simple payback technique is that it is based upon facts that are known today, not some future projections of what may or may not happen tomorrow. Also, since most of the energy alternatives that are usually considered for a building will have simple payback periods of several years or less, the relative influence of such items as inflation, escalation, interest, and taxes will be minimal on the results and conclusions.

When weighing alternatives that have payback periods of several years or longer, most building owners have their own unique set of financial and economic circumstances to consider. In addition, it may be beneficial to include the relative escalation in the cost of energy compared to the escalation in the cost of living, either higher or lower. For longer payback periods, this factor can become quite significant and must be evaluated very carefully. The degree of sophistication in this analysis will depend to a great extent upon the payback periods involved. However, it should be remembered that the energy cost calculations are usually not nearly as accurate as the economic calculations.

B. Life Cycle Costing

One of the most widely promoted economic methods today is that of life cycle costing, especially in government circles. Life cycle costing is a technique in which all of the costs of owning and operating the energy-related features of a building are taken into consideration over the life of the building or the systems. While the concept may have great appeal, it is most often used to promote and enhance more expensive energy-related alternatives that would not otherwise be considered reasonable, such as solar systems.

In considering life cycles, especially those in excess of 10 or 20 years, the most significant element in the life cycle cost becomes the inflation and escalation factors that are used, to a point where they overshadow the basis for the analysis—which are, hopefully; the facts that exist today. Thus, life cycle costing becomes a juggling of assumptions regarding things that may or may not happen in the future, which is made even worse by the compounding of those assumptions over a long period of time. It is possible to make a life cycle cost analysis show anything that the user wants to prove, simply by juggling the assumptions.

One example of a life cycle cost analysis for a conservative private taxpaying corporation is shown in Table 7-5. This analysis was made on six energy-related alternatives in the construction of a new building. Each alternative has a first cost and an annual energy cost associated with it. In the life cycle cost calculations, the cost of oil was assumed to escalate at 12% per year, while the cost of electricity was assumed to escalate at 8% per year, resulting in the life cycle costs shown. For each of the alternatives, the energy budget was calculated. Note that the system that has the lowest first cost does not have the highest energy consumption. Note also that the system that has the lowest life cycle does not have the lowest energy consumption. This shows that there is no direct relationship between life cycle cost and energy consumption.

Further, if another equally valid assumption were made, such as that the escalation in the cost of both oil and electricity would be 10% per year, then the life cycle cost would change rather substantially. The first cost would not change, nor would the annual energy cost, nor would the energy budget. In all probability, not only would the life cycle cost change substantially, but the order of the life cycle costs would also change, further distorting the relationship between energy consumption and life cycle cost.

Another example is given in Figure 7-11, which shows the data used by the U.S. government in connection with the procurement of a large new building for the Social Security Administration in Baltimore, Maryland in 1976. Prior to the bidding on first cost, each bidder was asked to submit details on the energy-related portions of the building that they were proposing. These data were evaluated by an independent firm on behalf of the government and life cycle energy costs were established. Each bidder was informed of the calculated life cycle energy cost and was told that the award of the contract would go to the firm whose sum total of first cost plus life cycle energy cost was the lowest.

The award of the contract was made to the U.S. Steel–Owens Corning Joint Venture in the amount of $37,500,000. It turned out that the low bidder not only had the lowest first cost but also had the lowest life cycle energy cost. This example shows the fallacy of the life cycle cost method because in the example the lowest life cycle cost coincided with the lowest first cost, which is contrary to the general thinking on life cycle costing.

Most often life cycle costing is used as a gimmick by the promoters of high-priced goodies that would not otherwise be considered. By making assumptions about what may happen in the future and using the life cycle cost technique, they make these high-priced items look more attractive than they otherwise would, using any other economic analysis technique.

While life cycle cost analyses can be considered for the purpose of making judgments on alternatives to be selected, they should not be used as the ultimate decision-making tool.

Table 7-5. Summary of Life Cycle Costs

SYSTEM NUMBER	FIRST COST $	ANNUAL ENERGY COST $	LIFE CYCLE COST $	ENERGY BUDGET	
				MBtu/gsf/yr	MJ/(m$^2 \cdot$yr)
1	Base	349,300	13,178,295	82.5	941
2	+75,000	345,209	13,139,615	79.0	901
3	+125,000	340,271	12,534,528	48.9	557
4	+325,000	335,165	12,720,649	45.6	520
5	−50,000	357,587	12,643,621	59.5	678
6	+75,000	338,111	12,916,086	70.9	808

INCREMENTAL SUMMARY OF LIFE CYCLE COSTS					
SYSTEM NUMBER	FIRST COST $	ANNUAL ENERGY COST $	LIFE CYCLE COST $	ENERGY BUDGET	
				MBtu/gsf/yr	MJ/(m$^2 \cdot$yr)
1	+50,000	+14,135	+643,767	82.5	941
2	+125,000	+10,044	+605,087	79.0	901
3	+175,000	+5,106	-0-	48.9	557
4	+375,000	-0-	+186,121	45.6	520
5	-0-	+22,422	+109,093	59.6	678
6	+125,000	+2,946	+381,558	70.9	808

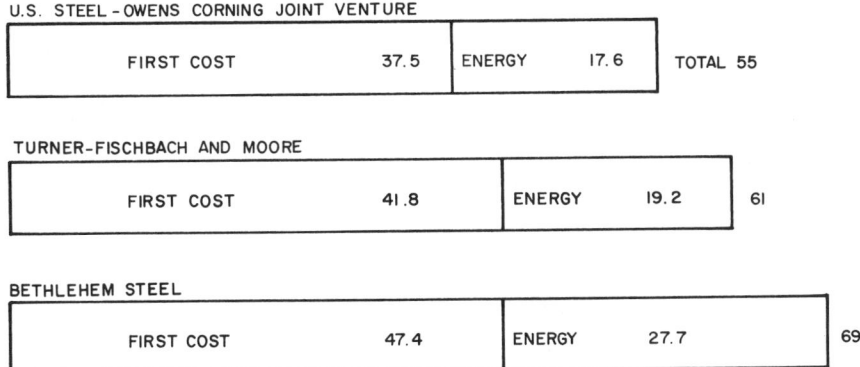

Figure 7-11. Life Cycle Costs, G.S.A.–Social Security Project, May 1976, in millions of dollars. Data used by the U.S. government in connection with the procurement of a large new building for the Social Security Administration in Baltimore, Maryland.

VII. CONCLUSION

The most significant concern in making energy and economic calculations becomes that of the professional judgment to be utilized in determining the degree of engineering precision necessary. This degree of precision will be different in almost every situation encountered. Generalized analyses and rules of thumb cannot be utilized for making these analyses. All too often they are incorrect. The world of buildings is made up of exceptions to the rule.

As one example, the Department of Energy in cooperation with the American Association of School Administrators selected ten elementary schools across the country [4], and hired several of the top consulting engineering firms in the country, giving them good fees to do detailed computerized energy audits in these schools. They were told to recommend anything and everything that had a payback of 12 years or less, which turns out to be a simple payback of about 20 years. That creates a climate for big capital investment. Then the Department of Energy funded all of the recommendations and had one of their laboratories instrument these buildings to measure how much energy was actually saved by implementing all these items.

In their audits, the engineers predicted that the overall heating energy savings would be 37%, but the actual measured heating energy saving was 15%. The engineers predicted that the overall electric energy savings would be 18%, but the actual measured electric energy savings was four-tenths of 1%, for all schools combined.

With a 20-year payback and no limit on capital investment, it is generally accepted that it is possible to save 37% of the heating energy and 18% of the electric energy in schools. Those are the kinds of numbers that people throw around from podiums in Congress, in Washington, and at trade association meetings. However, these people fail to look at what the meters show. In the report evaluating this project the longest chapter was on the subject of "excuses" for why the predicted savings were not achieved. Those excuses centered around the people that occupy the buildings because the formulas or computer programs that describe how people use buildings haven't yet been developed. It is necessary for engineers to use professional judgment to make responsible estimates that are realistic and can be achieved on the meter.

REFERENCES

1. *ASHRAE Handbook—1981 Fundamentals* (T. C. Elliott, ed.), American Society of Heating, Refrigerating and Air-Conditioning Engineers, Inc., Atlanta, GA. 1981, pp. 28.1–28.32.
2. *Engineering Weather Data,* U.S. Air Force Manual 88-29, Superintendent of Documents Stock Number 008-070-00420-8, Washington, D.C., 1978, pp. 3-1–3-427.
3. U.S. Department of Commerce, National Oceanic and Atmospheric Administration, Environmental

Data Service, National Climatic Center, Federal Building, Asheville, N.C. 28801.
4. J. Rudy, *Saving Schoolhouse Energy: Final Report,* Lawrence Berkeley Laboratory LBL-9106, Berkeley, Calif., 1979.

READING LIST

Energy Analysis:

A. W. Black III, "A Heretical View of Energy Programs or Is Bigger Really Better?," *ASHRAE Transactions,* Vol. 83, Part 2, pp. 300–311, 1977.

R. H. Howell and H. J. Sauer, Jr., *Bibliography on Available Computer Programs in the General Area of Heating, Refrigerating, Air Conditioning and Ventilating,* American Society of Heating, Refrigerating and Air Conditioning Engineers, Inc., Atlanta, GA 1980.

L. G. Spielvogel, "Computer Energy Analysis for Existing Buildings," *ASHRAE Journal,* Vol. 17, No. 8, pp. 40–41, 1975.

L. G. Spielvogel, "Comparisons of Energy Analysis Computer Programs," *ASHRAE Transactions,* Vol. 83, Part 2, pp. 293–299, 1977.

Utility Rates:

National Electric Rate Book, U.S. Department of Energy, Energy Information Administration, Washington, D.C. One book per state issued annually. Formerly available by subscription from the Superintendent of Documents, Washington, D.C.

8. Schoolhouse Energy Efficiency Demonstration: An Energy Audit Program*

Donna L. Rybiski
Public Affairs Department
Tenneco Inc., Houston, Texas

and

Milton Meckler, P.E.
President, The Meckler Group
Encino, California

		Page
I.	Introduction	89
II.	Initiating the Survey	92
III.	Findings	93
	Fuel Types	94
IV.	Energy Management	95
	The SEED Audit Process	97
	Summary	98
V.	Projected Savings	98
VI.	Training Energy Managers	99
VII.	Data Analysis	100
	Balance Sheet	101

Appendix A.	SEED School Energy Audit Checklist	105
Appendix B.	Energy Report Forms	143
Appendix C.	Energy Data Collection Forms	155
Appendix D.	Case Studies	167

I. INTRODUCTION

Schools present special needs and special opportunities. Industrial firms, in contrast to schools, have the economic incentive, the capital, and the engineering talent needed for energy management. Although a number of schools employ talented engineers and maintenance directors, few have sufficient staff to undertake new programs. Schools have been unable to pass on to taxpayers the higher energy costs of retrofitting expenditures. The alternative, a reduction in services, is generally considered unacceptable, although higher energy costs have caused a cut in some education programs.

The special opportunity refers to the school's unique ability to influence individuals and

---

*Portions of this chapter are reprinted from "Something Special from SEED: Energy Efficiency for Educators and Students," Tenneco Inc., Houston, Texas.

other institutions in the community. An earlier study found that schools can reduce energy consumption by at least 25% without major capital expenditures, and by 50% with capital expenditures. Our program was designed to prove the earlier study through demonstration school energy audits, and to communicate the findings to professional educators and patrons in order to encourage support for energy management in schools. Energy for education in grades kindergarten through high school now costs more than $3 billion per year, and at least $1 billion of that amount can be saved at today's prices. The potential is far greater, however. Schools with effective energy management programs can influence actions by others throughout the commercial–residential sector, multiplying the benefit of a school program.

The Schoolhouse Energy Efficiency Demonstration (SEED) was developed by Tenneco Inc. with the assistance of our education consultant, Dr. Shirley J. Hansen, director of energy programs for the American Association of School Administrators. The program's success can be traced to participation by nine national education groups which agreed to encourage and recommend the SEED program to their members: American Association of School Administrators, National Congress of Parents and Teachers, Association of School Business Officials, National School Boards Association, Education Commission of the States, Council for American Private Education, National Association of Elementary School Principals, National Association of Secondary School Principals, and National Education Association.

Programs were conducted in the public school districts of East Hartford, Connecticut; New Castle County, Delaware; DeKalb County, Georgia; Oak Lawn, Illinois; Lexington, Massachusetts; Livonia, Michigan; Manchester, New Hampshire; Lancaster, New York; Greenhills Forest Park, Ohio; West Irondequoit, New York; Haverford Township, Pennsylvania; Bethel Park, Pennsylvania; Pawtucket, Rhode Island; Newport News, Virginia; Garland, Texas; Roanoke County, Virginia; Houston, Texas; Racine, Wisconsin; as well as a school at Ft. Belvoir, Virginia, operated by Fairfax County, and the Baltimore Friends School.

The SEED program had two components: the technical activity, and the public education or motivational effort. First, the technical activity: We conducted comprehensive energy audits at 20 demonstration schools between May 1978 and April 1979. Each audit consisted of a fact-finding visit to the school, a pre-audit analysis of building plans and fuel records, a two-day thorough examination of the facility, and the writing of a complete energy study. All phases of the audits were conducted by a team headed by Tenneco's energy conservation consultant, Roger Rasbach of The Woodlands, Texas, an architectural designer. The audits were conducted under the direction of two professional engineers, Dr. Calvin M. Wolff of Houston and Milton Meckler, president of The Meckler Group, Encino, California, working in conjunction with Roger Rasbach Associates, The Woodlands, Texas.

On the second day of the audit, the professional engineer conducted an informal workshop for the host school district's maintenance staff, and others engaged in maintenance and energy management from nearby school districts and state and municipal governments.

Schools selected for the program were required to be representative of the region, so that recommendations could be transferred to the maximum extent. Because nearly 60% of schools in use today were built between 1955 and 1965, most demonstration schools were from that period, and had extensive areas of glass. Most were one- or two-story, with slab-on-grade foundations and uninsulated brick and block walls. The school in Oak Lawn, Illinois, was the only one with wall insulation. Roofs were usually flat with no insulation or limited rigid insulation combined with tar and gravel surfaces. Most buildings were rectangular with double-loaded corridors. One school was three stories, and one was a campus with five classroom buildings. Fifteen were elementary schools, three were junior high schools, one was a high school, and one, the private school, had kindergarten through high school. About half used oil and half used natural gas as the primary fuel, and one had supplemental electric baseboard heating. Five were air-conditioned. Most used either stream or hydronic heating systems, with unit ventilators or natural convection.

All school superintendents involved were aware of the need for energy efficiency. Some had advanced energy management programs, and others were in the early stages of program development. Resources of the districts varied considerably, although all had access to some assistance from state agencies. In almost every district, the key administrators had attended energy workshops. They were acquainted with the issue, and the well-publicized innovative new schools and expensive capital equipment for retrofitting. The need was for practical advice for low-cost, quick-fix solutions with the greatest return. We emphasized building modifications and changes in operating procedures that can be accomplished by the maintenance staff, and financed from the operating budget. Most recommendations have a payback period of two years or less. Major retrofitting (e.g., boilers and roofs) generally was advised only at the time of major repairs or replacement. Schools were advised to specify high-efficiency products when replacing fans, motors, pumps, kitchen appliances, and other energy consumers. Also, a switch to the new high-efficiency fluorescent lamps was recommended on a replacement basis.

The problems encountered most frequently, and the indicated solutions, meet the superintendents' request for guidance on quick-fix actions that provide immediate payback opportunities. For example, massive air movements caused excessive fuel consumption in 90% of the schools in the study. Such movement results from equipment that takes in too much cold fresh air and exhausts too much heated air, infiltration through cracks around doors and windows, open classroom doors and windows, and transmission through the large number of windows. Next, night and weekend temperatures in some cases were little different from daytime temperatures, and in most cases were too high. Fortunately, there are low-cost solutions to these problems, with the exception of heat transmission through windows.

The primary finding was that the average school can reduce its energy consumption by between 35 and 40% without major capital expenditures.

Since completion of our audits, the American Association of School Administrators (AASA) has published a study showing that the nation's schools have reduced energy consumption by nearly 30% since the 1974–75 school year. Btu's consumed per square foot of school space dropped from 148,113 in 1974–75 to 104,445 in the 1977–78 school year.

This leads to an obvious question: Have schools already achieved the savings we are predicting, or can schools that have cut consumption by nearly one-third now cut consumption another one-third? Although a reduction in energy consumption of more than one-half may sound overly optimistic or unrealistic at first, the potential seems to exist for many schools. The AASA report concerns progress through the 1977–78 school year, which is the base year for our study. Also, for the schools in our study with the best existing energy management programs, opportunities to reduce energy consumption in the range of 20 to 40% were recommended.

There are obvious reasons for the exceptional opportunity now facing schools in energy management. Schools constructed in the 1950s and 1960s had to meet the test of minimum first cost, with little consideration for operating costs. Administrators, faced with overcrowded classrooms, were under pressure to build schools as quickly and as cheaply as possible. Energy costs, in real terms, were constant or even declining during the school building boom that responded to the postwar baby boom. In recent years, the well-publicized financial difficulties of schools have created new problems. Schools have been forced to stretch the life of equipment, defer major repairs, and even defer regular maintenance.

Now that the opportunities in energy management have been identified, the task is to encourage support from school patrons. To do so, it is necessary to show energy management as an education issue. Again, AASA data can be used to make the point. The typical U.S. school district, according to AASA, allocates

85% of its budget for fixed costs, primarily compensation. The other 15%, called discretionary, goes for educational materials such as books and other operating costs such as energy. In the last five years, the portion of the discretionary budget that goes to pay energy costs has doubled, from 12.5% to 25%. Because total budgets have not increased proportionately, rising energy costs have replaced classroom materials.

In the SEED program, we have attempted to make this point to persons interested in quality education. Each of our audits received excellent newspaper and television news coverage. Articles, films, and presentations have been used to convey this message.

II. INITIATING THE SURVEY

For the study conducted by the American Association of School Administrators (AASA), a systematic random sample was drawn from its membership list stratified by district size. The ratio 1:4 for districts over 25,000 enrollment was used and 1:7 for districts under that figure. A total sample of 2,127 was drawn. The sample was matched to the universe of school systems (CIC list) and found to be generally representative of district sizes. (See Table 8-1.) The sample was also matched to the number of districts in the ten federal energy regions. (See Table 8-2.)

The survey instrument was developed in consultation with the Department of Energy and with the Educational Research Service which compiled and analyzed the data.

The instrument was mailed in January 1979, and a second mailing was sent out in March 1979. The response rate exceeded 50% at the April 13, 1979 deadline. Because of the complexity of the data requested, all responses were cleared by hand, and many verifying phone calls were made. Of the responses received, 629 were deemed usable. The usable response rate was thus 29.57%. While this was statistically sufficient to satisfy the conditions of the study, it is noted that responses by specific cells (i.e., certain fuel types in certain regions) are so low that caution is warranted in interpreting the findings. In certain instances the n was too low to treat the data at all. Where this is a concern, n is provided in the tables to assist the reader in assessing the data.

An analysis of the usable responses indicated that Region 5 (IL, IN, MI, MN, OH, and WI) was overrepresented and districts under 300 in size were underrepresented (also shown in Table 8-1).

The data were not adjusted for climate, but it is noted that the number of degree-days for the study's time frame, school year 1977–78 or calendar year 1978, was significantly higher than for 1973–74, the year upon which comparisons are based.

Table 8.1. Universe*/Sample/Respondent Breakdowns by Enrollment Groups.

ENROLLMENT GROUP	CIC UNIVERSE		SAMPLING RATIO	SAMPLE		RESPONDENTS	
				#	%	#	%
25,000 or more	413	2.8%	1:4	103	4.8%	33	5.2%
10,000 to 24,999	633	4.3	1:7	90	4.2	32	5.1
5,000 to 9,999	1,220	8.4	1:7	174	8.2	58	9.2
3,000 to 4,999	1,491	10.2	1:7	213	10.0	65	10.3
1,000 to 2,999	4,099	28.1	1:7	586	27.6	191	30.4
600 to 999	1,796	12.3	1:7	257	12.1	98	15.6
300 to 599	2,139	14.7	1:7	306	14.4	75	11.9
299 or less	2,784	19.1	1:7	398	18.7	59	9.4
Total	14,575	100.0		2,127	100.0	629**	97.1***

*Curriculum Information Center 1978–79 tape of all U.S. public school districts (enrollment based on fall 1977 figures).
**Response rate was therefore 29.6%.
***No enrollment data provided by 2.9% of respondents.

Table 8.2. Number/Percent of Local Public School Districts (Fall 1977) in the United States by DOE Energy Region.*

ENROLLMENT RANGE	U.S. TOTAL	DEPARTMENT OF ENERGY: ENERGY REGIONS									
		1	2	3	4	5	6	7	8	9	10
25,000 or more	187	3	8	20	47	26	29	8	7	34	5
10,000 to 24,999	530	37	49	61	78	91	59	19	22	92	22
5,000 to 9,999	1,104	87	135	133	210	221	152	46	18	114	47
2,500 to 4,999	2,067	165	254	257	343	548	171	74	59	134	71
1,000 to 2,999	3,463	253	405	248	331	1,130	368	290	99	206	133
600 to 999	1,864	121	169	18	52	538	312	320	97	140	97
300 to 599	2,323	157	161	7	33	465	507	473	236	158	126
Fewer than 300	4,296	367	141	1	5	373	709	1,189	822	393	296
Total in region	15,834	1,190	1,322	745	1,099	3,392	307	2,419	1,360	1,271	797
Percent of total U.S.	100.0	7.5	8.3	4.7	6.9	21.4	14.1	15.3	8.5	8.0	5.0
Respondents %		5.1	9.2	7.5	6.0	31.3	11.1	11.4	5.9	5.9	6.5

*Source of data: Table 4 (Number of operating local public school systems, by size of school system and state: United States, fall 1977), p. xix, *Education Directory Public School Systems 1977–78,* by Jeffrey W. Williams and Sallie L. Warf, National Center for Educational Statistics, 1978.

In other words, the reductions in consumption noted were made in a year when climatic demands on energy were much greater. Indeed, the weather bureau indicates it was one of two of the worst years in the last 50. The effects of school closings in the winter of 1977–78 due to fuel curtailment were also considered; but since the total number of pupil days lost was less than 1% of the operational demand, it was disregarded in the findings presented. While the number and percentage of electrically cooled schools are given, the schools are not counted in the total figures, as most of these buildings are also heated and thus would be counted twice.

III. FINDINGS

Table 8-3 indicated the median Btu's per square foot by total for 1977–78 and by the earlier findings of the Federal Energy Administration (FEA). It should be noted that while AASA's sample was randomly drawn, FEA's was a fortuitous or problematic sample, thus limiting the opportunity to make inferences. Nevertheless, the FEA national survey provides the only other available data on school energy consumption.

Consumption figures by region logically follow the respective severity of the climate. Figure 8-1 indicates the states comprising each federal energy region. However, the difference in energy reduction by region cannot be explained by known factors (i.e., cost of fuel, type and availability of fuel, or energy conservation programs at the state or regional level).

The study did not encompass the manner in which the reductions were achieved or the amount of money invested in energy-conserving measures. Knowledge of school energy operations outside the study suggests that savings

Table 8.3. Median Consumption All Fuels (in Btu's/sq ft).

	1972–73*	1977–78	% CHANGES 1972–1973 TO 1977–78
Total	161,312	104,445	35.25
Federal energy region			
1	176,710	114,999	34.92
2	185,999	109,687	41.03
3	166,069	111,241	33.01
4	112,013	75,604	32.50
5	184,948	110,488	40.26
6	132,603	77,907	41.25
7	163,245	117,766	27.86
8	182,640	120,103	34.24
9	131,342	66,523	49.35
10	162,761	85,978	47.17

*Federal Energy Administration's School Fuel Impact Survey.

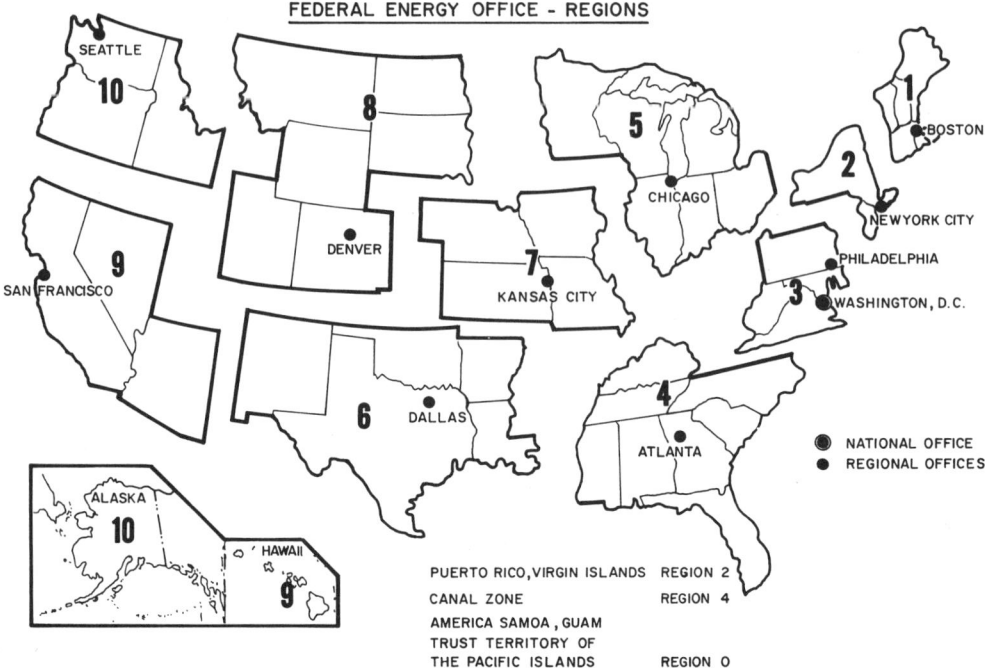

Figure 8-1. Federal energy office—regions.

can be primarily attributed to changes in operations, more energy-conscious maintenance, and implementation of low-cost retrofits. Further consideration of the median Btu's per square foot suggests that consumption is also a function of district size. Table 8-4 shows that districts over 5,000 consume 11.6% less than those under 5,000, a figure that suggests the benefits of on-staff expertise. The matter warrants closer scrutiny. If these findings are corroborated, the use of educational service agency resources to assist smaller districts could present a very cost-effective vehicle for training and technical assistance.

Fuel Types

One of the interesting facets of the findings was an analysis of types of fuel used by the schools, nationally and by regions of the country. No figures were available prior to the AASA study. Natural gas consumption on a national basis was surprisingly higher than anticipated at 55%. Table 8-5 depicts the fuel sources for the reporting buildings and the percentage they represent on a national basis.

The variations in fuel source by region were far greater than expected. Table 8-6, which shows fuel sources by region, reveals differences of considerable importance to decisions relative to energy in the schools. For example, the difference between Region 3's 31% and Region 6's .3% in oil consumption is significant to those wishing to reduce oil imports. Any analysis of the economic impact of gas deregulation on the schools should consider that the impact will be much greater in Regions 4, 5, 6 and 9, and that relatively little effect would be seen in Regions 1, 2, and 10. Coal strikes would hit Regions 3, 4, and 5 hardest.

Table 8-7 provides an analysis of fuel sources region by region. In Region 1, for ex-

Table 8.4. Median Btu/sq ft by District Enrollment.

	DISTRICT ENROLLMENT	
	Less than 4999	5000 or more
n	488	123
Median Btu/sq ft	107,294	94,826

Table 8.5. Fuel Sources, All Buildings.

FUEL TYPE	NO. OF BLDGS. REPORTED AND PERCENTAGE	
	N	%
Natural gas	5,999	54.89
Oil (all grades)	2,734	25.01
Electricity		
Htg/cooling	809	7.40
Htg only	962	8.80
Cooling only	(852)	
Propane	220	2.01
Coal	168	1.54
Butane	19	.002
Diesel	16	.0015
Total value	10,927	99.65

ample, oil supplies 79% of the schools' energy needs, while gas provides 11% and electricity 9%. Conversely, Region 6 relies on natural gas to serve 79% of its school needs, while oil is negligible and electricity contributes 18%.

The data also revealed that district size appears to influence some fuel usage. The larger districts use a disproportionate amount of natural gas, while small districts show greater reliance on electricity. Oil, on the other hand, seems to supply about 25% of the districts, regardless of size. Table 8-8 presents the percentage of certain fuels used relative to district size.

IV. ENERGY MANAGEMENT

It is impossible to limit the discussion of remedies for energy inefficiencies to the bricks and mortar of the physical facility. The school building and the people who use it are inexorably linked, for much of the energy that can be saved is saved through the vigilance and cooperation of individuals.

The most important elements in implementing an effective energy management program

Table 8.6. Percentage Consumption by Region of Specified Fuels (by School Building Count).

FED. EN. REGION		NATURAL GAS	OIL	FUEL SOURCE Htg/c	ELECTRICAL Htg/ONLY	Co/ONLY	COAL	PROPANE
1	n	34	236	13	16	9	-	4
	%	.6	8.6	2	2	1	0	2
2		81	293	9	37	13	-	1
		1	10.7	1	4	1	0	.4
3		255	854	97	58	125	53	1
		4	31	12	6	15	31	.4
4		1,072	248	182	220	205	36	99
		18	9	22	23	24	21	45
5		1,215	447	106	149	101	53	9
		20	16	13	15	12	31	4
6		1,370	9	206	104	168	-	25
		23	.3	25	11	20	0	11
7		456	164	51	45	33	6	42
		8	6	6	5	4	3	19
8		348	54	44	204	14	11	22
		6	2	5	21	2	6	10
9		1,044	45	54	73	165	-	13
		17	1.6	7	7	19	0	6
10		124	387	47	56	19	9	4
		2	14	6	6	2	5	2
Total		5,999	2,734	809	962	852	168	220

Table 8.7. Percentage of Fuel Sources, Federal Energy Region ($n < 10$ not treated, shown by X).

		FEDERAL ENERGY REGIONS									
		1	2	3	4	5	6	7	8	9	10
Fuel Type	n	34	81	255	1072	1215	1370	456	348	1044	124
Natural gas $n = 5999$%		11	19	19	58	61	79	60	51	84	19
Oil (all grades) $n = 2734$		236	290	854	248	447	9	164	54	45	387
		79	69	65	13	23	X	21	8	4	61
Electricity htg/cooling $n = 809$		13	9	97	182	106	206	51	44	54	47
		4	X	7	10	5	1	7	6	4	7
htg only $n = 962$		16	37	58	220	149	104	45	204	73	56
		5	9	4	12	7	6	6	30	6	9
cooled only $n = 852$		*									
Propane $n = 220$		4	1	1	99	9	25	42	22	13	4
		X	X	X	5	X	1	5	3	1	X
Coal $n = 168$		-	-	53	36	53	-	6	11	-	9
		X	X	4	2	3	X	X	1.6	X	X
Butane $n = 19$		-	-	-	-	-	16	-	-	3	-
		X	X	X	X	X	.9	X	X	X	X
Diesel $n = 16$		-	-	-	-	-	-	-	-	5	11
		X	X	X	X	X	X	X	X	X	2
All Fuels Total	$n = 10929$										

*No total figures are given for electrically cooled buildings because most of these buildings are also heated and thus would be counted twice.

are the commitment and teamwork of the people involved. School personnel are highly committed to provide quality education for students. They are equally committed to performing the other, supportive tasks necessary to ensure that quality education is provided. In recent years a new activity has been added to the long list of tasks for superintendents—energy management. School administrators have come to embrace energy management as a valuable tool to assist them in holding down operating expenditures and channeling money into quality education services.

The commitment begins at the top, with the school board and superintendent. They begin by making the school community aware of problems relative to energy and motivating them to assist in solving those problems.

Everyone can have a role—from the school board to the principal, from the superintendent to parent–teacher groups, from the maintenance engineer to students and teachers in every classroom. Energy conservation goals should be set, and teamwork should be used because it is the best approach to goal accomplishment.

Energy conservation teams appointed by the superintendent usually consist of an energy coordinator, the business official, the maintenance engineer, a principal, a PTA member, a teacher, and a student representative. In many school districts there are both district-wide teams and teams for each school.

The role of the team is to set out goals for energy conservation and to develop a plan to reach the established goals. Individual members of the team are then responsible for implementation of certain aspects of the plan. Typical roles might include the following:

School board and superintendent	Establish policy
Energy coordinator	Coordinate the implementation of all energy conservation activities

Table 8.8. Fuel Source by District Size; Number of Buildings and Percentages.

Fuel type	LESS THAN 600 n = 419		600 TO 4,999 n = 2,996		5,000 OR MORE n = 7,353	
	n	%	n	%	n	%
Natural gas n = 5,999	154	37	1,446	48	4,327	59
Oil (all grades) n = 2,734	103	25	732	24	1,853	25
Electricity						
htg/cooling n = 809	20	5	254	8	528	7
htg only n = 962	72	17	341	11	524	7
(cooling only) n = 852	(19)		(153)		(671)	
Propane n = 220	47	11	138	5	28	.4
Coal n = 168	11	3	78	3	77	1
Butane n = 19	9	2	-		10	.1
Diesel n = 16	3	.7	7	.2	6	.08

Business official
- Report to superintendent and school board
- Keep accurate records of energy expenditures and projected savings
- Monitor billings closely for an early detection of newly developing energy inefficiencies

Maintenance engineer
- Keep all mechanical and electrical equipment in good operating condition
- Make necessary modification to equipment and operating procedures to ensure energy conservation

Principal
- Set the tone of energy conservation in the school
- Post and enforce energy conservation guidelines and operating procedures for the school

Teacher
- Implement and monitor energy conservation in the classroom
- Motivate students to learn about energy conservation

Students
- Assist in implementing and monitoring energy conservation activities

The SEED Audit Process

In addition to identifying problems and opportunities, the energy audit and subsequent data analysis define the framework for an effective energy management program. The data analysis provides realistic quantifiable estimates of where energy and money may be saved. It points out the areas of greatest loss and therefore the areas of greatest energy savings opportunities. In this manner, this process assists in setting priorities for energy conservation activities.

The SEED energy audit process consists of three major steps: the pre-audit, the full audit survey, and a presentation of the results. The pre-audit is the data collection and preliminary analysis phase of the audit process. At this time, monthly fuel and power consumption records and billing schedules for at least the past year or two are assembled. Weather data for recent years are collected from the local weather station or the National Oceanic and Atmospheric Administration. Architectural drawings, including modifications and additions, are inspected to compute square footage, percentage of wall area, amount of window area, and the insulating properties of construction materials. Mechanical and electrical specifications are examined to determine power demands. A careful examination of the data collected will point out potential areas of energy inefficiencies. At this time, a boiler efficiency test is made to determine how much of the fuel consumed is actually used to heat the building.

Once the anomalies between the as-designed and actual energy consumption rates are defined, the full on-site audit inspection is conducted, with inspectors looking for opportunities to correct those anomalies and save energy.

In addition, the audit team interviews the principal and maintenance personnel to determine building use and number of occupants. These interviews are extremely valuable not only for the information gathered but also for conveying the importance of energy conservation to those who operate the building.

Once the audit survey and data analysis are complete, a written report is prepared. The report clearly and precisely enumerates the problems and lists alternative recommendations. To assist the school administrator in justifying expenditures, cost estimates are provided for each proposed solution. The length of time required to recover the initial cost (payback period) is included as an additional tool to justify expenditures. Payback periods should be used with caution, as they are only a good yardstick for determining relative costs.

The recommendations suggest priorities based upon the amount of energy saved, cost estimates, payback period, and the effect of each recommendation on other problems and/or proposed solutions.

Summary

In summary, there is a proper sequence, according to SEED, in which the remedies should be implemented:

1. Tighten the building envelope. This includes insulating windows, weatherstripping, caulking, and reducing ventilation and exhausts.
2. Lower the building temperature, especially the night and weekend settings.
3. Modify the mechanical systems.

The lighting modifications can be implemented at any point in the energy management program. (See appendices to this chapter for energy audit checklist, data collection forms, and energy report forms.)

V. PROJECTED SAVINGS

The results of the 20 audits have been compiled in Table 8-9. The principal finding of the SEED 20-school study is that the average school can reduce its energy consumption by 48.6%, without the use of expensive capital equipment or exotic technology. To the extent that the selected schools are representative, there is a significant opportunity for school energy conservation.

In the audits, the engineers first identified sources of heat gain and loss and quantified each. This information, averaged for the 20 schools, is presented in the energy consumption profile, Table 8-10.

The profile identifies the percent of heat lost through the most common circumstances, and shows the amount of heat gained from sources other than the furnace. It also divides electrical consumption into several categories. This analysis led to identification of low-cost opportunities for energy efficiency. With one or two

Table 8.9. SEED Summary of Projected Savings.

	PERCENT OF FUEL	PERCENT OF ELECTRICITY	PERCENT OF TOTAL	PAYBACK PERIOD (MONTHS)
East Hartford, Connecticut	41.0	40.0	40.0	4
Newark, Delaware	52.3	23.6	36.3	4
Decatur, Georgia	75.0	16.3	34.4	17
Oak Lawn, Illinois	69.0	32.5	47.2	5
Baltimore, Maryland	58.2	30.5	41.2	19
Lexington, Massachusetts	48.5	34.2	42.3	18
Livonia, Michigan	56.4	24.4	44.0	10
Manchester, New Hampshire	65.4	32.3	48.0	8
Lancaster, New York	53.7	45.2	50.8	12
West Irondequoit, New York	51.5	30.8	45.8	9
Cincinnati, Ohio	63.6	49.1	55.2	13
Bethel Park, Pennsylvania	65.0	31.3	51.3	14
Havertown, Pennsylvania	75.8	64.9	67.6	18
Pawtucket, Rhode Island	75.0	21.0	63.0	9
Garland, Texas	59.5	57.6	58.0	7
Houston, Texas	51.0	44.6	46.3	5
Ft. Belvoir, Virginia	85.0	29.0	60.7	12
Newport News, Virginia	62.5	21.3	43.4	12
Salem, Virginia	58.2	59.8	55.2	21
Racine, Wisconsin	61.8	9.9	41.5	14
AVERAGE	**61.4**	**34.9**	**48.6**	**11.6**

exceptions, recommendations were limited to off-the-shelf materials and proven procedures.

Table 8-11 identifies projected savings by activity. This table is the best support for the overall conclusion that energy consumption in the selected schools can be cut nearly in half. Under heating fuel, five of the first six items, other than window insulation, account for more than half the projected savings. This is possible through such inexpensive and interrelated actions as night thermostat setback, tightening of the building envelope, and control of excessive air movements.

VI. TRAINING ENERGY MANAGERS

The technical workshop, a vital part of the SEED program, was developed to accomplish three objectives: to explain the interrelationships of the energy-consuming aspects of the building; to illustrate where and how much energy is being lost and at what cost; and to demonstrate ways to conserve energy. The audience was comprised primarily of nonprofessional maintenance engineers and school business officials. These people are dedicated, conscientious, and experienced in sound, practical methods of school maintenance and operation. They are the "front line of defense" for any effective energy management program.

As the SEED program proceeded, the form and substance of the technical workshop were modified to meet the needs of the audience. One of the major purposes of the workshop was to share with the audience our findings and experiences in establishing energy management programs. We too learned from the SEED audit process, in particular from the audiences of the technical workshops. For example, in Livonia, Michigan, we learned that placing pegboard over the outside of fresh air inlets is an excellent way to reduce the amount of outside air entering the classroom. We encouraged the attendees to speak out, to describe their problems, and to identify the solutions they had developed. Problems most frequently discussed related to energy losses due to transmission, infiltration, ventilation, lighting, and heating and cooling systems. This real exchange of ideas and information was best accomplished in a small, informal setting wherein people were willing to become active

Table 8.10. SEED Energy Consumption Profile.

HEATING FUEL LOSSES	PERCENT OF LOSS	PERCENT OF TOTAL
Transmission		
Roof	20.6	
Windows and doors	35.6	
Walls	20.4	
Floors	5.6	
		82.2
Ventilation		20.7
Infiltration		
Windows	9.8	
Doors	7.2	
		17.0
Potable hot water		7.6
System losses		2.2
Total heating losses		129.7

HEAT GAIN FROM SOURCES OTHER THAN HEATING FUEL	PER CENT OF GAIN	
Electric Power	15.3	
Occupants	7.2	
Daylight	7.2	
Total heating gains		29.7
Heating losses minus gains		100.0

ALLOCATION OF ELECTRIC POWER CONSUMPTION		
Fluorescent lighting		33.8
Incandescent lighting		19.8
Exterior lighting		3.2
Ventilation		11.6
Hydronic pumping		15.3
Refrigeration		2.2
Other kitchen		5.5
Miscellaneous		8.6
Total electric power consumption		**100.0**

Table 8.11. SEED Projected Savings by Activity.

HEATING FUEL	PERCENT OF SAVINGS	PAYBACK PERIOD (MONTHS)
Caulk and seal windows, skylights and air conditioning window units	6.4	19
Weatherstrip and deactivate doors	4.9	2.5
Insulate windows and skylights	32.5	24
Reduce fresh air intake	7.3	14
Reduce ventilation rates	8.1	2.7
Tune thermostat system	30.7	3.1
Adjust and modify heating system	6.3	18
Install independent water heaters	2.1	65
Lower hot water temperature	1.7	7.6
ELECTRIC POWER		
Deactivate lights near windows	25.4	4
Use high efficiency fluorescent lamps and ballasts	15.0	4
Replace incandescent lamps	33.6	25
Reduce hydronic flow rate	12.7	20
Reduce ventilation rate	4.7	10
Implement energy saving kitchen procedures	2.0	12
Miscellaneous	6.6	8

participants. Also of great importance to us was the identification of barriers to conservation (i.e., why certain recommendations we make can't be carried out in some schools). The dialogue in the workshops identified a number of barriers.

SEED, through its technical workshops, attempted to show that maintenance personnel are capable of implementing effective energy management programs. Nothing new or revolutionary is needed. Recommendations, presented in the technical workshops, included only simple, quick-fix, low-cost energy conservation measures. They included a quantification of the amount of energy and money saved. The recommendations were designed to assist schools in developing energy management programs by establishing priorities for implementation of energy-saving measures.

VII. DATA ANALYSIS

After fully defining the problems relative to energy inefficiencies and collecting all the data to describe a building's energy usage, the energy manager needs to know how to analyze the data. Three sets of calculations are given to help the energy manager in the analysis: (1) "rules of thumb," guidelines to determine where energy dollars are going; (2) a balance sheet that allocates energy loss and cost to specific problem areas; and (3) recommendations for problem solutions that are quantified.

"Rules of thumb" are general guidelines that put a price tag on things that use or misuse energy. These formulas can apply to any school in the district, and were derived to support the SEED audit process. They are easily understandable and provide emphasis for energy saving concepts.

For example, a 1° lowering of the night thermostat on a boiler has the same effect as a 3.5° reduction during the day. Why? Because the school building is normally unoccupied 3.5 times longer than it is occupied. Therefore, the energy manager can readily see that daytime temperature settings are not as important as night settings.

Other "rules of thumb" will provide answers to the following questions:

How much does a MMBtu (one million Btu's) of heating cost?

$$\frac{\text{(Fuel costs—\$ per unit)}}{\text{(MMBtu per unit)} \times \text{(boiler efficiency)}} = \frac{\$}{\text{MMBtu}}$$

How much is the cost of heating by electricity?

$$\frac{\$/\text{kWh} \times 1 \text{ million}}{3{,}412 \text{ Btu/kWh}} = \frac{\$}{\text{MMBtu}}$$

How much money is lost by transmission through windows?

$$\frac{\text{\# degree-days} \times 24 \text{ hr} \times 1.4 \text{ Btu/ft}^2/\text{hr}/°F}{1 \text{ million}}$$
$$\times \$/\text{MMBtu} = \$ \text{ lost/ft}^2/\text{yr}$$

How much does heat contributed by occupants cost per year?

$$.38 \text{ MMBtu/person/yr} \times \$/\text{MMBtu}$$
$$= \$ \text{ gained/person/yr}$$

How much money does heat gained from sunlight cost per year?

$$.1 \text{ MMBtu/ft}^2/\text{yr} \times \$/\text{MMBtu}$$
$$= \$ \text{ gained/ft}^2/\text{yr}$$

How much do losses from infiltration cost?

$$\frac{1.08 \text{ Btu}/°F \text{ hr cfm} \times \text{\# degree-days} \times 24 \text{ hr}}{1 \text{ million}} \times \$/\text{MMBtu}$$
$$= \$ \text{ lost/yr/cfm}$$

How much do losses from ventilation cost?

$$\frac{5.4 \times \text{\# degree-days}}{1 \text{ million}} \times \$/\text{MMBtu}$$
$$= \$ \text{ lost/kWh}$$

How much do gains from electric power contribute?

$$\frac{90° \text{ of annual electric power consumption} \times 3{,}412 \text{ Btu/kWh}}{1 \text{ million}}$$
$$\times \$\text{MMBtu} = \$ \text{ gained/kWh}$$

Balance Sheet

A balance sheet, much like an accountant's debit and credit sheet, illustrates where energy is lost or gained and how much it costs. The balance sheet is the basis for recommendations, which are in turn evaluated against cost of implementation to determine payback period. This identification of energy inefficiencies helps the energy manager determine where the greatest energy savings can occur. Anomalies (imbalance) in the balance sheet indicate that some things were not properly accounted for, or that there are some heretofore unknown consumers of energy, indicating a further search. It was these anomalies that prompted our "midnight raids." It is the allocation of dollars to energy loss that can assist the energy manager in calculating the time required to recover capital expenditures. This feature, the payback period, helps the energy manager convince those who control school budgets to make the expenditures necessary to implement energy conservation measures. When the energy manager combines the identification of large energy inefficiencies with the cost and payback period of conservation activities, an energy management plan can be defined.

Balance sheets can be prepared for energy expenditures for heating and electricity. An example of a condensed heating balance sheet is given below:

	LOSSES			GAINS	
	MMBtu	$		MMBtu	$
Transmission			Electric power	2,000	6,000
Windows	8,000	24,000	Occupants	1,000	3,000
Others	10,000	30,000	Daylight	500	1,500
Ventilation	2,000	6,000		3,500	10,500
Infiltration	1,500	4,500	Net use, calc.	18,000	54,000
Total	21,500	64,500	Actual use	20,000	60,000
			Discrepancy	10%	

CONDENSED HEATING BALANCE SHEET

An energy manager can analyze this example and readily determine that the greatest opportunity for energy savings exists in the windows. Therefore, it will be necessary to figure out ways to eliminate heat transmission through windows.

Once general "rules of thumb" have been applied and a balance sheet prepared, the energy manager is in a position to define alternative means of correcting the various problems of energy loss, allocate a cost, and define an energy management program.

In the SEED technical workshops, recommendations were made that were specific to the particular school building that had been audited. In addition, general recommendations were made for most commonly found problems so that energy managers could apply them to other school situations.

The general approach to improving the energy efficiency of a building is to tighten the building envelope, that is, to insulate windows and seal cracks, holes, and other openings where energy can be wasted and heat lost. The result of "tightening up" the building is that the demand for energy drops. Then, one can make the necessary internal (temperature control) adjustments that will save even more energy.

In order to reduce heat loss due to transmission, infiltration, and ventilation—to tighten up the building envelope—the energy manager should consider the following actions:

1. Close and seal operable windows that are not needed.
2. Ensure that windows remaining operable seal well when closed.
3. Caulk and weatherstrip around windows and doors.
4. Insulate the upper two-thirds of windows with a translucent window covering.
5. Deactivate unused doors by issuing administrative procedures.
6. Keep doors to corridors closed.
7. Place pegboard over the outside of fresh air inlets.
8. Ensure that dampers on unit ventilators and roof-top exhaust fans are working properly.
9. Disconnect unnecessary roof-top fans.
10. Place remaining fans on timers.
11. On hydronic heating systems, reduce pump flow and power by shaving pump impellers.
12. Plant trees and shrubs as a windbreak.

Once the cracks in the building envelope have been reduced, the energy manager can make the necessary internal adjustments. The boiler provides the greatest opportunities for energy savings. Because the building now leaks less heat, the night setback temperature on the boiler can be turned back. The energy manager should experiment with temperature settings and determine the lowest night and weekend thermostat settings permissible for that particular building. In most cases, it will be from 45°F to 55°F, resulting in time-averaged night–weekend temperatures of around 60°F.

Room thermostats can probably be readjusted. The energy manager should make certain that all thermostats are calibrated properly. Keep in mind that the daytime temperatures are not as important as nighttime settings. This fact makes it easier for the energy manager to keep teachers and students happy and warm and still be able to implement energy savings.

APPENDIX A. SEED SCHOOL ENERGY AUDIT CHECKLIST

This section is designed as a guide to provide an energy management team with a convenient, organized listing of areas of energy waste and inefficiency most commonly found in school buildings.

While conducting a tour through the building, use each checklist INDICATOR in the order it appears here to help quickly identify conditions of energy misuse. If TRUE is checked, the INDICATOR implies good energy management; no remedial action is necessary. If, however, the INDICATOR is FALSE, then remedial action is required.

The first remedial action taken should be the LOWER COST CORRECTIONS, operational and maintenance opportunities (O & M's) suggested in the center. Where applicable, these suggestions should be assigned to specific personnel and implemented immediately. By themselves they can reduce energy consumption by 30 to 50% in buildings where no improvements in energy efficiency have yet been made.

Finally, consider each applicable suggestion under HIGHER COST CORRECTIONS. These are redesign, retrofit, or energy conservation measures (ECM's), and will usually require capital expenditures for implementation. Professional architects and/or engineers should be consulted prior to their implementation. Payback and life cycle cost analysis should also be conducted before any of these decisions are made in order to determine which will save the most energy dollars.

A. HUMAN SYSTEMS

PATTERNS OF USE INDICATOR	LOWER COST CORRECTIONS (O & M's)	HIGHER COST CORRECTIONS (ECM's)
#1. The change of seasons has been reflected in thermostat settings. TRUE ☐ FALSE ☐ NOTES: _____	☐ Adjust thermostat settings so that room temperatures of occupied areas are 68°F in heating season and 78°F during cooling season.	☐ Replace existing thermostat with a thermostat which has a separate setting for cooling and a separate setting for heating, or use one thermostat to control heating and one thermostat to control cooling.

PATTERNS OF USE INDICATOR	LOWER COST CORRECTIONS (O & M's)	HIGHER COST CORRECTIONS (ECM's)
#2. Location of all thermostats provides for moderate temperature fluctuations. TRUE ☐ FALSE ☐ NOTES: _____	☐ Move thermostats from areas next to windows, doors, heating or cooling units to areas which reflect temperatures of conditioned spaces more accurately. ☐ If thermostat cannot be moved, protect it from the source creating fluctuation.	☐ Relocate thermostat in HVAC return system, where possible.

A. HUMAN SYSTEMS

PATTERNS OF USE INDICATOR	LOWER COST CORRECTIONS (O & M's)	HIGHER COST CORRECTIONS (ECM's)
#3. Areas which are unoccupied or minimally used have modified heating/cooling temperature settings. TRUE ☐ FALSE ☐ NOTES: _____	☐ Turn off heating system or close heat register if nothing in space can freeze. ☐ If area has its own thermostat, reduce heat setting to 55°F. ☐ Turn off cooling system or close registers. ☐ Use infra-red spot heaters in large spaces with low occupancy.	☐ Reduce heating/cooling demands of unoccupied areas with appropriate control systems.

PATTERNS OF USE INDICATOR	LOWER COST CORRECTIONS (O & M's)	HIGHER COST CORRECTIONS (ECM's)
#4. Activities, including custodial services, not normally considered part of the school day, are scheduled to reflect wise energy management. TRUE ☐ FALSE ☐ NOTES: _____	☐ Reschedule activities to accommodate partial reduction of energy demand. ☐ Consolidate activities to reduce building usage. ☐ Reschedule cleaning and maintenance activities during daylight working hours where possible. ☐ Where rescheduling is not possible, use lights and equipment only in those areas where individuals are working.	☐ Consider installing an automated energy control system that will predetermine the environment of occupied spaces.

A. HUMAN SYSTEMS

PATTERNS OF USE INDICATOR	LOWER COST CORRECTIONS (O & M's)	HIGHER COST CORRECTIONS (ECM's)
#5. Temperatures have been adjusted to reflect temporary usage of lobbies, corridors, vestibules and other public places. TRUE ☐ FALSE ☐ NOTES: _____	☐ Reduce heat supply to these areas. ☐ Disconnect or turn off electric heaters in these areas unless it creates a freeze-up problem. ☐ Discontinue air conditioning of these areas.	None practical

PATTERNS OF USE INDICATOR	LOWER COST CORRECTIONS (O & M's)	HIGHER COST CORRECTIONS (ECM's)
#6. Building temperatures have been reduced to reflect unoccupied periods. TRUE ☐ FALSE ☐ NOTES: _____	☐ Reduce thermostat settings to 50°F when building is unoccupied during heating season. Begin reduction of heating during last hour of occupancy. ☐ Turn off all air conditioning units when building is unoccupied. ☐ Experiment with heating "turn-on" times to determine length of pre-heat period required for satisfactory comfort levels of occupants. ☐ Install appropriate automatic controls; e.g., time clocks.	None practical

A. HUMAN SYSTEMS

OCCUPANT IMPACT INDICATOR	LOWER COST CORRECTIONS (O & M's)	HIGHER COST CORRECTIONS (ECM's)
#1. Thermostats are locked to eliminate unauthorized adjustment. TRUE ☐ FALSE ☐ NOTES: _____	☐ Install locking screws. ☐ Post signs displaying proper thermostat setting. ☐ Replace defective thermostat with non-adjustable pre-set thermostats. ☐ Install tamper-proof covers.	☐ Relocate thermostat in HVAC return system, where possible.

OCCUPANT IMPACT INDICATOR	LOWER COST CORRECTIONS (O & M's)	HIGHER COST CORRECTIONS (ECM's)
#2. Staff utilizes blinds, curtains and other window-covering devices to reduce energy usage. TRUE ☐ FALSE ☐ NOTES: _____	☐ Instruct staff on techniques of reducing energy consumption with blinds, curtains and other window-covering devices. ☐ Inform staff to use natural lighting and solar heat gain when appropriate. ☐ Repair damaged shading devices. ☐ Install devices where needed.	☐ Install reflective or heat absorbing films in areas with excessive heat gain in summer.

110 3/ENERGY ANALYSIS TECHNIQUES AND STRATEGIES

A. HUMAN SYSTEMS

OCCUPANT IMPACT INDICATOR	LOWER COST CORRECTIONS (O & M's)	HIGHER COST CORRECTIONS (ECM's)
#3. The school provides a suitable inservice training program for building staff in the importance and techniques of wise energy use. TRUE ☐ FALSE ☐ NOTES: _____	☐ Appoint interested individuals to develop and implement an energy management program for staff. ☐ Extend the training program to include students through appropriate instruction. ☐ Remind staff and students of good energy conservation techniques through the utilization of Energy Conservation Plaques.	None practical.

OCCUPANT IMPACT INDICATOR	LOWER COST CORRECTIONS (O & M's)	HIGHER COST CORRECTIONS (ECM's)
#4. Occupants turn off lights in unoccupied areas. TRUE ☐ FALSE ☐ NOTES: _____	☐ Instruct staff on importance of conservation of electrical energy. ☐ Utilize only necessary lighting in large areas such as libraries, gymnasiums, auditoriums and cafeterias. ☐ Color-code switches to indicate minimum levels of light needed for normal activities.	☐ Install automatic timing switches in areas used for short periods, such as storerooms and janitors' closets. ☐ Rewire switches so that one switch does not control all fixtures in multiple task areas.

A. HUMAN SYSTEMS

OCCUPANT IMPACT INDICATOR	LOWER COST CORRECTIONS (O & M's)	HIGHER COST CORRECTIONS (ECM's)
#5. Doors and windows remain closed while building is being heated or cooled. TRUE ☐ FALSE ☐ NOTES: _____	☐ Instruct staff to keep doors and windows closed while building air is being conditioned. ☐ Adjust door mechanism for faster closing. ☐ Remove device which allows doors to be locked open. ☐ Permanently seal unnecessary operable windows.	None practical.

B. STRUCTURAL SYSTEMS

DOORS INDICATORS	LOWER COST CORRECTIONS (O & M's)	HIGHER COST CORRECTIONS (ECM's)
#1. All exterior doors are aligned properly, fit tightly and operate efficiently. TRUE ☐ FALSE ☐ NOTES: _____ _____ _____ _____ _____ _____ _____	☐ Realign doors that do not close properly. ☐ Replace or readjust automatic door closing. ☐ Repair or replace threshholds and/or gaskets. ☐ Repair or replace weatherstripping.	☐ Consider installing vestibule doors at heavily used entrances. ☐ Install mechanisms which automatically close doors to unconditioned spaces. ☐ Consider installing removable center posts with proper gaskets for double doors with large gaps. ☐ Consider installing wind screens to protect exterior doors from direct blasts of prevailing winds. ☐ Insulate large exterior doors.

B. STRUCTURAL SYSTEMS

WINDOWS INDICATORS	LOWER COST CORRECTIONS (O & M's)	HIGHER COST CORRECTIONS (ECM's)
#1. All windows on exterior walls are aligned properly, fit tightly and operate effectively. TRUE ☐ FALSE ☐ NOTES: _____	☐ Realign windows that do not close properly. ☐ Permanently seal those which cannot be properly aligned. ☐ Caulk and weatherstrip windows. ☐ Replace any broken or cracked windows.	☐ Consider replacing faulty or non-essential windows with walls or insulated panels. ☐ Consider replacing deteriorated or badly fitting frames with new energy efficient models. ☐ Install wall or insulated panel if window is not essential.

WINDOWS INDICATOR	LOWER COST CORRECTIONS (O & M's)	HIGHER COST CORRECTIONS (ECM's)
#2. Exterior walls have little excessive glassed areas. TRUE ☐ FALSE ☐ NOTES: _____	☐ Insulate the upper portion of each window.	☐ Consider replacing windows with walls or insulated panels. ☐ Install double pane windows if cost-effective. ☐ Investigate other products such as reflective or heat absorbing film. (Be sure to anticipate maintenance problems.) ☐ Use thermopane windows (utilizing the same casings) when replacing windows.

B. STRUCTURAL SYSTEMS

WALLS
INDICATOR

LOWER COST CORRECTIONS
(O&M's)

HIGHER COST CORRECTIONS
(ECM's)

#1. Penetrations in exterior walls and the joints where different wall materials join have been properly caulked.

☐ Ensure that all areas of potential air infiltration are sealed, using quality caulking materials.

☐ Cover all window cooling units when not in use to prevent air leakage through the units.

None practical

TRUE ☐   FALSE ☐

NOTES: _____
_____
_____
_____
_____
_____
_____
_____

WALLS
INDICATOR

LOWER COST CORRECTIONS
(O & M's)

HIGHER COST CORRECTIONS
(ECM's)

#2. Wall insulation is adequate for local weather conditions.

None practical

☐ Consider adding insulation to increase the R-factor, when remodeling or when replacing segments of the exterior wall.

TRUE ☐   FALSE ☐

NOTES: _____
_____
_____
_____
_____
_____
_____
_____

B. STRUCTURAL SYSTEMS

ROOFS INDICATOR	LOWER COST CORRECTIONS (O & M's)	HIGHER COST CORRECTIONS (ECM's)
1. Roof is adequately insulated for local weather conditions. TRUE ☐ FALSE ☐ NOTES: _____	☐ Ensure that the vapor barrier faces the conditioned space.	☐ Bring the current insulation level up to standard recommended for your climate. ☐ Replace any water-damaged insulation.

ROOFS INDICATOR	LOWER COST CORRECTIONS (O & M's)	HIGHER COST CORRECTIONS (ECM's)
#2. Skylights have been modified to reflect wise energy management. TRUE ☐ FALSE ☐ NOTES: _____	☐ Caulk and seal around entire skylight. ☐ Be certain glass or plastic is in good condition to prevent air infiltration and heat loss. ☐ Install an additional glass or plastic barrier to reduce heat loss. ☐ Insulate skylight with a translucent insulating material.	☐ Replace skylight with insulating materials.

C. LIGHTING SYSTEMS

INTERIOR INDICATOR	LOWER COST CORRECTIONS (O & M's)	HIGHER COST CORRECTIONS (ECM's)
#1. Incandescent lamps are used rarely except for decorative purposes. TRUE ☐ FALSE ☐ NOTES: _____	☐ Replace burned-out incandescent lamps with lower wattage types, where possible. ☐ Consider replacing incandescent lamps with lower-wattage, self-ballasting mercury vapor bulbs in large areas such as gymnasiums, cafeterias and auditoriums. (Note: Be sure at least one incandescent bulb is left in to provide a safety light should the mercury vapor lamps be accidentally turned off.) ☐ Discontinue use of extended service lamps in locations where they can be replaced easily.	☐ Substitute energy conserving lamps for non-decorative incandescent lamps. The former provides more light for less wattage at lower cost.

INTERIOR INDICATOR	LOWER COST CORRECTIONS (O & M's)	HIGHER COST CORRECTIONS (ECM's)
#2. Wherever fluorescent lamps have been removed, the ballasts have been properly disconnected. TRUE ☐ FALSE ☐ NOTES: _____	☐ Remove or disconnect ballasts. (Note: Although they appear inactive, ballasts consume significant amounts of electricity even though the lamps have been removed.)	☐ Investigate replacing unnecessary tubes with models which draw a small current, yet provide a uniform lighting effect.

C. LIGHTING SYSTEMS

INTERIOR INDICATOR	LOWER COST CORRECTIONS (O & M's)	HIGHER COST CORRECTIONS (ECM's)
#3. The entire school has been re-lamped with energy efficient fluorescent tubes, and ballasts have been replaced as needed with energy conserving types. TRUE ☐ FALSE ☐ NOTES: _____	☐ Replace defective fluorescent tubes with more efficient and lower wattage types. ☐ Replace burned-out ballasts with more efficient, lower wattage models.	☐ Install more efficient, lower wattage fluorescent tubes in all fixtures.

INTERIOR INDICATOR	LOWER COST CORRECTIONS (O & M's)	HIGHER COST CORRECTIONS (ECM's)
#4. Proper delamping, the removal of unnecessary fluorescent tubes and incandescent bulbs, has been completed. TRUE ☐ FALSE ☐ NOTES: _____	☐ Do not replace defective fluorescent tubes in areas where delamping is feasible. In four-lamp fixtures allow two lamps to remain, disconnecting appropriate ballasts. ☐ Do not replace burned-out incandescent bulbs in areas where delamping is feasible.	☐ Consider lowering fixtures in order to increase illumination levels on task areas. Higher illumination levels will allow a delamping program. ☐ Consider removing entire fixtures in areas where illumination is excessive. ☐ Install three-tube fixtures which allow three levels of illumination for varying tasks, when remodeling.

C. LIGHTING SYSTEMS

INTERIOR INDICATOR	LOWER COST CORRECTIONS (O & M's)	HIGHER COST CORRECTIONS (ECM's)
#5. Routine cleaning of lamps, tubes and fixtures is part of the energy management program. TRUE ☐ FALSE ☐ NOTES: _____	☐ Clean lamps, tubes and fixtures regularly. ☐ Replace yellowed, cracked or defective light diffusers as needed.	None practical

INTERIOR INDICATOR	LOWER COST CORRECTIONS (O & M's)	HIGHER COST CORRECTIONS (ECM's)
#6. Natural lighting is optimized. TRUE ☐ FALSE ☐ NOTES: _____	☐ Clean walls regularly. ☐ Repaint with light reflective, non-glossy colors to enhance illumination. ☐ Clean windows on a routine basis. ☐ Use drapes, blinds and curtains to increase natural light gain.	None practical

C. LIGHTING SYSTEMS

EXTERIOR INDICATOR	LOWER COST CORRECTIONS (O & M's)	HIGHER COST CORRECTIONS (ECM's)
#1. Defective exterior lamps and/or fixtures have been replaced with more energy efficient types. TRUE ☐ FALSE ☐ NOTES: _____	☐ Replace 150-watt security flood lights with 75-watt lamps. ☐ Consider blacking-out building and grounds during unoccupied night time periods. (Note: Some school districts have had success with this program.) ☐ Consider using remotely monitored intrusion alarms.	☐ Replace defective exterior lamps with mercury vapor, metal halide or high-pressure sodium lamps.

EXTERIOR INDICATOR	LOWER COST CORRECTIONS (O & M's)	HIGHER COST CORRECTIONS (ECM's)
#2. Exterior lighting is automatically controlled. TRUE ☐ FALSE ☐ NOTES: _____	☐ Reduce hours of operation of exterior lights. ☐ Ensure proper functioning of photo-cell controls.	☐ Install automatic lighting controls if needed. ☐ Modify the control system to include photocell "turn-on" and time clock "turn-off".

C. LIGHTING SYSTEMS

EXTERIOR INDICATOR	**LOWER COST CORRECTIONS** (O & M's)	**HIGHER COST CORRECTIONS** (ECM's)
#3. Playing field lighting system has been upgraded to save energy dollars. TRUE ☐ FALSE ☐ NOTES: _____	☐ Clean lighting fixtures regularly. ☐ Maintain and refinish lighting reflectors, as required, to obtain maximum illumination. ☐ Use only lights required during night practice periods.	☐ Replace incandescent lamps with more efficient and effective models such as high-pressure sodium or metal halides. ☐ Install separate meter for playing field lights to reduce total electrical demand, if applicable. (Note: This procedure will not save energy but can save dollars.)

D. MECHANICAL SYSTEMS

HEATING INDICATOR	**LOWER COST CORRECTIONS** (O & M's)	**HIGHER COST CORRECTIONS** (ECM's)
#1. Multiple boilers or heaters have been modified to prevent simultaneous firing. TRUE ☐ FALSE ☐ NOTES: _____	☐ Adjust boiler or furnace controls so that unit #2 will not fire until unit #1 can no longer satisfy demand.	☐ Install automatic staging controls, if needed.
#2. Stack temperature is in normal range as verified by routine flue gas analysis. TRUE ☐ FALSE ☐ NOTES: _____	☐ Perform regular flue gas testing to ensure proper air to fuel ratio and make necessary corrections. (Example: Clean air intake filters; check that spuds and nozzles are properly sized and not clogged; ensure that fuel pressures are not excessive, and there is an adequate supply of combustion air.) ☐ Reduce the boiler's firing rate.	☐ Purchase kit for flue gas analysis, or digital flue gas analyzer.

D. MECHANICAL SYSTEMS

HEATING INDICATOR	LOWER COST CORRECTIONS (O & M's)	HIGHER COST CORRECTIONS (ECM's)
#3. Thermostat settings accurately reflect room temperatures. TRUE ☐ FALSE ☐ NOTES: _____	☐ Recalibrate thermostats and controllers. ☐ Bleed and clean pneumatic lines, if applicable. ☐ Clean contacts of electrical control system, if applicable. ☐ Ensure that control valves and dampers are modulated properly and that heat distribution to space is unobstructed.	☐ Investigate new types of thermostatic control devices for application in specific situations. Install suitable type.

D. MECHANICAL SYSTEMS

HEATING INDICATOR	LOWER COST CORRECTIONS (O & M's)	HIGHER COST CORRECTIONS (ECM's)
#4. Automatic controls schedule heating water temperature in accordance with outdoor temperature. TRUE ☐ FALSE ☐ NOTES: _____	☐ Check temperature of hot water in heating system. If it appears excessive during periods of mild weather: a. Ensure that reset controls are functioning properly. b. Experiment with hot water temperature reduction when an acceptable comfort level is attained. ☐ Turn off boiler, pumps or heat source during summer. ☐ If the building has been properly caulked, sealed, weatherstripped and insulated, turn the boilers off at night. Exercise caution if the outside temperature is expected to go below freezing. In addition, make certain that the boiler will tolerate being turned off.	☐ Install temperature controls which automatically program water temperature according to outdoor temperature and will turn off heating unit when outside temperature reaches 60°F.

D. MECHANICAL SYSTEMS

HEATING INDICATOR	**LOWER COST CORRECTIONS** (O & M's)	**HIGHER COST CORRECTIONS** (ECM's)
#5. Heating pilot lights are scheduled to be turned off during summer. TRUE ☐ FALSE ☐ NOTES _____	☐ Turn pilot(s) off on prescheduled date. ☐ Post a reminder in boiler/furnace room of pilot reactivation date.	☐ Replace pilot lights with new electronic ignition models whenever feasible.
#6. Steam radiators and other steam equipment are maintained regularly for proper functioning. TRUE ☐ FALSE ☐ NOTES: _____	☐ Replace defective air vent valves. ☐ Clean thermostatic control valves on radiators. Replace if necessary. ☐ Clean or replace bellows element in defective thermostatic traps. ☐ Check the temperature on the down stream side of steam traps. Unless the pipe is moderately hot (as hot as a water pipe), it is malfunctioning. Take appropriate action.	☐ Reclaim steam trap heat losses by diverting the steam-vacuum system exhaust through a potable hot water pre-heat tank.

D. MECHANICAL SYSTEMS

HEATING INDICATOR	**LOWER COST CORRECTIONS** (O & M's)	**HIGHER COST CORRECTIONS** (ECM's)
#7. Insulation on hot water pipes has been inspected, is adequate and is in good condition. TRUE ☐ FALSE ☐ NOTES: _____	☐ Replace damaged or missing insulation.	☐ Install additional pipe insulation in accordance with good energy conservation practices.

HEATING INDICATOR	**LOWER COST CORRECTIONS** (O & M's)	**HIGHER COST CORRECTIONS** (ECM's)
#8. Oil burner is operating efficiently without excessive smoke or sooting. TRUE ☐ FALSE ☐ NOTES: _____	☐ Check for proper oil pressure. Adjust as required. ☐ Confirm that oil is at proper temperature and free-flowing. ☐ Inspect burner nozzles for cleanliness and correct spray angles. ☐ Verify proper air to fuel ratio by routine flue gas analysis. ☐ Reduce firing rate.	☐ Purchase vent for flue gas analysis or a digital flue gas analyzer.

D. MECHANICAL SYSTEMS

HEATING INDICATOR	LOWER COST CORRECTIONS (O & M's)	HIGHER COST CORRECTIONS (ECM's)
#9. Routine maintenance is employed to insure efficient operating of heating units. TRUE ☐ FALSE ☐ NOTES: _____	Boilers: ☐ Remove soot from tubes during routine maintenance. ☐ Remove scale deposits, sediments and precipitates on water-side surfaces. (Note: Rear portion of boiler is most susceptible to scale formation.) ☐ Repair all damaged or worn boiler insulation, refractory, brickwork and boiler casings. Furnaces: ☐ Repair or replace solenoid valve if fire does not cut off immediately when unit shuts down. Both: ☐ Reset hot water temperature. Limit switch to a higher setting if burner short cycles. ☐ Reduce firing rate.	☐ Replace dangerous or ineffective units with more efficient models.

D. MECHANICAL SYSTEMS

HEATING INDICATOR	**LOWER COST CORRECTIONS** (O & M's)	**HIGHER COST CORRECTIONS** (ECM's)
#10. Hot water/steam radiation units are operating efficiently. TRUE ☐ FALSE ☐ NOTES: _____	☐ Open air vents and bleed off air until water appears. ☐ Repair or replace faulty thermostats. ☐ Check pneumatic lines for proper functioning. ☐ Replace faulty valves. ☐ Check water pumps for proper operation. Make necessary adjustments. ☐ Check boiler for correct operating temperature. Correct as required. ☐ Remove objects which may be obstructing heating units.	☐ Clean heating elements.

D. MECHANICAL SYSTEMS

VENTILATION INDICATOR

#1. A quantity of outdoor air which does not exceed code requirements is used to ventilate the building.

TRUE ☐   FALSE ☐

NOTES: _____
_____
_____
_____
_____
_____
_____

LOWER COST CORRECTIONS (O & M's)

☐ Reduce outdoor air quantity to the minimum allowed by codes by adjusting appropriate dampers.

☐ Be sure that outdoor air dampers are closed when building is unoccupied.

☐ Check all outdoor dampers for defective seals and for proper closure. Repair, if needed.

☐ Place pegboard over the outside of the fresh air inlet.

HIGHER COST CORRECTIONS (ECM's)

☐ Replace defective or damaged dampers with new opposed-blade models.

☐ Install automatic controls to close dampers when building is unoccupied.

VENTILATION INDICATOR

#2. The ventilation systems are programmed to use natural cooling whenever possible.

TRUE ☐   FALSE ☐

NOTES: _____
_____
_____
_____
_____
_____
_____
_____

LOWER COST CORRECTIONS (O & M's)

☐ Utilize outside air for cooling rather than refrigeration units whenever possible.

☐ Be sure the economizer cycle is operating properly.

HIGHER COST CORRECTIONS (ECM's)

☐ Install an economizer cycle with enthalpy control.

D. MECHANICAL SYSTEMS

VENTILATION INDICATOR	LOWER COST CORRECTIONS (O & M's)	HIGHER COST CORRECTIONS (ECM's)
#3. The building's exhaust system functions in accordance with occupancy patterns. TRUE ☐ FALSE ☐ NOTES: _____	☐ Disconnect unnecessary exhaust fans and cover grill to prevent conditioned air loss. ☐ Re-wire special exhaust fans to operate only when room is occupied. ☐ Schedule all other exhaust fans to operate only when needed. ☐ Program exhaust fans with time clocks or other controls.	☐ Install variable speed motors (when replacing exhaust fans) to modulate fan speed, allowing only required ventilation. ☐ Modify all exterior exhaust ducts with controlled or gravity dampers.
#4. Return, outdoor air and exhaust dampers are sequencing properly. TRUE ☐ FALSE ☐ NOTES: _____	☐ Check damper linkage. Adjust for proper closure. ☐ Readjust indicators to indicate damper positions.	☐ Replace defective or damaged damper with new opposed-blade models.

D. MECHANICAL SYSTEMS

VENTILATION INDICATOR

#5. Temperature of the air entering through the ducts during the heating season feels comfortably warm.

TRUE ☐ FALSE ☐

NOTES: _____

LOWER COST CORRECTIONS (O & M's)

☐ Reduce air volume to eliminate "draft effect".

☐ Raise supply temperature to 65°F in perimeter zones (60°F in interior zones) during heating season only.

HIGHER COST CORRECTIONS (ECM's)

None practical

VENTILATION INDICATOR

#6. The air flow is well-balanced and consistent throughout the building.

TRUE ☐ FALSE ☐

NOTES: _____

LOWER COST CORRECTIONS (O & M's)

☐ Check air filters. Clean or replace regularly.

☐ Remove objects obstructing air flow; clean diffusers, registers and grilles.

☐ Balance air flow system, if required.

HIGHER COST CORRECTIONS (ECM's)

None practical

D. MECHANICAL SYSTEMS

AIR CONDITIONING INDICATOR	LOWER COST CORRECTIONS (O & M's)	HIGHER COST CORRECTIONS (ECM's)
#1. Thermostat settings accurately reflect room temperatures. TRUE ☐ FALSE ☐ NOTES: _____	☐ Restrict outdoor air intake when not using economizer cycle. ☐ Insure that control dampers and valves (especially the economizer cycle) are working properly. ☐ Calibrate thermostats and controllers. ☐ Clean thermostatic controls.	☐ Investigate new types of thermostatic control devices for application in specific situations. Install suitable type.

AIR CONDITIONING INDICATOR	LOWER COST CORRECTIONS (O & M's)	HIGHER COST CORRECTIONS (ECM's)
#2. Zone temperatures are maintained without the use of reheat coils. TRUE ☐ FALSE ☐ NOTES: _____	☐ Determine if temperatures remain in comfort zone when boilers are shut down during cooling.	☐ Convert to variable air volume system, if practical.

132 3/ENERGY ANALYSIS TECHNIQUES AND STRATEGIES

D. MECHANICAL SYSTEMS

AIR CONDITIONING INDICATOR	**LOWER COST CORRECTIONS** (O & M's)	**HIGHER COST CORRECTIONS** (ECM's)
#3. Multiple air conditioning compressors have been modified to prevent simultaneous start-up.	☐ Adjust compressor controls so that unit #2 will not start-up until unit #1 can no longer satisfy cooling demand.	☐ Install automatic staging controls, if needed.

TRUE ☐   FALSE ☐

NOTES: _____
_____
_____
_____
_____
_____
_____

AIR CONDITIONING INDICATOR	**LOWER COST CORRECTIONS** (O & M's)	**HIGHER COST CORRECTIONS** (ECM's)
#4. Insulation on cooling lines, pipes and ducts has been inspected, is adequate and is in good condition.	☐ Replace damaged or missing insulation.	☐ Install additional insulation on all delivery lines and ducts in accordance with good energy conservation practices.

TRUE ☐   FALSE ☐

NOTES: _____
_____
_____
_____
_____
_____
_____

D. MECHANICAL SYSTEMS

AIR CONDITIONING INDICATOR	LOWER COST CORRECTIONS (O & M's)	HIGHER COST CORRECTIONS (ECM's)
#5. Cool air in adequate volume is discharged into space. TRUE # FALSE ☐ NOTES: _____	☐ Clean condenser and evaporator coils, fins and tubes. Remove if necessary. ☐ Clean or replace air filters. ☐ Ensure that fire and balancing dampers are open and in correct positions. ☐ Verify that fan is rotating in proper direction.	None practical

AIR CONDITIONING INDICATOR	LOWER COST CORRECTIONS (O & M's)	HIGHER COST CORRECTIONS (ECM's)
#6. Refrigerated air is at recommended temperature levels. TRUE ☐ FALSE ☐ NOTES: _____	☐ Remove and clean the strainer if frost or sweat is visible at the strainer outlet. ☐ Inspect and clean all coils (including dehumidification coils) regularly. ☐ Clean dirty condensers. Dirty condensors increase system pressure, causing a decrease in system efficiency. ☐ Repair or adjust defective compressor valves. These can be identified by high discharge temperatures.	None practical.

D. MECHANICAL SYSTEMS

AIR CONDITIONING INDICATOR	LOWER COST CORRECTIONS (O & M's)	HIGHER COST CORRECTIONS (ECM's)
#7. Evidence indicates that chilled water piping, valves and fittings are intact.	☐ Repair all leaks. ☐ Check valves and fittings. Replace as required.	None practical.

TRUE ☐   FALSE ☐

NOTES: _____
_____
_____
_____
_____
_____
_____
_____

D. MECHANICAL SYSTEMS

AIR CONDITIONING
INDICATOR

LOWER COST CORRECTIONS
(O & M's)

HIGHER COST CORRECTIONS
(ECM's)

#8. Refrigeration compressor is operating efficiently.

TRUE ☐   FALSE ☐

NOTES: _____
_____
_____
_____
_____
_____
_____

☐ Check refrigerant charge level. Adjust to equipment specifications. Repair leaks.

☐ Check for faulty or fused electrical control circuits. Repair or replace as required.

☐ Ensure that liquid line solenoid valve is not leaking or stuck open.

☐ Inspect all compressor valves. Take corrective action if needed.

☐ Clean evaporation and condensor coils. Clean liquid line strainer if clogged.

☐ Reset high/low pressure control differential settings if needed.

None practical.

E. SPECIAL SYSTEMS

WATER INDICATOR

#1. Hot water storage tanks, piping and heaters are operating efficiently.

TRUE ☐   FALSE ☐

NOTES: _____
_____
_____
_____
_____
_____
_____

LOWER COST CORRECTIONS (O & M's)

☐ Reset heater thermostat to 105°F-115°F or to local code requirement. This may require installing a "booster" for the kitchen.

☐ Turn off recirculating pump(s) when building is unoccupied.

☐ Repair all leaks.

☐ Replace damaged or missing insulation.

HIGHER COST CORRECTIONS (ECM's)

☐ Install adequate insulation on all lines and tanks containing hot water.

☐ Consider installing smaller domestic water heaters to maintain desired temperature in storage tanks.

☐ Install separate gas fired water heater for use in cafeteria and kitchen.

WATER INDICATOR

#2. Heating cycle on electric water heater is restricted to low electrical demand periods.

TRUE ☐   FALSE ☐

NOTES: _____
_____
_____
_____
_____
_____
_____

LOWER COST CORRECTIONS (O & M's)

☐ Utilize heater's "vacation cycle" during extended vacation periods.

☐ Use a time clock or automatic controls to restrict the duty cycle.

HIGHER COST CORRECTIONS (ECM's)

None practical

E. SPECIAL SYSTEMS

WATER INDICATOR	LOWER COST CORRECTIONS (O & M's)	HIGHER COST CORRECTIONS (ECM's)
#3. Devices which restrict hot water usage have been utilized. TRUE ☐ FALSE ☐ NOTES: _____	☐ Install inexpensive flow restrictors in lines or faucets.	☐ Install mixing valves that pre-determine water temperature. ☐ Re-plumb locker room showers to include a master valve to control shower time. ☐ Replace standard faucets with spring-loaded shut-off types. ☐ Consider installing, if practical, a solar water heater to assist in meeting hot water needs.

WATER INDICATOR	LOWER COST CORRECTIONS (O & M's)	HIGHER COST CORRECTIONS (ECM's)
#4. Swimming pool is energy efficient. TRUE ☐ FALSE ☐ NOTES: _____	☐ Maintain water temperature in maximum range of 80°F-84°F. ☐ Clean filters as required. ☐ Ensure that pool heating equipment is maintained regularly. ☐ Use a flue gas analysis to maintain proper air-to-fuel ratio. ☐ Experiment with ventilation/exhaust system to attain minimum air displacement and still meet code. ☐ Use a buoyant plastic sheet to cover water when not in use.	☐ Consider installing enclosure on outdoor pools. ☐ Investigate solar water heaters to maintain pool temperature.

E. SPECIAL SYSTEMS

KITCHEN CAFETERIA INDICATOR	LOWER COST CORRECTIONS (O & M's)	HIGHER COST CORRECTIONS (ECM's)
#1. Exhaust hoods and fans operate efficiently and only when needed. TRUE ☐ FALSE ☐ NOTES: _____	☐ Program operating of exhaust fan. ☐ Ensure that fans and hoods are properly sized to prevent excessive ventilation of other areas of building. Reduce fan speed if necessary.	☐ Investigate heat recovery options of exhausted air.

KITCHEN CAFETERIA INDICATOR	LOWER COST CORRECTIONS (O & M's)	HIGHER COST CORRECTIONS (ECM's)
#2. Food preparation equipment is used wisely. TRUE ☐ FALSE ☐ NOTES: _____	☐ Turn on electrical equipment only when needed; keep it off when not required. ☐ Clean refrigeration coils regularly. Be sure coils have sufficient air circulation space. ☐ Move refrigerator and/or freezer away from any heat source. ☐ Check (and replace, as required) gaskets on refrigerator, freezer and oven doors for proper seal. ☐ Keep refrigerators full.	☐ Install newer energy efficient models when replacing damaged or defective equipment.

E. SPECIAL SYSTEMS

CAFETERIA INDICATOR	LOWER COST CORRECTIONS (O & M's)	HIGHER COST CORRECTIONS (ECM's)
#3. Food service staff is cognizant of good energy conservation techniques. TRUE ☐ FALSE ☐ NOTES: _____	☐ Operate dishwasher only with full-loads. ☐ Train employees to conserve hot water. ☐ Cook with lids in place on pots and kettles. ☐ Thaw frozen foods in refrigerated compartments. ☐ Use fans to cool people not heat sources. ☐ Do not allow refrigerator/freezer doors to remain open. ☐ Avoid preheating ovens. (Studies indicate preheating is unnecessary even for baked goods, if food is allowed to remain in oven several minutes after oven is turned off.)	None practical.

E. SPECIAL SYSTEMS

LAUNDRY INDICATOR	LOWER COST CORRECTIONS (O & M's)	HIGHER COST CORRECTIONS (ECM's)
#1. Laundry functions are carried out in an energy efficient manner. TRUE ☐ FALSE ☐ NOTES: _____	☐ Train staff to keep lint filters in dryers and the exhaust hoods clean. ☐ Reschedule laundry operation to avoid electrical peak demand hours. ☐ Wash and dry full loads only. ☐ Develop concise operating instruction for each piece of equipment. Post Energy Conservation Plaques. ☐ Reduce washing and drying cycles. Studies indicate that accepted washing/drying times are excessive for most items.	☐ Investigate heat recovery options.

E. SPECIAL SYSTEMS

**OFFICE MACHINES
ELECTRICAL EQUIPMENT**
INDICATOR

#1. Operation of electrical devices is carefully monitored.

TRUE ☐   FALSE ☐

NOTES: _____
_____
_____
_____
_____
_____
_____
_____

LOWER COST CORRECTIONS
(O & M's)

☐ Turn on office machines only when needed, e.g., copiers, coffee urns, typewriters.

☐ Use special high-demand equipment such as kilns and electric welders in low demand periods.

HIGHER COST CORRECTIONS
(ECM's)

☐ Install a demand limiter to prevent excessive electrical load.

APPENDIX B. ENERGY REPORT FORMS

The following pages are reporting forms, designed in compliance with requirements of DOE rule-making of April 2, 1979.

The following forms are included:

1. Auditor Certification
2. Schoolhouse Energy Usage Forms
3. The energy data collection forms (according to fuel type)
4. Evaluation of Potential for Energy Conservation Measures
5. Evaluation of Potential for Solar and Renewable Resource Materials

Special instructions are included to clarify information elements and to facilitate completion of the report.

AUDITOR CERTIFICATION

1. I, _____ , am qualified to perform energy audits by:
   (print)

   ☐ successful completion of State training course of

   _____
   (date)

   ☐ training and experience equivalent to successful completion of State training course:

   _____
   (detail or date of State waiver of requirement)

2. I am employed to operate this building. ☐ yes   ☐ no

3. My financial interests in connection with this energy audit are:
   ☐ no outside interests
   ☐ own, have stock in, or employed by
      ☐ a consulting firm
      ☐ an equipment manufacturer
      ☐ an energy supplier

4. The audit has been conducted in conformance with the requirement of the State of _____ and 10 CFR Part 450, Subpart E.

   _____   _____
   (Signature)   (Date)

SCHOOLHOUSE ENERGY EFFICIENCY DEMONSTRATION: AN ENERGY AUDIT PROGRAM

SCHOOLHOUSE ENERGY USAGE FORMS

Select a year which you consider energy usage to have been reasonable and normal. For purposes of this audit the year selected will be known as your "base year".

Using utility bills from your base year and from the most recent months of the current year, complete the following forms. If your bills use units other than those specified on the forms, use the conversion factors listed below to obtain the desired units.

Heating and cooling degree days can be obtained from your utility or your local weather bureau.

CONVERSION FACTORS

To Change From:	To:	Multiply By:
Electricity:		
Kilowatt Hours (KWH)	Btu's	3,413+/11,600#
Watts	Kilowatts	.001
Natural Gas:		
Thousands of Cubic Feet (MCF)	Cubic Feet	1,000
Hundreds of Cubic Feet (CCF)	Btu's	103,000
Therms	Btu's	100,000
Oil:		
Gallons of #2	Btu's	138,690
Gallons of #6	Btu's	149,690
Propane:		
Gallons	Btu's	95,475
Coal:		
Short tons	Btu's	24,500,000
Purchased Steam:		
Pounds	Btu's	1000+/1390#

+ School Conversion Factor
# Point of Generation Conversion Factor
(for federal reporting)

NATURAL GAS

	A	B		C		D		E		F		G	
M o n t h	Reading Date	Gas Used CCF		Gas Cost		Gas Cost Adjustment		Total Cost (C + D = E)		$/CCF (E ÷ B = F)		Heating Degree Days	
	FROM / TO	CURRENT	BASE	CURRENT	BASE	CURRENT	BASE	CURRENT	BASE	CURRENT	BASE	CURRENT	BASE
Jan													
Feb													
Mar													
Apr													
May													
Jun													
Jul													
Aug													
Sep													
Oct													
Nov													
Dec													

	TOTAL		
	BTU Conversion Factor	x 103,000	x 103,000
	Total BTU's		

NATURAL GAS _____ CCF/YR. Current Year _____
Natural Gas Rate No. _____ Base Year _____
Building _____

ELECTRICITY

Month	A Reading Date		B KWH Used		C Measured Demand KW		D $ Cost		E (FCA) Fuel Cost Adjustment ($)		F Total Cost (D + E = F)		G $/KWH (F ÷ B = G)	
	FROM	TO	CURRENT	BASE	CURRENT	BASE	CURRENT	BASE	CURRENT	BASE	CURRENT	BASE	CURRENT	BASE
Jan														
Feb														
Mar														
Apr														
May														
Jun														
Jul														
Aug														
Sep														
Oct														
Nov														
Dec														
Column B Totals														
On-Site BTU Conversion Factor			x 3,413	x 3,413										
Total BTU's														
Column B Totals														
Point of Generation BTU Conversion Factor			x 11,600	x 11,600										
Total BTU's														

ELECTRICITY _____ KWH/YR.   Current Year _____

Electricity Rate No _____   Base Year _____

Building _____

OIL/PROPANE

Month	A Reading Date		B Fuel Used (Gallons)		C $ Cost		D $/Gallon (C÷B=D)		E Heating Degree Days		F Fuel Used Per Degree Day (B÷E=F)	
	FROM	TO	CURRENT	BASE	CURRENT	BASE	CURRENT	BASE	CURRENT	BASE	CURRENT	BASE
Jan												
Feb												
Mar												
Apr												
May												
Jun												
Jul												
Aug												
Sep												
Oct												
Nov												
Dec												
TOTAL												

Conversion Factor:		
#2 Oil	x 138,690	x 138,690
#6 Oil	x 149,690	x 149,690
Propane	x 95,475	x 95,475
Other	x	x
Total BTU's		

OIL ☐ #2 Oil ☐ #6 Oil _____ Gal./Yr.
PROPANE _____ Gal./Yr. Other _____ Gal./Yr.

CURRENT YEAR _____ BASE YEAR _____

STEAM

Month	A Reading Date		B Steam Used (1,000 of lbs.)		C $ Cost		D $ per 1,000 lbs. (C÷B=D)		E Heating Degree Days		F Fuel Used Per Degree Day (B÷E=F)	
	FROM	TO	CURRENT	BASE	CURRENT	BASE	CURRENT	BASE	CURRENT	BASE	CURRENT	BASE
Jan												
Feb												
Mar												
Apr												
May												
Jun												
Jul												
Aug												
Sep												
Oct												
Nov												
Dec												
Column B Totals												
On-Site BTU Conversion Factor			x 1,000	x 1,000								
Total BTU's												
Column B Totals												
Point of Generation BTU Conversion Factor			x 1,390	x 1,390								
Total BTU's												

STEAM _____ LBS./YRS.

CURRENT YEAR _____

BASE YEAR _____

EVALUATION OF POTENTIAL FOR ENERGY CONSERVATION MEASURES

The Department of Energy has developed a Relative Importance Factor (RIF) for the five items listed: The RIF, ranging from 15 to 35 is assigned to each of the five items. Within each item, conditions are described and a Weighting Factor (WF) assigned to each condition. The evaluation of the potential of the building for energy conservation measures is based on the sum of the products of the RIF's and WF's. The higher this combined value, the greater the potential for energy savings. Determine the Weighting Factor as follows:

A. HVAC System Type (RIF: 35)
Note the type of HVAC system found in your building.

	WF
☐ Reheat or Dual Duct	1.0
☐ Multizone or Induction Units	.9
☐ Rooftop Units, Wall Units, or Unit Ventilators	.8
☐ Fancoil, VAV, or Heat and Vent System	.7
☐ Radiation, Unit Heaters (no fan systems)	.6

B. Outside Air (RIF:20)
Note the amount of outside air introduced through your ventilation systems. Check the ventilation system for outside air percentage.

	WF
☐ 76 to 100% Outside Air	1.0
☐ 51 to 75% Outside Air	.9
☐ 26 to 50% Outside Air	.8
☐ 10 to 25% Outside Air	.7
☐ Infiltration, toilet exhausts only	.6

C. Building Envelope (RIF:15)
Percentage of glass area can be estimated by dividing the glass area in a typical wall by the total wall area. The degree of infiltration can be determined by noting how tight-fitting the outside doors and windows are in their frames; i.e., tight fitting doors and windows, low infiltration; loose fitting, high infiltration.

	WF
☐ Bldgs over 40% glass and large infiltration	1.0
☐ Bldgs over 40% glass	.9
☐ Bldgs with large infiltration	.8
☐ Bldgs under 40% glass	.7
☐ Bldgs with low infiltration	.6
☐ Bldgs under 15% glass	.5

EVALUATION OF POTENTIAL FOR ENERGY CONSERVATION MEASURES

D. Lighting (RIF:15)
Add the wattage of all lamps in the building and divide by the gross floor area of the building.

	WF
☐ Lighting over 3 watts/SqFt	1.0
☐ Lighting 2 to 3 watts/SqFt	.9
☐ Lighting 1 to 2 watts/SqFt	.8
☐ Lighting reduced by changes in switching	.7
☐ Lighting levels that cannot be reduced	.6

E. Fan Energy (RIF:15)
To determine square feet per fan horsepower (HP), divide building gross floor area by total HP of all HVAC and ventilating fans in the building. HP rating can be found on nameplates of pumps and motors in your air handling systems.

	WF
☐ Under 200 SqFt per fan HP	1.0
☐ 200-600 SqFt per fan HP	.9
☐ 601-1000 SqFt per fan HP	.8
☐ 1001-1500 SqFt per fan HP	.7
☐ 1501-2000 SqFt per fan HP	.6
☐ Over 2000 SqFt per fan HP	.5

Complete the following table to determine the energy conservation measure potential index:

	RIF	x WF =	EVALUATION
a. HVAC System Type	35		
b. Ratio Outside Air	20		
c. Bldg. Envelope — % Glass and Infiltration	15		
d. Lighting Levels	15		
e. Fan Energy	15		
Energy conservation measure potential index:			

EVALUATION OF POTENTIAL FOR SOLAR AND RENEWABLE RESOURCE MATERIALS

This evaluation is to be completed in the same manner as the Energy Conservation Measures Evaluation in the previous section.

A. Available Insolation (RIF:30)
Available insolation is a function of geographic location and site characteristics. Determine average annual horizontal insolation on a horizontal surface from information provided by the State or from the National Weather Service data for your location. Observe whether the building is shaded or unshaded (a building whose roof and south-facing wall are approximately more than half-shaded for more than approximately four hours per day — should be considered "shaded".) If the building itself is shaded, note whether there is open, unshaded land available adjacent to the building site.

	WF
☐ Unshaded and 1300 BTU/SqFt or more	1.0
☐ Unshaded, less than 1300 BTU/SqFt	.5
☐ Open land and 1300 BTU/SqFt or more	1.0
☐ Open land, less than 1300 BTU/SqFt	.5
☐ Shaded, 1300 BTU/SqFt or more	.2
☐ Shaded, less than 1300 BTU/SqFt	.1

B. Wind Speed (RIF:30)
Determine average monthy wind speed, using data supplied by the State, obtained from the National Weather Service or local records. Note whether there are natural or man-made barriers in the direction of prevailing winds.

	WF
☐ Greater than 15 mph, no obstructions	1.0
☐ Between 10-15 mph, no obstructions	.5
☐ Greater than 15 mph, some obstructions	.7
☐ Between 10-15 mph, some obstructions	.3
☐ Less than 10 mph	.2

C. Fuel Used (RIF:20)
Note the fuel used for space and water heating

	WF
☐ All electric	1.0
☐ Oil or gas heat	.8
☐ Coal heat	.4
☐ Oil or gas heat, hot water	.4
☐ Coal heat, hot water	.2

EVALUATION OF POTENTIAL FOR SOLAR AND RENEWABLE RESOURCE MATERIALS

D. Wall Characteristics (RIF:20)
Determine the glass area of the south-facing walls as a percentage of the total wall area, noting the construction material.

	WF
☐ Over 75% glass; masonry	1.0
☐ Over 75% glass; aluminum or metal	.7
☐ Over 75% glass; wood or other material	.6
☐ 25%-75% glass; masonry	.7
☐ 25%-75 glass; aluminum or metal	.6
☐ 25%-75% glass; wood or other material	.4
☐ Under 25% glass; masonry	.5
☐ Under 25% glass; aluminum or metal	.3
☐ Under 25% glass; wood or other material	.2

E. Building Characteristics (RIF:10)
A "favorable" building is one which is compact (i.e., square or rectangular), in which the equipment is in one location on the roof or adjacent to the south-facing wall.

A building which is "fair" would be other than compact (i.e., E-shaped, L-shaped, etc.) but in which the equipment is in one location on the roof or adjacent to the south-facing wall, or a building which is compact but in which the equipment is within 5 floors of the roof or 50 feet of the south-facing wall.

A building which is not compact and in which the equipment is located beyond five floors of the roof and 50 feet of the south-facing wall is to be characterized as "moderate".

A high irregular building or one in which equipment is in scattered location — most of which are more than five floors from the roof or 50 feet from the south-facing wall — is characterized as "poor".

	WF
☐ Favorable	1.0
☐ Fair	.8
☐ Moderate	.5
☐ Poor	.2

EVALUATION OF POTENTIAL FOR SOLAR AND RENEWABLE RESOURCE MATERIALS

F. Roof Characteristics (RIF:10)

Characterize the building as "favorable" if the roof is flat or pitched nearly to the south, if the roofing is built-up, shingled or otherwise sufficiently durable to withstand mounting and maintaining solar collectors, if the structural members are strong enough to support additional weight and the roof area is free of obstructions.

A "fair" rating would be given a building meeting the above conditions except that the roof pitch is only approximately in the direction of south or where there are roof obstructions.

Describe a building as "moderate" if the roof pitch is only approximately toward the south and there are roof obstructions.

A building which meets none of these conditions is characterized as "poor".

	WF
☐ Favorable	1.0
☐ Fair	.8
☐ Moderate	.5
☐ Poor	.2

Complete the following table to determine the solar and renewable resources potential index:

	RIF	x WF =	EVALUATION
a. Insolation Available	30		
b. Wind Speed	30		
c. Fuel Used	20		
d. Wall Characteristics	20		
e. Building Characteristics	10		
f. Roof Characteristics	10		
Solar and renewable resource measure potential index:			

APPENDIX C. ENERGY DATA COLLECTION FORMS

This section organizes important school data according to the following categories:

A. General Administrative Information
B. Human Systems
C. Structural Systems
D. Energy Systems
    A. Lighting
    B. Mechanical
    C. Special Systems
E. Energy Consumption Summary
F. Solar and Renewable Resource Potential

A. GENERAL ADMINISTRATIVE INFORMATION

1. Name of Building or Complex: _____
   Owner: _____
   Public _____ Private _____
   Non-Profit _____ Indian Tribe _____
2. Building Category:
   ☐ Elementary   ☐ Vocational
   ☐ Secondary   ☐ LEA Admin.
   ☐ Junior College   ☐ Other, Specify
   ☐ College or Univ.   _____
3. Building Address: _____ City _____
   State _____ Zip _____ Telephone Number _____
4. Year constructed: _____ Year of last major addition or modification: _____
5. Principal/Manager: _____ Telephone Number: _____
   Head Custodian/Operator: _____ Telephone Number: _____
6. Energy Management Coordinator designated: ☐ Yes   ☐ No
7. Anticipated building modifications: _____
   _____
   _____

8. Previous energy audit completed: ☐ Yes   ☐ No   Specify, _____
   _____

9. Conservation Measures (retrofit) already implemented or under consideration:
   Yes   No   Specify project, cost and expected energy savings: _____
   _____
   _____

10. Previous architectural/engineering studies: ☐ Yes   ☐ No   Specify: _____
    _____

SCHOOLHOUSE ENERGY EFFICIENCY DEMONSTRATION: AN ENERGY AUDIT PROGRAM 157

B. HUMAN SYSTEMS

1. Complete occupancy schedule. If the school operates on a seasonal schedule, or has other periods of at least a week's duration when the building is only partially occupied, the number of weeks partial use by calendar quarter should be entered, along with the approximate percentage of total gross square feet in use during such periods.
2. Note U.S. heating and cooling zone in which building is located (see maps below) and climate information unique to your location. Call your utility or local weather bureau.

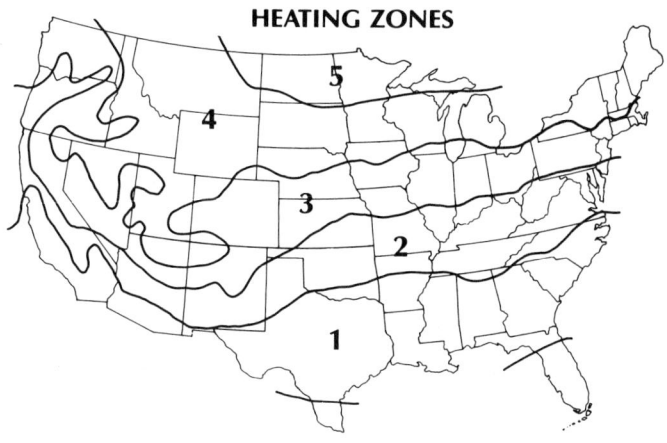

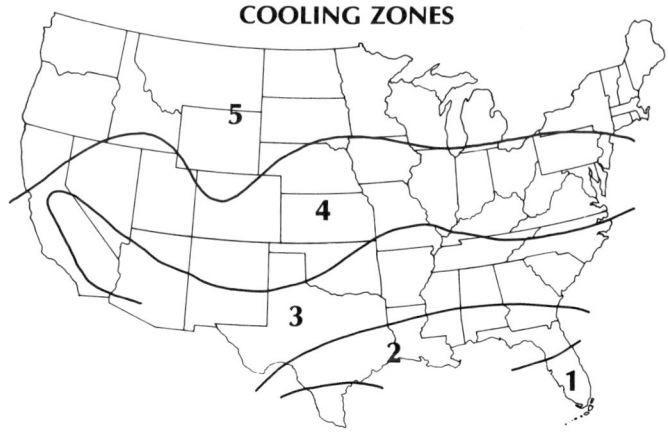

B. HUMAN SYSTEMS

Day(s)	Time Period	Average Occupancy	or	% gsf Occupied	No. of Hours	Weeks/Year
Mon-Fri	day					
	evening					
	night					
Saturday	day					
	evening					
	night					
Sunday	day					
	evening					
	night					

Quarterly Partial Usage:

Quarter	Weeks	% Gross Square Feet (gsf)
1st		
2nd		
3rd		
4th		

2. Building Location:
U.S. Heating Zone # _____ U.S. Cooling Zone # _____
Annual Heating Degree Days _____ Annual Cooling Degree Days _____
(Base 65°F)

SCHOOLHOUSE ENERGY EFFICIENCY DEMONSTRATION: AN ENERGY AUDIT PROGRAM 159

C. STRUCTURAL SYSTEMS

1. To calculate gross square feet (gsf) multiply the outside dimensions or measure from the centerline of common walls and multiply by the number of floors. If the building has wings, or the number of floors varies in one part from another, divide the building into sections, calculate the area of each section, and total. (Deduct from the total area any parking garages or other areas which are neither heated nor cooled.)
2-3. Note type and thickness of insulation material in roof, walls and floors. Note if there is none.
4. Sketch position of facility on site.
5. Briefly describe general building conditions. Indicate any structural flaws or deterioration, if applicable. If building is in good condition, so state.

1. Gross Floor Area: _____ Gross _____ Sq. Ft.
2. Insulation Type: Roof _____ Wall _____ Floor _____
3. Insulation Thickness: Roof _____ Wall _____ Floor _____
4. What is orientation of building on site? (Draw sketch.)

```
      N
      |
W ----+---- E
      |
      S
```

5. Description of general building conditions:

_____
_____
_____

160 3/ENERGY ANALYSIS TECHNIQUES AND STRATEGIES

D. ENERGY SYSTEMS

A. Lighting

1-2. For each lighting type, fluorescent and incandescent, note the percentage of gross square feet of the building illuminated. Include an estimate of average usage in hours per week and hours per year.
   3. Determine the total wattage presently used to illuminate the building's interior and divide by the gross floor area to compute an average lighting level in watts per square foot.
   4. Indicate potential for lower lighting levels: classrooms, corridors, offices, gymnasium(s), auditorium(s), library, cafeteria, shops, others.

1. Fluorescent:
   Percentage of Gross Sq.Ft: _____ % Usage: _____ Hr/Wk _____ Hr/Yr
2. Incandescent:
   Percentage of Gross Sq.Ft: _____ % Usage: _____ Hr/Wk _____ Hr/Yr
3. Average Interior Lighting Level: _____ Watts/SqFt
4. Potential for lower lighting levels (check box if yes):
   ☐ Classrooms   ☐ Auditoriums   ☐ Other (Specify)
   ☐ Corridors   ☐ Library   _____
   ☐ Offices   ☐ Cafeteria   _____
   ☐ Gymnasiums   ☐ Shop(s)

D. ENERGY SYSTEMS

B. Mechanical

1-10. Check the type and capacity of HVAC systems found in your school. If knowledge of the system is not available, obtain the information from the mechanical engineer, blueprints, specifications or nameplates. Total the cubic feet per minute (CFM) of air that the air systems supply to the building. Note what percentage of outside air is used. Note the heating and cooling capacities and fan horsepower.
11. Note the principal fuels used by the heating and cooling systems.
12-13. Larger buildings tend to have boilers or purchase hot water or steam. Small buildings tend to have unitary direct fired equipment. Determine your system type. Estimate the number of hours per day that the heating plant operates during winter. Examine the cooling equipment to determine system(s) type. Estimate the number of hours per day that the cooling system(s) operates during summer.

System Type	Total CFM	Minimum % Outside Air	Capacity BTU/hr	Fan Horsepower
1. Terminal Reheat				
2. Multizone				
3. Dual duct				
4. Variable Air Volume				
5. Induction				
6. Fan Coil		N/A		
7. Heat Pump				
8. Air Exhaust			N/A	N/A
9. Radiation	N/A	N/A		N/A
10. Other (Specify) _____				

11. Heating System

_____
Fuel Type

12. Systems Types: (Check √)
____ Boilers
____ Purchased water or steam
____ Unitary Direct Fired
____ Furnaces
____ Package Equipment

13. Operation Profile:
____ hrs/weekday
____ hrs/Saturday
____ hrs/Sunday

Cooling System

_____
Fuel Type

System Types: (Check √)
____ Absorption
____ Electric Drive
____ Steam Turbine Drive
____ Water Cooled Packaged Unit
____ Air Cooled Packaged Unit

Operation Profile:
____ hrs/weekday
____ hrs/Saturday
____ hrs/Sunday

D. ENERGY SYSTEMS

C. Special Systems
1. Indicate water heating source.
2. Complete data for swimming pool, if applicable.
3. Are laundry services provided? Enter fuel type.
4. Are food services provided?
5. List type and number of office equipment.
6. List kind and wattage of special high electrical demand equipment; e.g., kilns, electric welders.

1. Domestic Water Heated By: Electricity _____ Natural Gas _____
   Other _____
   (Specify)
2. Swimming Pool:
   ☐ Indoor   ☐ Outdoor
   Heating Fuel: _____
   Normal Water Temperature: _____ °F
3. Laundry Services ☐ Yes   ☐ No
   Fuel Type: Washers _____ Dryers _____
4. Food Services ☐ Yes   ☐ No
   Major cooking fuel: _____
5. Office Machines (kind and number of each)

   _____ #_____
   _____ #_____
   _____ #_____
   _____ #_____

6. High Demand Electrical Equipment (kind and wattage)

   _____ _____ watts
   _____ _____ watts
   _____ _____ watts
   _____ _____ watts

E. ENERGY CONSUMPTION SUMMARY

1. Complete fuel use summary for base year (or last 12 months if no base year has been established), using utility records and Energy Usage Forms in Section 2. (If no past records have been kept, call your utility.) Multiply by the conversion factors (as required by the *Federal Register,* Section 450.42(11), April 2, 1979) and enter the results in Column D as annual BTU consumption.

   Transfer annual cost for each fuel from the appropriate Energy Usage Form to Column E.

   Compute consumption in BTU's per gross square foot per year by dividing the total of the entries in Column D by (C-1) gross square feet.

   Compute energy dollars per gross square foot per year by dividing the total of the entries in Column E by gsf. (Obtain gross floor area from previous chart (C-1) and energy costs from Energy Usage Forms.)

   Notes: • No. 2 oil should include other distillate fuel oils.
   • No. 6 oil should include other residual fuel oils.
   • Use a standard engineering manual or conversion factors provided by the State for other fuels (Row 8, Column C).

2. Based on your past year's utility bills, complete peak electrical demand data. For buildings or complexes over 200,000 gsf or if the electric rate contains a demand charge, determine if demand is recorded. Note times at which typical peaks occur during daily operation. Also note whether demand fluctuates on a seasonal basis, indicating month in which the highest demand occurs.

3. *If data is available,* indicate the fuel used by each of the major energy using systems listed and the annual consumption of each. For #5 "special", indicate special purpose facilities (e.g., food service, laundry) which use significant amounts of energy, fuel type used and annual usage.

E. ENERGY CONSUMPTION SUMMARY

A	B	C	D	E	F	G
Fuel	Previous 12 Month Totals	Conversion Factor	BTU's Consumed	Annual Cost	BTU's/Gross SqFt/Year	S/Gross SqFt/Yr
1. Electricity	KWH x	11,600 =				
2. Natural Gas	CCF x	103,000 =				
3. #2 Oil	gallons x	138,690 =				
4. #6 Oil	gallons x	149,690 =				
5. Steam	pounds x	1,390 =				
6. Coal	tons x	24,500,000 =				
7. Propane	gallons x	95,475 =				
8. Other, Specify	x	=				
9. TOTALS	N/A	N/A				

2. **Peak Electrical demand:**
   Daily: _____ KW   Annual: _____ KW
   Time: _____   Month: _____

3. **Fuel Use by Major Energy-Using Systems:**

SYSTEM	FUEL TYPE	ANNUAL USE
Heating		
Cooling		
Hot Water		
Lighting		
Special, specify		

F. SOLAR AND RENEWABLE RESOURCE POTENTIAL

1. Check characteristics of adjacent property.
2. Indicate nature of location.
3. Indicate building shape as square, rectangular, E-shaped, H-shaped, L-shaped or attach a rough sketch of the configuration. Note whether roof and southern wall are shaded or unshaded.
4. Note roof design. For the orientation of a pitched roof, indicate the *compass direction* of a line perpendicular to the ridgeline in the direction of the down slope. Note presence of roof obstructions such as chimneys, space conditioning equipment, water towers, mechanical rooms and stairwells. Identify the principal structural material of the roof; e.g., steel, concrete, or wood structural components. Also identify the type of roofing such as shingle, slate or built-up.
5. Indicate structure composition of southern facing wall. Determine percentage of wall area covered by glass.
6. Check type and location of space and water heating equipment.
7. Using information from the National Weather Service or your State's energy office enter monthly average solar insolation and wind speeds.
8. Note any special conditions or characteristics related to potential for solar or other renewable resource application.

APPENDIX D. CASE STUDIES

BRIARLAKE ELEMENTARY SCHOOL DECATUR, GEORGIA

Description: The school, built in 1964, is a single story T-shaped building with 32,407 sq. ft. The walls are uninsulated with extensive window area, 39% of the exterior wall. The school is situated on a hill but has enough trees to buffer the building from prevailing winds.

Energy Budget: $18,539   Gas: $5,724   Electricity: $12,815

Prior to the energy audit, Briarlake School used 17.6 BTU/sq. ft./degree day. If the SEED recommendations are implemented, the consumption can be reduced to approximately 4.4 BTU/sq. ft./degree day.

Recommendations	Cost	Annual Savings
TIGHTEN THE BUILDING ENVELOPE		
• Weatherstrip doors	$ 606	$ 273
• Caulk and regasket windows	1,900	540
• Place pegboard over fresh air inlets	260	1,211
• Insulate upper portion of windows	4,902	1,501
PROCEDURAL CHANGES		
• Tune thermostat system	600	1,600
• Cease using kiln to heat kitchen	—	60
• Install Energy Conservation Plaque	20	—
MODIFY EQUIPMENT		
• Reduce hydronic pumping rate	150	425
• Install high efficiency fluorescent lamps	100	345
• Put water heater on timer	154	350
• Install fresh air inlet for heat pump	300	464
TOTAL	**$8,992**	**$6,769**

ROY L. CLARK ELEMENTARY SCHOOL OAK LAWN, ILLINOIS

Description: The school, built in 1964 with additions in 1967 and 1971, is a single story, L-shaped structure. The foundation is an uninsulated slab-on-grade and totals 46,000 sq. ft. The walls are insulated and are built of brick and block totaling 15,653 sq. ft. with only 7% single glazed windows. The windows are shaded by a 3 ft. overhang. The roof is also well insulated. However, the school site is completely devoid of buffering vegetation.

Energy Budget: $12,012   **Gas:** $6,012   **Electricity:** $6,000

If all SEED recommendations are fully implemented, Clark Elementary can reduce its energy consumption as much as 47%.

Recommendations	Cost	Annual Savings
TIGHTEN THE BUILDING ENVELOPE		
• Reduce fresh air intake	$ 280	$2,000
• Insulate upper portion of windows	525	1,870
PROCEDURAL CHANGES		
• Tune thermostat system	260	200
• Deactivate unnecessary lamps	—	1,577
• Adjust back draft dampers	—	15
• Reduce temperature on water heaters	—	75
• Install Energy Conservation Plaques	20	—
• Institute housekeeping measures	45	245
MODIFY EQUIPMENT		
• Install high efficiency fluorescent lamps	1,637	777
• Install timers on exterior lights	100	400
• Place timers on roof top exhaust fans	500	400
TOTAL	**$3,367**	**$7,559**

HILLSIDE JUNIOR HIGH SCHOOL MANCHESTER, NEW HAMPSHIRE

Description: The school, constructed in 1965, is a three story structure built on pier-on-beam and slab-on-grade with 116,500 sq. ft. of floor space. A bomb shelter is located under the slab-on-grade portion of the school. The walls are of brick and block cavity construction with only 8.5% window area. Only the roof is insulated. The structure is built into the side of a moderately sloping hill and has no landscaping other than low shrubs.

Energy Budget: $33,961 **Heating Oil: $13,878** **Electricity: $20,083**

This school used only 6.1 BTU/sq. ft./degree day due primarily to the small amount of window area. However, as a result of the energy audit the projected reduction in the rate of energy consumption is 4.1 BTU/sq. ft./degree day.

Recommendations	Cost	Annual Savings
TIGHTEN THE BUILDING ENVELOPE		
• Seal unused shop door	$ 5	$ 20
• Place pegboard over fresh air inlets	240	760
• Insulate upper portion of windows	1,640	1,050
• Place dampers on roof-top exhaust	7,000	3,910
PROCEDURAL CHANGES		
• Tune thermostat system	—	3,370
• Deactivate unnecessary classroom lights	—	1,167
• Reduce temperature on water heater	—	140
• Shut down boilers when possible	1,000	300
• Savings due to reduction in system operating time	—	800
• Install Energy Conservation Plaque	200	—
• Institute housekeeping measures	20	104
MODIFY EQUIPMENT		
• Reduce hydronic pumping rates	2,500	2,800
• Install high efficiency fluorescent lamps	154	618
• Install independent water heater	1,250	500
• Place timers on exhaust fans	300	200
TOTAL	**$14,309**	**$15,739**

IROQUOIS MIDDLE SCHOOL WEST IRONDEQUOIT, NEW YORK

Description: The school, constructed in 1948 with an addition in 1962, is a three level, T-shaped structure. The basement foundation is an uninsulated slab-on-grade with 24,456 sq. ft. The building has a total floor space of 65,345 sq. ft. The walls are uninsulated and composed of brick and tile totaling 22,478 sq. ft. of which 23% is single glazed, operable windows. The site is surrounded by a residential area with little significant buffering vegetation.

Energy Budget: $21,765   Gas: $11,240   Electricity: $10,525

During the 1977-1978 school year, Iroquois Middle School consumed energy at a rate of 9.2 BTU/sq. ft./degree day. If the SEED recommendations are fully implemented, the consumption rate could be reduced to 5.4 BTU/sq. ft./degree day.

Recommendations	Cost	Annual Savings
TIGHTEN THE BUILDING ENVELOPE		
• Weatherstrip doors	$ 350	$ 85
• Seal and caulk windows	92	153
• Insulate upper portion of windows	610	1,890
PROCEDURAL CHANGES		
• Tune thermostat system	460	510
• Deactivate unnecessary lamps	—	1,500
• Deactivate portable freezer when not in use	—	200
• Reduce temperature on water heater	—	60
• Institute housekeeping measures	—	130
MODIFY EQUIPMENT		
• Install high efficiency fluorescent lamps	533	450
• Install multivapor lamps in gym	1,200	528
• Insulate boiler manifold	200	100
• Insulate expansion tank	122	38
• Replace gasket on ice cream freezer	15	50
• Replace kettle	500	320
• Install dampers and timers on rooftop exhaust fans	175	830
TOTAL	**$4,257**	**$6,844**

MONTCLAIR ELEMENTARY SCHOOL GARLAND, TEXAS

Description: The school, constructed in 1968 and retrofitted for air-conditioning in 1973, is a 31,000 sq. ft. single story, U-shaped structure made of brick and clay tile on a slab-on-grade foundation. The total wall area is 12,462 sq. ft. of which 34% is single glazed, operable windows. The windows are shaded by 5-ft. overhangs. The roof is insulated only through use of an acoustical tile ceiling. There was no buffering landscaping.

Energy Budget: $13,161   Gas: $3,483   Electricity: $9,678

If the following recommendations are fully implemented, Montclair Elementary could reduce its energy consumption as much as 58%.

Recommendations	Cost	Annual Savings
TIGHTEN THE BUILDING ENVELOPE		
• Plant a windbreak	$1,200	$ 200
• Install and seal partition plenum	170	2,200
PROCEDURAL CHANGES		
• Turn off heaters at night	—	1,800
• Lower temperature on refrigerator	—	100
• Seal economizer	50	300
MODIFY EQUIPMENT		
• Install high efficiency fluorescent lamps	703	1,673
• Reduce compressor tonnage rating	—	300
• Use high EER compressor	2,100	1,000
• Insulate water heater	40	72
TOTAL	**$4,263**	**$7,645**

9. HVAC System Analysis

Kermit S. Harmon, Jr., P.E.

*President, Texas Energy Engineers, Inc.
Houston, Texas**

		Page
I.	Energy Accounting	172
II.	General Analysis	174
	A. Facility Familiarization	174
	B. System Familiarization	174
	C. System/Facility Interaction Analysis	175
III.	Energy Audit Review	176
	A. Review Energy Consumption for Suspected Energy-Abusing System	176
	B. Perform Efficiency Tests on Suspect Equipment and System Functions	176
	C. Establish Levels for All Energy-Consuming Categories	176
IV.	Preliminary Retrofit Opportunities List	176
	A. Order of Cost/Benefit Ratio	176
V.	Temperature Control Systems	177
VI.	Equipment Efficiency Check	181
	A. Boilers	181
	B. Chillers	182
	C. Heat Exchangers	182
	D. Pumps	183
	E. Fans	183
	F. Motors	184
	G. Cooling Towers	185

I. ENERGY ACCOUNTING

A too frequently committed error in energy analysis stems from neglecting the universal law of energy conservation that states that energy can neither be created nor destroyed. The total amount of energy in the universe remains constant. By accepting this law, we can astutely analyze how a given quantity or constant flow of energy is utilized.

A helpful analogy is to think of energy as a bank account: you cannot draw out more cash than the account possesses.

The closely related and nearly synonymous terms "work" and "energy" should not be confused. The "work" concept may be considered as forming a crucial link between early Newtonian physics and the modern term "energy."

The nineteenth century term "energy" has been defined as the capacity to do work, and, as such, "energy" may be present in many different forms: we may have electrical, mechanical, thermal, magnetic, chemical, or nuclear

---

*Current affiliation: Texas Energy Engineers, Inc., Houston, Texas.

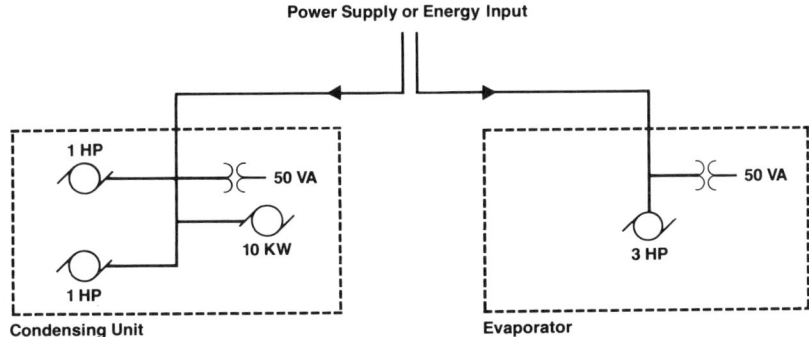

Figure 9-1. One-line electrical diagram.

"energy." The closely related term "power" is the rate of doing work, or the rate at which energy is transformed from one form to another.

During the process of energy analysis, many forms of energy transformation and transmission may be realized, yet the energy will not cease to exist. Only the energy's form will change.

Refer to Figure 9-1 for a simple hypothetical example containing the following elements of a typical HVAC system:

1. 10-kW sealed hermetic compressor
2. 3-hp fan motor (2.75-bhp load)
3. Two 1-hp condenser fan motors (1.85-bhp total load)
4. Two 50-VA control transformers (35-VA load each)

Assume that motor efficiency and drive efficiency are 0.85 and 0.95 respectively, for the evaporator fan and the condenser fan (items 2 and 3). The loads in kilowatts of this equipment may be expressed as follows:

1. Compressor = 10.00
2. Evaporator fan $\dfrac{2.75 \text{ hp}}{0.85 \times 0.95} \times \dfrac{0.746 \text{ kW}}{\text{hp}}$ = 2.54
3. Condenser fan $\dfrac{1.85}{0.85 \times 0.95} \times \dfrac{0.746 \text{ kW}}{\text{hp}}$ = 1.71
4. $2 \times 35 \text{ VA} \times 0.90 \text{ PF} \times \dfrac{\text{kW}}{1{,}000 \text{ W}}$ = 0.06

Total power input = 14.31 kW

If this system provides ten tons of cooling, the system energy efficiency ratio (EER) can be obtained by dividing ten tons by 14.31 kW, as follows:

$$\text{EER} = \dfrac{10 \text{ tons}}{14.31 \text{ kW}} \times \dfrac{12{,}000 \text{ Btu}}{\text{ton-hr}} \times \dfrac{\text{kW}}{1{,}000 \text{ watt}} = 8.39 \text{ Btu/watt-hr}$$

The 10 or 120,000 Btu (126.7 × 10⁶ ton-hr joules) of cooling is not energy from the system, but rather the *absence of energy* within the building. Heat energy was removed from the building by pumping the heat to the outdoors with mechanical refrigeration. So what happened to the energy purchased from the power company if it was not the 120,000 Btu of cooling? The 14.31 kWh of purchased electrical power is equivalent to 48,840 Btu (51.5 × 10⁶ J) of thermal energy (3413 Btu/kWh). This conversion rate is for *boundary* energy, not *raw source* energy. Raw source energy would be approximately three times as great, owing to generation, transmission, and distribution system losses. What actually happens is that the 120,000 Btu (126.7 × 10⁶ J) of heat pumped from inside the building, plus the 48,840 Btu (51.5 × 10⁶ J) of converted electrical energy, are transferred to the outdoor air by the condensing unit coil. Figure 9-2 illus-

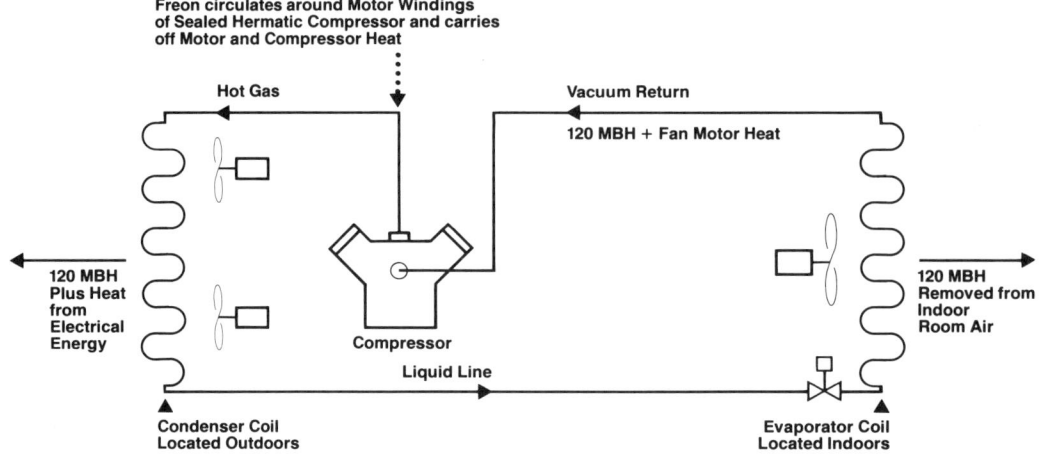

Figure 9-2. Refrigerant piping diagram.

trates how almost all of the 168,840 Btu (178.2 × 10⁶ J) total is transferred to the outdoor air by the condenser coil. A portion of the heat from the sealed, hermetic, compressor casing and hot gas piping and all of the heat from the condenser fan motors are transferred directly to the outdoor air by conduction, convection, and radiation.

The important point is that *all* energy considered should be accounted for. The fact that the system receives energy from more than one source [48,840 Btu (51.5 × 10⁶ J) from the power company and 120,000 Btu (126.7 × 10⁶ J) from room air] should not confuse the basic issue. The system eventually transfers the 168,840 Btu (178.2 × 10⁶ J) total to the outdoor air.

Of what value is such accurate energy accounting? Such energy analysis has caused the discovery of many energy-conserving opportunities, heat reclamation systems being probably the most significant discovery. The 168,840 Btu (178.2 × 10⁶ J) of otherwise lost heat, if entirely reclaimed, can heat 579 gallons (2.19 m³) of domestic water from 70°F (21°C) to 105°F (41°C). If the cooling load is for a space that requires winter cooling, such as interior computer rooms, the reclaimed heat can be used to heat the building perimeter spaces.

Please note that reclaimed heat is free, and is replacing an electric resistant heat equivalent of 50 kWh of boundary energy or 150 kWh of raw source energy. If purchased at 5¢ per kWh, 50 kW of resistant heat would cost $2.50 per hour. Reclaimed heat use, therefore, could easily amount to substantial energy cost savings over a year's operation.

II. GENERAL ANALYSIS

A. Facility Familiarization

When analyzing any HVAC system, first become familiar with the facility that the system serves. Review exterior elevations, floor plans, the roof, and each type of space within the facility. Find out how the facility is used, the number of employees usually present, employee tasks, and normal working hours.

Ask sufficient questions to identify the better retrofitting opportunities. For example, through questioning you may find that one of the vice presidents works ten hours every Sunday and operates the 500-ton central plant at a cost of $25.00 per hour to keep his office cool. In this case, a separate direct expansion (DX) system might be installed in the vice president's office and pay for itself with the first month's energy savings.

B. System Familiarization

After facility familiarization, become thoroughly acquainted with the facility's HVAC systems. Think of them as energy systems, and list them in categories such as unitary DX, split DX, central chilled water, two-pipe, four-pipe, and so on. Note the distribution patterns of all HVAC power wiring, piping, ductwork, and supply and return air passages. Think of such routes as avenues for energy transmission to and from the listed energy systems. Observe the air path in such detail as to learn exactly how air enters and leaves each environmental space.

Become knowledgeable about the HVAC system controls. The two basic categories of controls are manual and automatic. Manual controls require an operator to make observations, to make a decision based on these observations, and to manually adjust the controls to create the desired environmental results. Many exhaust systems, pumping systems, air-handling units, and similar types of equipment are manually controlled. All automatic controls have a form of master manual control; this may be only the circuit breaker or fusible switch that feeds power to the control system or the HVAC system.

Automatic controls, when energized automatically, respond to such devices as thermostats, pressure sensors, or flow switches. Most automatic temperature control devices are reset manually like typical room thermostats. However, some control devices are reset from a sensor signal, a combination of sensor signals such as a discriminator control circuit, or a totalizer control circuit. Time clock or computer controls may further automate the control system, thus reducing or eliminating manual tasks. Obviously, computer-made decisions must be better than manually made decisions if energy is to be saved.

With this information in mind, review system installation drawings, temperature control drawings, and the actual systems. Note control locations and how they really work. In particular, note room thermostat locations, since these thermostats are the final points of temperature control. Also, note what device, such as damper, valve, or electric reheat coil, the room thermostat triggers to achieve final room temperature.

Barring major complications, the foregoing effort should be completed within a day or two for a 100,000-sq-ft facility. Avoid being delayed by unimportant details as you become acquainted with the overall facility and its HVAC systems. Later in the analysis, you will want to focus in greater detail on areas where opportunities appear to exist.

C. System/Facility Interaction Analysis

After becoming familiar with the facility and its HVAC systems, analyze how well the HVAC systems respond to facility requirements. Interview operators, take detailed notes of their daily activities, review their daily log sheets, and review occupant complaint logs. Complaint call logs are very important; complaints may frequently be from energy-abusive areas that are overheated or overcooled.

The key question is: Does the system interact with the facility so that the system fulfills occupant comfort needs without being forced into an energy-abusive mode of operation? Related questions are:

- Do some occupants use portable heaters?
- Can computer rooms or other areas used 24 hours per day be operated without operating the entire central plant?
- Do any occupants open windows?
- Is extensive makeup air required by a starved exhaust hood that keeps the building under a vacuum?
- Must the heating boiler be operated all summer to provide a small amount of domestic hot water?
- Are fans computer-cycled for demand control instead of being slowed down and fitted with smaller motors with better power factors to continuously save energy?
- Has lighting been reduced to save energy without corresponding reductions in fan horsepower?

Such questions are numerous and will continue to be refined as we gain experience in analyzing HVAC systems for retrofit opportunities.

III. ENERGY AUDIT REVIEW

A. Review Energy Consumption for Suspected Energy-Abusing System

Although most energy audits do not investigate HVAC systems in as much detail as would most design consultants, audit data can be most informative. For certain types of buildings, air-conditioning energy usage can be estimated by its absence from energy bills during winter months, and heating energy usage can be estimated by its absence from energy bills during summer months. Cooling and heating energy usage can be even more accurately broken down when a fossil fuel is used for heating while electrical energy is used for cooling. It is easy to establish fairly accurate energy consumption estimates for lighting, fan motors, and pumps; this can be done by verifying their loads and multiplying them by known operating hours. The numerous notes contained in some energy audits may also provide ideas for retrofitting energy-consuming areas.

B. Perform Efficiency Tests on Suspect Equipment and System Functions

Completely thorough efficiency tests on equipment and systems are costly, and may not be necessary in some cases. In most cases, the fossil fuel boiler is not operating efficiently. The primary reason for testing the boiler would be to prove to the owner that efficiency upgrading could pay for itself within two years. Chiller efficiencies can, in some cases, be calculated from logged data obtained from chiller instrumentation and water balance and service reports. Chiller efficiency is very low if most of the operating hours are at low load, particularly if load is often added to keep the chillers on line. Fan efficiencies should be checked by calculating the air transport factor as defined in ASHRAE 90-75. Deteriorated cooling towers offer visual evidence of water chilling plant efficiency loss.

C. Establish Levels for All Energy-Consuming Categories

Although considerable data have been published on energy budgets, the breakdowns have often been in percentages. Unfortunately, percentages are practically meaningless if only one category of energy consumption changes significantly.

Computer-modeling programs probably provide the best reference for a breakdown of energy use by categories. However, there is still a need to establish the levels of accuracy of some programs by modeling buildings that are equipped with meters to break down the energy.

It is proposed that levels be established for the energy use categories against what experience has shown can be achieved through conscientious design effort. For example, internal loads can be reduced through high-efficiency luminaries and efficient office or processing equipment. Use generous-sized ductwork and piping to minimize fan and pump horsepower. The known achievable loads of these items, when calculated against operating hours, will provide realistic energy quantities against which to compare estimates of the existing systems.

IV. PRELIMINARY RETROFIT OPPORTUNITIES LIST

A. Order of Cost/Benefit Ratio

At this point, sufficient familiarity with the facility and HVAC systems has been gained. Now, an idea-generating exercise can prove beneficial. This can be an excellent group exercise, and it may include individuals familiar with the facility as well as individuals who do not know the facility but are known to have good retrofit ideas. To gain maximum benefit from the exercise, each participant must feel free to suggest all types of ideas, even those that may seem impractical. A seemingly impractical idea suggested by one participant may be complementary to an idea from another participant.

After completing the retrofitting list, review the list and delete items that are too impractical. This might be accomplished by a group vote on each item. Arrange the remaining items in descending order of cost/benefit ratio, with those items that have the best cost/benefit ratio listed first. It is not necessary to determine accurate cost/benefits at this time; the cost/benefit of ideas can be estimated by group matrix evaluation. For example, each group member can give each idea a cost/benefit value on a scale of one to ten, and the total number of points received for each item would establish that item's appraisal value for priority listing.

Select the highest 5, 10, or 20 items, or as many as appropriate, for pursuing an accurate evaluation. Conceptual design documents for each of the selected retrofit ideas will be required for pricing. The conceptual design documents may be freehand schematic drawings or written descriptions for some items, but they must be comprehensive enough that estimates may be obtained from a qualified contractor.

Both energy savings and resulting cost savings should be calculated for each retrofit item to determine the actual cost/benefit ratio. These calculations will require careful and astute judgment. Be practical; use shortcuts and approximations for calculating savings from items that represent low capital cost but could cost a lot of engineering hours if highly accurate calculations were produced. For example, it may be possible to install an adjustable time delay for $150.00 on outside air dampers so that the dampers remain closed during startup. It could easily cost over $200.00 in engineering time to take the weather data for each day of a typical year and manually calculate the savings or model the operation on a computer. However, a *reasonably accurate estimate* could be calculated in about 15 minutes by selecting a few typical weather conditions throughout the year. Energy-saving calculations for retrofit items that require considerable capital investment should be investigated selectively. Do not hesitate to invest greater engineering effort and computer time *if necessary* for accuracy.

The cost/benefit ratio may be expressed as simple payback period according to the following formula:

$$\text{Simple payback period} = \frac{\text{Capital cost of retrofit item}}{\text{First-year savings in energy cost}}$$

If energy cost is escalating at a greater rate than interest on investments, the *actual payback period* will be shorter than the period as calculated above.

A simple payback period of five years represents a 20% annual rate of return on the first year of the investment. The rate of return on investment will increase as energy rates increase; so such investments are obviously excellent for those owners who want an investment that exceeds inflation rates.

V. TEMPERATURE CONTROL SYSTEMS

A. Of all subsystems, elements, and components comprising an HVAC system, the temperature control systems generally offer more energy cost-reduction opportunities than are found anywhere else. Conversely, temperature control system malfunction is frequently the greatest energy abuser. For this reason, thoroughly inspect the system to determine if the controls are operating as intended. This inspection will also better acquaint you with the system and should reveal control system retrofit opportunities. The number of items that have been wrongly installed, or that have been tampered with, and remain undiscovered for lengthy periods, would shock some of the "theoretical" geniuses who engineered them. Tampering is often the result of an unsuccessful attempt to remedy a malfunctioning control. Obviously, the retrofit program should include correcting all defective controls.

B. After thoroughly inspecting and testing the temperature control system, develop a list of possible control enhancements. Some enhancements may have already been included in the previously discussed preliminary list. Evaluate the list and process the items in a manner similar to that used for the preliminary items retrofit list.

The following typical list is by no means comprehensive, but it should help you get started:

1. *Outside Air Closed on Startup.* Outside air dampers on HVAC systems with automatic temperature controls have routinely been programmed to open upon evaporator fan startup and to close upon fan shutoff. Imagine the energy abuse during the fifties and sixties when it was common to operate large central systems 24 hours per day. Now it is common to operate systems in office buildings for 12 hours per work day, although the occupants may use the building for no more than 9 hours per day. Early system startup helps bring building temperature under control before office hours. The outside air load during warmup or pulldown is wasted energy. Startup might be delayed by half an hour if outside air dampers remain closed, reducing the load, until office occupancy hours. Some operators wastefully operate cooling or heating systems in the more climatically severe afternoons or on holidays for a few hours, without occupants in the building, to more easily bring the building under temperature control the next morning, again with outside air dampers open. Time delay devices, time clocks, or remote manual overrides can be installed to control outside air flow on an "as needed basis." [Since the manuscript for this chapter was drafted, the city of Houston, Texas has passed an ordinance prohibiting use of outside air in unoccupied buildings.]

2. *Temperature Scheduling.* If HVAC system fluids temperatures can be limited to minimum deviation from final control point objectives, considerable energy can be saved. In other words, keep all temperatures as neutral as possible and yet deviate enough from neutral to permit the systems to do their job. The less the magnitude of the temperature differentials is, the less the rate of heat transfer. Savings are increased by diminishing undesirable gains and losses such as those through insulated walls of buildings, pipes, ducts, or storage tanks. Also, mechanical refrigeration efficiencies improve at lower heat pressures.

Supply Air. Supply air temperatures can be reset by outdoor air temperature, or by zone thermostats that have the greatest demand for cooling. Control circuits that reset from a thermostat having the greatest demand are commonly referred to as discriminator control circuits. Each air system must be analyzed individually to see how supply air temperature resets can be achieved to save energy.

For example, on a hot and cold deck multizone air unit, it may be practical to control the cold deck temperature reset with a discriminator control circuit and to control the hot deck temperature reset by outdoor air temperature. In this case, heating water might not flow to the warm deck coil when the outdoor air temperature is 60°F (16°C), and yet it might begin flowing to the coil as the outdoor air temperature drops below 60°F. The graduation could be such that the hot deck temperature uniformly rises from 80°F (27°C) to 120°F (49°C) as the outdoor air temperature decreases from 60°F to 10°F (-12°C). The savings thereby achieved on the multizone air unit are primarily due to minimizing the mixing of cooled air and heated air for temperature control. Savings continue to be realized even for zones in full-heat or full-cool mode, because of the inevitable air mixing that occurs from normal damper leakage.

An average double duct system can be viewed as a multizone unit. The hot deck and cold deck are extended by the hot duct and cold duct. The zone mixing dampers are located in double duct mixing boxes.

In variable air volume (VAV) systems such as the Varitrane or Mammouth's FM System, normally constant-temperature supply air is dumped into the return air plenum by terminal units as room thermostats become satisfied. Return air can therefore become even colder than room temperature, thus causing energy abuse by excessive temperature gain from the plenum's warm walls and roof. Supply air temperature can be reset by either outdoor air temperature or return air temperature. Additional benefit is gained by better air movement at low loads within the conditioned rooms. These types of VAV systems do not throttle the fan; therefore, fan horsepower savings are not lost from supply air temperature reset.

In VAV systems such as the Barber Coleman Jetronic (single inlet induction boxes), resetting supply air temperature is detrimental to the fan horsepower savings achieved through throttling the supply air. However, savings in reheat energy may be much greater than would otherwise have been saved in fan horsepower. Furthermore, it may be necessary to increase supply air temperature because of lighting load reduction. Air movement within conditioned rooms can become too low as lighting energy is reduced to approximately 2 watts per square foot. Reset control for this type of VAV system can be accomplished with the discriminator control circuit.

Obviously, supply air temperature reset and the discriminator control circuit would be effective on systems that are zoned by reheat.

Heated and Chilled Water. Generally, heated and chilled water temperatures can be effectively reset on a schedule based on outdoor air temperature. The reset schedule should be field-adjustable so that optimum values can be determined by trial and error. For maximum energy conservation, chilled water temperatures should never be lower than necessary to provide air temperatures that will meet cooling demands. Before concluding that increased chilled water is the best option, determine what additional savings might be achieved in fan horsepower by slowing down fans and installing smaller fan motors. Where lighting loads have been greatly reduced, it probably will be possible to increase chilled water temperature in addition to reducing supply air flow. If automatic temperature reset based on outdoor air dry bulb, wet bulb, or enthalpy is deferred because of cost, implementation of seasonal manual reset can also save energy.

Condenser Water. In past years, most systems were designed to control condenser water at 85°F (29°C). Although systems are designed for 85°F (29°C), in most cases it is unnecessary to operate at this temperature. The chiller manufacturer should be consulted to determine the recommended low limit for condenser water. Setting condenser water at the lower limit can expend more energy in cooling tower fan horsepower, but the savings in reduced chiller power will generally exceed the penalty. In a 1974 study a major HVAC equipment manufacturer indicated savings sufficient to buy a new chiller within 20 years.

In areas where very low condenser water temperatures are achievable for several months of the year, and the chiller must operate at very light loads, the "free cooling" cycle should be considered. This cycle is a chiller modification, made available in recent years by major manufacturers. It enables a chiller to operate with the compressor shut down.

3. *Additional Time Clocks.* Adding time clocks to automatically turn "on" and "off" various items of equipment or entire systems can be an effective and inexpensive means of energy-conserving automation. A variety of reliable time clocks have such features as 24-hour carryover for power failures and skip-a-day for automatic shutdown on weekends. Clocks with 24-hour dials can generally be programmed much more accurately than clocks with seven-day dials. Minicomputers now becoming available at low cost can perform all time clock functions, gather data, and monitor equipment. Careful analysis is required to ensure that time clocks are applied in the most effective manner possible. Connecting time clocks in control circuits to prevent operation of manually started equipment during unauthorized periods is a good application. Where time clocks are easily accessible, the manual override switch should be key-operated. The computerized-type switch uses a password to accomplish this same function.

4. *Dead Band Thermostats.* Dead-band thermostats which can be directly substituted for existing thermostats are becoming available. The dead-band thermostat, as the name implies, permits the building's temperature to drift within a preselected dead-band range without calling for heating or cooling. For example, a thermostat with a 5°F (3°C) dead-band could be set to call for cooling at 76°F (24°C) and heating at 71°F (22°C). The throttling range, if completely outside the dead-band, permits greater overshoot; i.e., temperature would drop to 69.5°F (21°C) and rise to 77.5°F (25°C) with 1.5°F (1°C) overshoot.

An alternative to the dead-band thermostat

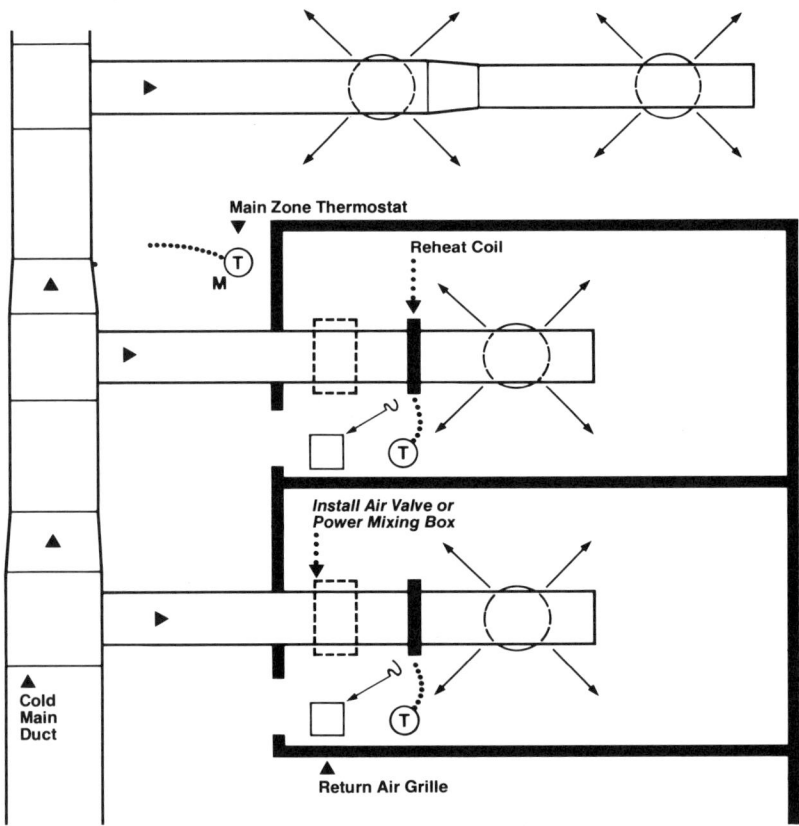

Figure 9-3. Partial air conditioning plan.

is a thermostat that throttles within the semi-dead-band. A thermostat with this characteristic would have only a 1°F or less true deadband, and would modulate (throttle) heating or cooling in proportion to room temperature deviation from set point. If the room temperature ever reached the limits, 71°F (22°C) or 76°F (24°C), the thermostat would (at that point) be calling for full heating or cooling. This type of thermostat will save energy and, at the same time, provide greater human comfort.

5. *Reheat Coils.* Reheat coils are often used for zone splitting to provide individual room control on multizone, VAV, or even double duct systems. The main zone thermostat is located in the room that has the predominant cooling requirement, and the reheat coil thermostats are located in the rooms that would tend to overcool. (Refer to Figure 9-3.)

In other cases, all thermostats on the zone would control reheat coils, and the zone duct temperature would be reset through a discriminator circuit from the room that has the greatest demand for cooling. In either case, energy-conserving modifications can be made to the systems.

The first step is to explore whether or not the heating coils can be deactivated or eliminated completely by rebalancing the air, relocating the main zone thermostat, and similar measures. The next step is to determine whether the coil can be replaced with an air valve to provide VAV control. If neither of the foregoing remedies can be successfully employed, install a powered mixing box ahead of the heating coil and sequence the mixing box damper with the heating coil valve as shown in Figure 9-4. The damper fully closes to the cold deck, and warm air from the heated return air is reclaimed before any new energy is used. This arrangement can also improve a room that is

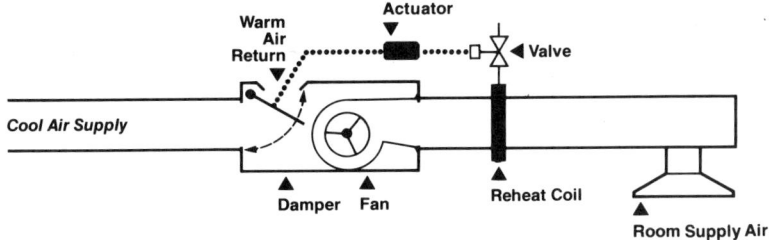

Figure 9-4. Schematic detail of reheat retrofit.

at times starved for cool air. Consider the extent of volume variation imposed on the central fan which does not have variable volume control. Where air valves or powered mixing boxes will control only a small portion of the total air, the system may tolerate the modification without a noticeable difference in performance elsewhere in the system. The diversity in load being reflected in the air requirements may also help other areas that at times have an air deficiency.

VI. EQUIPMENT EFFICIENCY CHECK

A. Boilers

The text here relates to fuel boilers, since electric boilers are nearly 100% efficient in transferring heat (boundary, not raw source energy) within the boiler. Precise boiler efficiencies can be determined only by laboratory test under fixed conditions. However, approximate combustion efficiency can be determined by flue gas analysis. By recording flue gas temperature and percentage of carbon dioxide (CO_2) and referring to a chart or table for the fuel in use, approximate combustion efficiency can be found. The formula is as follows:

Combustion efficiency
$$= \frac{\text{Input} - \text{Flue loss}}{\text{Input}} \times 100\%$$

Refer to the operating log to identify an increase in flue loss by a rise in flue temperature. This can result from scale buildup on heat transfer surfaces. The objective is to approach complete (stoichiometric) combustion by minimizing excess air (thereby minimizing oxygen, O_2, in the flue gas) without an increase of carbon monoxide (CO) in the flue gas.

Overall efficiency is always less than combustion efficiency owing to radiation loss from the boiler outside surface. The formula is as follows:

Overall efficiency
$$= \frac{\text{Input} - \text{Flue loss} - \text{Radiation loss}}{\text{Input}} \times 100\%$$

Overall efficiency may also be expressed as follows:

$$\text{Overall efficiency} = \frac{\text{Gross output}}{\text{Input}} \times 100\%$$

Determine an approximate value for overall efficiency by measuring input and output over a specific time period under a stable load condition. Measure input from the gas or oil flowmeter after turning off other loads served by the flowmeter. Water flow rate and temperature difference in supply and return are obtainable. For steam with 100% condensate return, measure the condensate return flow rate and temperature.

If the boiler is not already equipped with a modern draft burner, probably over 40% of the purchased fuel is being wasted.

When retrofitting a boiler, explore the use of equipment that will reduce excess air, preheat incoming air, increase flame temperature, increase firebox turbulence, and maintain boiler cleanliness. Only ⅛ inch (3 mm) of soot buildup on heat transfer surfaces can decrease heat transfer by up to 25%. Also, explore flue gas heat recovery by means of waste heat boilers. Claims are being made that less than one-year payback is common for boiler retrofitting.

Fire tube boilers can be retrofitted with devices, installed within the tubes, to increase turbulence and uniformly distribute fire flow through the tubes. Consult boiler manufacturers before making this alteration.

B. Chillers

Conduct all tests of reciprocating liquid chillers for rating or verification of rating in accordance with ASHRAE Standard 30-77. Derive or verify centrifugal or screw-type liquid chiller ratings tests in accordance with ARI Standard 550-77.

Such precision testing is costly, and such a degree of accuracy is not necessary to determine if an opportunity exists for chiller retrofit or replacement. Condensing temperature or temperature lift, load, and fouling affect chiller efficiency, most often referred to as kilowatts per ton or kilowatt-hours per ton-hour. These units can be converted to coefficient of performance (COP) or energy efficiency ratio (EER) as follows:

First, invert the expression of kW/ton to tons/kW. Then multiply the results by units of conversion as follows:

$$COP = \frac{tons}{kW} \times \frac{12,000 \text{ Btu}}{ton\text{-}hr}$$
$$\times \frac{kWh}{3,413 \text{ Btu}} = \frac{Btu \text{ cooling}}{Btu \text{ input energy}}$$

$$EER = \frac{tons}{kW} \times \frac{12,000 \text{ Btu}}{ton\text{-}hr}$$
$$\times \frac{kW}{1,000 \text{ watts}} = \frac{Btu \text{ cooling}}{watt\text{-}hr}$$

Note that EER = COP × 3.413, and that input energy is boundary energy. Additional credit should be given to absorption chillers that utilize raw source energy.

Review the operating log sheets to determine which chiller load exists for the greatest number of operating hours. Perform efficiency tests during the time when this predominant load is relatively stable. Use either a kilowatt strip chart recorder or a watt-hour meter on the line side of the chiller starter. An adjustment in readings will be necessary if some operating chiller accessories are not fed through the metered feeder. Verify flowmeter and thermometer accuracy on the chilled water circuit, and then record readings throughout the test duration. The following readings will be used for a sample calculation:

Test period = 2 hr
Energy consumed = 300 kWh (1.08 × 10^9 J)
Flow rate = 450 gpm (0.0284 m³/s)
Entering water temperature = 55°F (13°C)
Leaving water temperature = 45°F (7°C)

$$\text{Ton-hours} = \frac{450 \text{ gal}}{\text{min.}} \times \frac{8.33\#}{\text{gal}} \times \frac{60 \text{ min.}}{\text{hr}}$$
$$\times \frac{\text{Btu}}{\#°F} \times \frac{(55-45)°F \times 2 \text{ hr}}{12,000 \text{ Btu/ton}}$$
$$= 375 \text{ ton-hr}$$

$$\text{Performance} = \frac{300 \text{ kWh}}{375 \text{ ton-hr}} = \frac{.80 \text{ kW}}{\text{ton}}$$

Check this number against the manufacturer's published performance curves. If the number reasonably agrees with the curves, you probably have a clean condenser and the equipment is performing satisfactorily.

If the chiller is old and manufacturer performance data are unavailable, a comparison should be made against a likely replacement chiller. A payback calculation may reveal that a new, more efficient chiller will pay for itself within five years.

C. Heat Exchangers

Air-to-air, water-to-water, water-to-air, refrigerant-to-air, and refrigerant-to-water heat exchangers lose operating efficiency by becoming clogged and from film buildup on heat transfer surfaces. Unfortunately, cleaning is not always easy, and a thorough visual inspection may be difficult. For this reason, filters, strainers, and water treatment systems have a very important role.

Cooling coils can clog completely, even when filtered, if there is enough dirt buildup to support mold and bacteria growth. It is not unusual to find leaves and debris clogging up to one-half the condenser tubes in a wooded apartment complex where someone has removed the strainer basket.

Cleaning a heat exchanger is much easier than performing detailed heat transfer calculations to predict performance compared to that for which the heat exchanger was designed. However, you can check actual performance against published manufacturer data for similar, if not identical, equipment. If entering and leaving temperatures for the heat exchanger low side and high side have been logged periodically, deterioration in performance will be apparent.

Heat exchangers can be retrofitted with better instrumentation to monitor performance, better water treatment systems, filters, and even an automatic brush cleaning system, which may be a necessity when brackish water is used for condensing purposes. Heat exchangers can often be supplemented by additional heat exchangers to upgrade operational efficiency.

D. Pumps

When analyzing pump efficiency, first analyze the water circuits to determine if pumping requirements can be reduced. On systems where three-way valves have been used extensively, replacing such valves with two-way control valves will cause water volume variation according to block load demand. This often means more than a 20% reduction in peak water flow. If the three-way valves are relatively new, explore throttling down or completely shutting off the bypass circuit balancing valves.

After evaluating the water circuits and determining existing flow and heat requirements, compare pump, drive, and motor in-service efficiency against replacement candidates. Be certain to calculate energy *input* difference when determining payback, rather than the difference in brake horsepower energy required. (Overall efficiency = pump efficiency × drive efficiency × motor efficiency.) Always consider a change of impeller, motor, or both, as a potentially viable alternative for improved efficiency.

Explore variable speed pump drives, since the region of best efficiency, like the system curve, follows a parabolic curve (head versus flow) to zero as pump speed decreases. Inverters (SCR variable frequency) now available can efficiently vary squirrel-cage induction motor speed. Impeller diameter reduction is commonly employed to cause a pump to operate at a specified point, although pump efficiency usually decreases with appreciable reduction of impeller diameter. A false head (overpressure, as with balancing valves) is also used to cause a pump to operate at a specified flow rate. This practice can often appear to yield a higher efficiency on the pump performance curve than that of the trimmed impeller, but the apparent benefit can be deceiving. The efficiency indicated on the curve may be higher, but since the pump does more work to overcome the increased heat, more energy is used in this case than a trimmed impeller would use. Commonly used alternatives to the variable speed drive for limiting overpressure are:

1. Multiple pumps operating in parallel
2. Multiple pumps operating in series
3. Multispeed pumps

E. Fans

Much of the philosophy that applies to pumps also applies to fans. Explore all opportunities for reducing head and flow requirements (static pressure and cubic feet per minute). The energy required may be reduced, and some nuisance noise problems may disappear.

Excessive static pressure, like pump overpressure, wastes energy. Review the complete air circuits to determine where constrictions exist; these may be modified to relieve excessive static pressure. Return air passage area may be inadequate. Sound attenuators may be misused so as to add more noise than they attenuate, through increased static pressure.

Excessive use of balancing dampers will create unnecessarily high system static pressure. For a well-designed air supply system, only minimal adjustments are required by balancing dampers. However, system loads always change, and the existing system may be in serious need of rebalancing. But rebalancing may be inadequate without redesign of the ductwork. Load changes in various zones can be drastic in many buildings that have had lighting reduced to conserve energy and satisfy thermal lighting codes. Obviously, rebalancing must include reducing air quantities to areas having reduced loads, and slowing down the central fan as required. Sealing leaking ductwork should not be overlooked; it can save considerable fan horsepower in some projects.

A number of items must be considered to achieve optimum fan efficiency. (Overall efficiency = fan efficiency × drive efficiency × motor efficiency.) If the fan blades and scroll are clean, fan efficiency should be as reflected on manufacturer performance curves. Check the belts for wear, alignment, and tension. Check the bearings to ensure a minimum of drive loss. With the fan cleaned and slowed down and the drive assembly serviced, the motor may be oversized for the required duty. Most motors operating under light load lose efficiency and have a low power factor. Low overall power factors create additional charges on most utility rate structures. Replacement with a smaller motor may be a good investment.

F. Motors

Thoroughly analyze and minimize all system elements, including drive losses that contribute to motor loads, before seriously analyzing the motor. Additionally, determine and quantify the resulting specific load requirements in brake horsepower (BHP).

Where loads are constant, simply select a motor that can carry the given load with the least quantity of input energy. It is very important to think of selecting the most efficient motor in these precise terms. For example, by installing a new, high-efficiency, T-frame, squirrel-cage, polyphase, induction motor that is oversized for the application, you can end up with a less efficient installation than one using a less efficient motor correctly sized for the load.

Where the load will vary, as with VAV systems or variable water flow systems, the minimum load must also be determined and quantified in brake horsepower. However, even this load may be insufficient for complete evaluation if the efficiency is very low at minimum load, but the motor has only a few operating

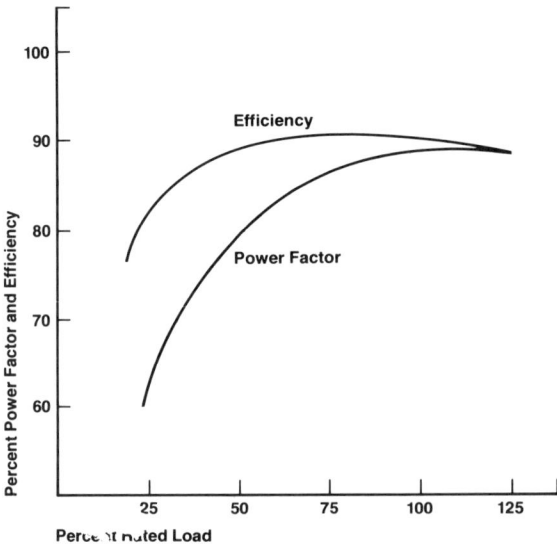

Figure 9-5. Variation of efficiency and power factor against load for a typical induction motor.

hours at that point. It is often necessary to determine the number of operating hours at various loads. Computer modeling is ideally suited for developing such load profiles.

Figure 9-5 illustrates how efficiency and power factor of a typical induction motor vary with load. Use the performance curves developed from testing the specific motor under consideration, since there can be a wide variation in characteristics for different sizes, types, and manufacturers. Use the performance curves in combination with the load profile to calculate difference in energy consumption, operating cost, and payback.

The power factor is important not only because of increased I^2R energy loss in electrical distribution systems, but also because of energy cost penalties levied in most utility company rate structures. These penalties should be included in the payback analysis mentioned in the preceding paragraph. Remember also that the power factor can be effectively corrected with capacitors for the existing motor. Furthermore, do not overlook an opportunity to relocate or exchange the function of existing motors within a large facility. Motors oversized for their load may be ideally suited for replacing larger motors that are also oversized.

The preceding discussion on motor selection has been intentionally limited to energy considerations. It assumes that other considerations such as torque, speed, duty, and current characteristics will also be taken into account. It is important to note that unusual service conditions may result in the consumption of additional energy. The usual NEMA service conditions for polyphase induction and synchronous motors are:

1. Ambient temperature, 0°C to 40°C
2. Altitude, under 3,300 feet
3. Installation on rigid mounting surface
4. Installation in locations that do not interfere with ventilation
5. Voltage tolerance, ± 10%
6. Frequency tolerance, ± 5%

Be cautious about replacing old U-frame motors with the newer, more common, T-frame motors. Studies funded by the now defunct Federal Energy Administration revealed that 1955/56 motors below ten horsepower were considerably more efficient than 1975 standard commercial motors. The U-frame motors are still available from some manufacturers, and they should be considered, since, in some applications, they are more efficient than many of the new "high efficiency" motors. U-frame motors are rugged and easily repairable.

You may also find that totally enclosed, fan-cooled (TEFC) motors generally have 2 to 3% better efficiency than open drip-proof type motors. It is also important to specify a cast iron main frame and end bells for efficient replacement motors. Higher-speed induction motors are inherently more efficient than most other types. Motors for pumps, fans, and compressors that start unloaded can accept lower locked rotor torque than other types can, which also allows for higher designed-in efficiency.

For small fractional horsepower motors, such as for direct drive fans, consider permanent split-capacitor motors; they are inherently more efficient than shaded-pole fan motors. The capacitor-start/capacitor-run motor is inherently more efficient than either capacitor-start or split-phase motors. Ball bearings with rolling friction are also inherently more efficient than sleeve bearings with sliding friction.

G. Cooling Towers

Cooling towers, can be classified into two basic categories. The first category involves direct contact between heated water and atmosphere; this category is normally called cooling tower. The second category involves indirect contact between heated fluid and atmosphere; this category is called closed-circuit fluid cooler, evaporative condenser, or closed-circuit cooling tower.

The discussion here will be limited to cooling towers, since closed-circuit fluid coolers are less efficient in cooling range and, therefore, should be used only where other benefits may be realized. For example, the closed-circuit fluid cooler is an alternative to the double bundle chiller application in a heat reclaim system.

The direct contact category of cooling towers can be further subclassified into gravity atmospheric towers, atmospheric spray towers, ejector towers, chimney towers (hyperbolic), ponds, spray ponds, spray module ponds and channels, induced-draft counterflow (single or double), and forced-draft counterflow.

Gravity atmospheric towers or ponds might be considered the most efficient, since they require the least amount of operating energy. However, since either would present a space problem on almost any modern project, we must limit tower size considerations to available space. Using available space as a constant, we may then analyze first cost against owning and operating costs. Consider the possibility of locating the new cooling tower in a location other than the space occupied by the existing cooling tower. Additional space will provide the opportunity to consider a more efficient tower. Two related axioms are well worth repeating:

1. The greater the effective volume, the better the performance, other things being equal.
2. The greater the wetted surface, the greater the cooling, other things being equal.

In calculating owning and operating costs, be sure to include water consumption and maintenance, since these items can vary widely between cooling tower solutions. The energy consumption difference between induced-draft and forced-draft types can also be considerable. Two-speed motors used for condenser water temperature control can also save energy.

Do not overlook the possibility of refurbishing the existing cooling tower as a viable alternative if the basin and a major portion of the tower structure are in good condition. Many existing induced-draft cooling towers can be refurbished by patching, painting, and adding new spray heads. Such cooling towers can be packed with new high-efficiency fill to realize a 20% increase in original tower cooling capacity.

Part 4
Commercial Retrofitting Case Studies

10. A Computer Simulation Approach to HVAC System Retrofitting*

Milton Meckler, P.E.
President, The Meckler Group
Encino, California

	Page
I. Introduction	189
II. Practical Considerations	189
III. Computer Run Breakdowns	203

I. INTRODUCTION

Experience has shown that the use factor is more important than extremes in local weather in determining the energy consumption of a building. Consequently, retrofitting strategies that concentrate on use factors can often be the most successful strategies for reducing energy consumption. The studies described herein illustrate this principle in a variety of ways, while simultaneously providing useful techniques for energy reduction in a range of office buildings. Information from a compilation of building studies conducted under the auspices of the FEA is augmented here by data from an in-depth study of a typical office building in the Midwest.

This latter study was conducted to investigate operations and energy consumption of the installed HVAC systems within a 19-story office building. Our objective was to define areas of potential energy savings, to evaluate the use of additional building automated control systems or subsystems, and to establish and rank those energy conservation measures that appear to be cost-effective based on decision criteria provided by management.

II. PRACTICAL CONSIDERATIONS

For any given building, energy consumption is often the result of two (principle) factors:

1. Basic energy requirements created by the building structure and its occupancy.
2. The major building systems used to satisfy these basic requirements and their programmed usage.

The building energy consuming systems, principally the HVAC system, can often consume amounts several times greater than their basic functional requirement. Thus, a key objective of this study was to investigate both the basic building energy requirement and the effect of the installed systems, to determine the economic and technical feasibility of obtaining attractive energy paybacks through implementation of the practical energy conservation measures.

---

*Portions of this chapter are reprinted with permission from "A Computer Simulation Approach to HVAC System Retrofitting," *Specifying Engineer,* January 1981, pp. 61–65.

The building in question was completed in November 1963. Principal floors 2 through 18 are served from a high pressure, high velocity, dual duct air distribution system supplied from one of four mechanical fan rooms. The fan rooms received chilled water and 124 psi (861.25 kPa) steam from respective main distribution loops emanating from the town power plant.

A thorough inspection of the building mechanical and associated electrical equipment was made. Verification was made of existing equipment types, locations, and characteristics. These were compared with information on available-record drawings. In addition, existing mechanical equipment and/or occupancy schedules were established through interviews with building maintenance personnel, and with power plant personnel.

A review of operating records indicated the building was consuming far more energy per unit area than were comparable office buildings in the Oklahoma location. Therefore, it appeared to be a prime candidate for meaningful energy use reduction and associated operating cost savings.

Nine profiles, characterizing the principal building and equipment usage on an hourly basis, and plotted as a percentage of the building's respective percentage daily peak requirement, were derived from records and interviews. Following analysis of this information, various energy conservation measures were developed and incorporated as part of an overall energy conservation strategy. Proprietary computer programs developed by the Trane Company and Johnson Controls Company were employed to estimate probable energy usage for each of the various ECM alternates. From this information, operating and energy cost savings were estimated based on applying actual billing criteria.

Because activities within the building could not be interrupted, and major demolition could not be tolerated, we focused on those energy conservation measures that could be accommodated on an incremental basis, with minimum interference to occupied building areas.

A number of energy conservation measures (ECM) were developed and evaluated. From these, several alternates offered attractive energy savings and could be readily implemented readily.

Three intial energy conservation measures (ECM) were examined:

ECM Alternate #1: Modifications to include installation of a mixed air, temperature-controlled economizer cycle.

ECM Alternate #2: Modification to include conversion of existing dual duct to variable air volume/dual duct system.

ECM Alternate #3: Same as ECM Alternate #2 except that modifications would also include an enthalpy-controlled economizer cycle.

Each of the ECM alternates was analyzed and compared with the existing dual duct systems. ECM Alternate #2 was found to provide an estimated annual operating cost of approximately $130,000. This was roughly a 22% cost savings as compared with existing operations for all building HVAC systems.

In the initial analysis, it was determined that use of so-called free cooling, employing outside air in lieu of running mechanical refrigeration equipment, would not be beneficial, because of the high cost of installing new return air fans, enlarging and extending existing ductwork, providing new openings, adding controls, etc., for each of the four main systems. Additional operating costs were also anticipated. Figures 10-1, 10-2, and 10-3 plot estimated monthly electrical, gas, and water usage for a representative year, based on implementation of ECM numbers 1, 2, and 3.

A cursory review of Table 10-1 reveals the principal reasons net operating savings are not realized with an economizer cycle. In comparing computer runs 1 and 2, notice that although the combined operating cost of the refrigeration chiller and its associated cooling tower auxiliaries is estimated to save $9,042 in associated electrical and water costs, additional boiler operating costs, associated primarily with preheating of outside air on the hot deck side, amount to $14,866. This trend continues in the case of the variable air volume/dual duct system conversion, although

Table 10-1. Operating Summary of Computer Simulation of ECM Study Alternates.

	ELECTRICITY($)			
	BASE* COMPUTER RUN #1	ECM ALT. #1 COMPUTER RUN #2	ECM ALT. #2 COMPUTER RUN #3	ECM ALT. #3 COMPUTER RUN #4
Chillers	32,432	24,815	17,186	13,601
Pumps	15,941	15,941	15,924	15,924
Main supply fans	41,745	44,490	23,896	23,896
Auxiliary fans	5,434	4,658	4,478	4,183
Controls	250	250	250	250
Lights and misc.	60,916	60,916	60,916	60,916
Total electricity	156,717	151,070	122,650	118,770
	GAS($)			
Boiler	6,831	21,681	6,180	10,481
	WATER($)			
Make-up cooling tower	2,668	1,948	1,315	1,018
Make-up boiler	7	23	7	11
Total water	2,675	1,971	1,322	1,029
Grand total	166,223	174,722	130,152	130,280

*Modified manually to account for variable volume operating features during heating season.
Source: Reprinted from *Specifying Engineer*, January 1981, with their permission.

the cost difference with the economizer cycle was found to be negligible. This is due to the advantages of enthalpy versus mixed air temperature control.

Having determined that the conversion to variable air volume/dual duct operation resulted in maximizing major building HVAC systems operating savings, it was decided next to explore further some optimizing measures that might provide additional operating savings. These were:

ECM Alternate #4: Installation of a solar compensated ambient hot/cold deck temperature reset control.

ECM Alternate #5: Installation of zone temperature sensed hot/cold deck. This temperature reset control alternate was offered as a substitution for Alternate #4.

ECM Alternate #6: Installation of central optimal main fan system(s) start/stop controller.

Refer to Table 10-2 for a summary of life cycle analysis studies covering promising ECM Alternates #2, 4, 5, and 6. (See Figures 10-4 and 10-5 for ECM Alternates #5 and 6.)

All the ECM's satisfy the five-year maximum payout criteria established by company management. Although ECM Alternate #5 requires approximately seven times the capital investment of ECM #4, the calculated payouts are approximately the same, slightly less than one year. Subsequent substantially greater annual cost savings would be provided by ECM Alternate #5. Savings estimated from implementation of ECM Alternate #5 were established by a proprietary computer

Table 10-2. Summary of Life Cycle Analysis Results, Major ECM Alternate Candidates.

FINANCIAL CRITERIA	ECM ALTERNATE NO.			
	#2	#4	#5	#6
Investment ($)	71,040	3,800	25,920	16,000
Payback* years	2.7	0.5	0.7	1.6
Ranking	4	1	2	3

*Based on an assumed 6¼% (or ½% over current prime) interest rate, compounded annually opportunity cost for investment monies.
Source: Reprinted from *Specifying Engineer*, January 1981, with their permission.

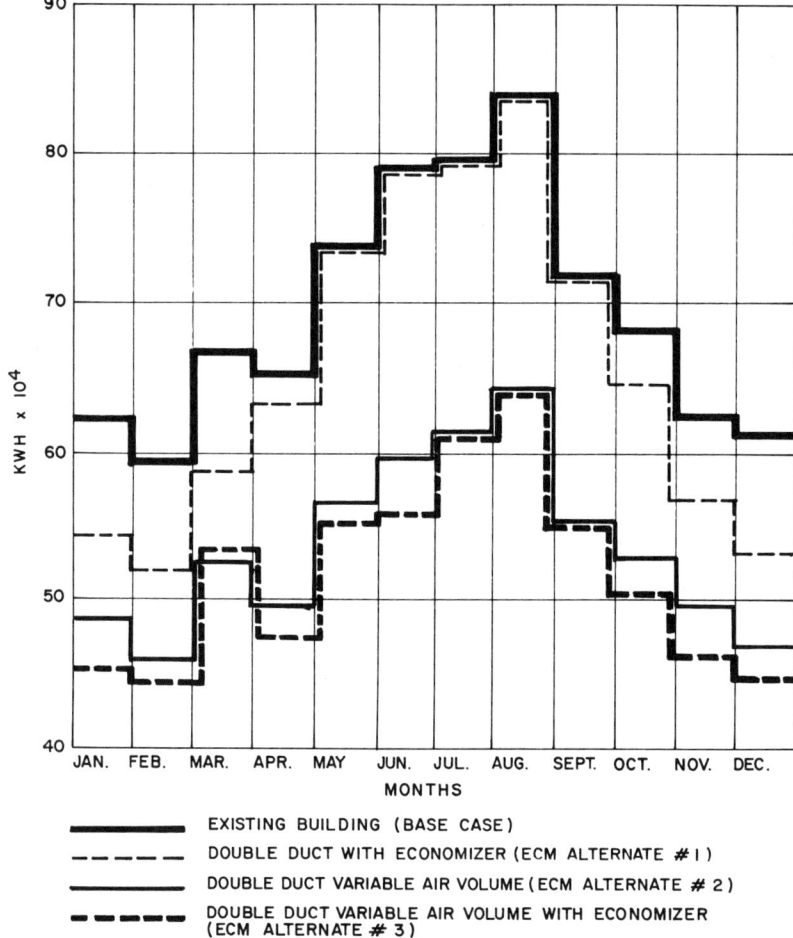

Figure 10-1. Comparison of alternatives: electrical consumption. (Reprinted from *Specifying Engineer*, January 1981, with their permission.)

simulation program developed by Johnson Controls Co. However, no such computer program exists for simulation of ECM Alternate #4; the estimated savings had to be approximated from the results of the ECM Alternate #5 computer run. Therefore, they are believed to be less accurate than computer results.

Additional instrument air requirements necessary to accommodate ECM Alternates #2, 5, and 6, were estimated to be 10 SCFM (4.72 l/s). Subsequent discussion with building central plant operating personnel revealed adequate existing compressor capacity to handle this additional load without materially affecting its current duty cycle. Auxiliary compressor standby capacity was also available, if required.

In calculating the savings realized through implementation of each of the energy conservation measure alternates, the cost of gas, electricity, and water were escalated in accordance with criteria supplied by management. These anticipated costs are summarized in Table 10-3. A 6¾% interest rate was used in computing the payout for the estimated investment for each of the proposed energy conservation measures. It represented a value 0.5% above the prime rate of February 23, 1977. Other pertinent operating data are outlined in Figures 10-6 through 10-16.

Modifications to accomplish each of the promising ECM alternates described were then developed and separately costed.

It also became necessary to modify the out-

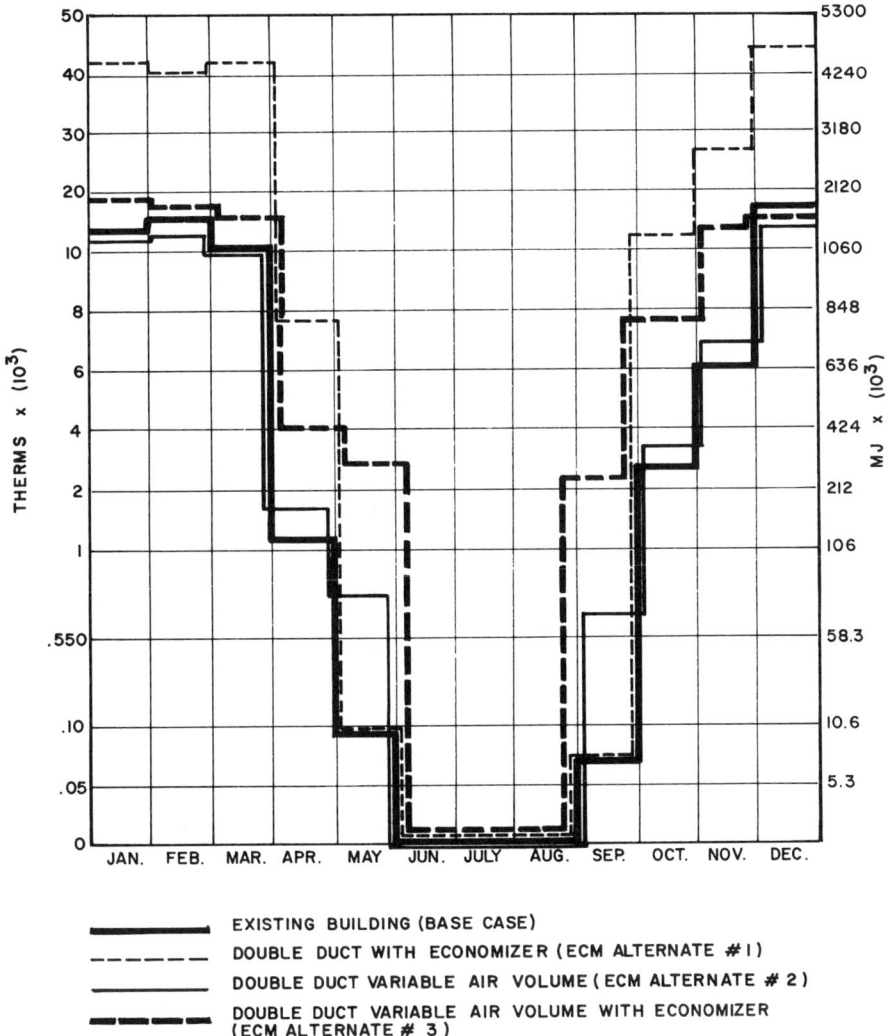

Figure 10-2. Comparison of alternatives: natural gas consumption. (Reprinted from *Specifying Engineer*, January 1981, with their permission.)

Table 10-3. Summary of Annual Energy Savings.

	JOHNSON CONTROLS COMPUTER RESULTS				
		COOLING OPERATING		TOTAL	
ECM ALTERNATE NO. SIMULATED	HEATING OPERATIONS MMBtu (kJ × 10⁶)	MMBtu (kJ × 10⁶)	AIR-SUPPLY FANS kWh	MMBtu (kJ × 10⁶)	kWh
#5*	15,027 (15,853)	6.540 (6,900)	906,903	21,567 (22,753)	906,903
#6	1,565 (1,651)	kWh 457,230	224,990	1,565	682,220

*ECM Alternate #4 estimated conservatively at 20% of all above listed savings for ECM Alternate #5.
Source: Reprinted from *Specifying Engineer*, January 1981, with their permission.

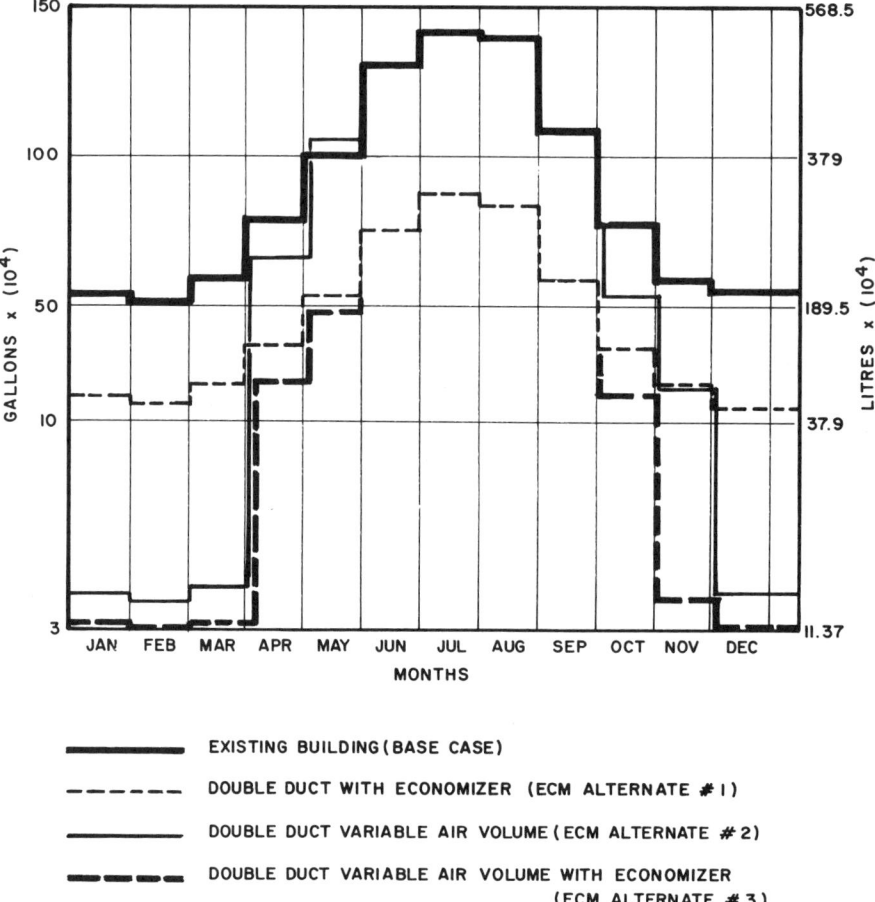

Figure 10-3. Comparison of alternatives: water consumption. (Reprinted from *Specifying Engineer*, January 1981, with their permission.)

put results of the TRACE simulation of the four main existing (base case) dual duct systems because we wished to reflect current variable fan drive operations, which were then controlled in accordance with an outdoor air temperature reset schedule.

It was recommended that ECM Alternates #2, 5, and 6 be adopted in the near future as an attractive investment opportunity resulting in significant operating and energy cost savings. ECM Alternates #2, 5, and 6 required a total investment of approximately $112,960. Payout averaged 1.7 years, based on criteria listed in Table 10-2.

ECM Alternate #2 involves control modifications to permit the existing mixing boxes to function as a variable air volume (VAV)/dual duct distribution system, and modification of the fan systems in each of the four principal built-up air handling systems serving floors 2 through 18 to convert from variable speed (winter) to conventional VAV fan operations. As formerly operated, each separate zone thermostat modulated its respective mixing box hot deck actuator, and the existing air flow controller operated its respective mixing box cold deck actuator.

ECM Alternate #5 was designed to accomplish the same result described under more precise control. As in the case of ECM Alternate #2, it was desirable to reduce at any time the differential temperature between the hot and cold decks to the lowest possible value.

ECM Alternate #6 was designed to delay

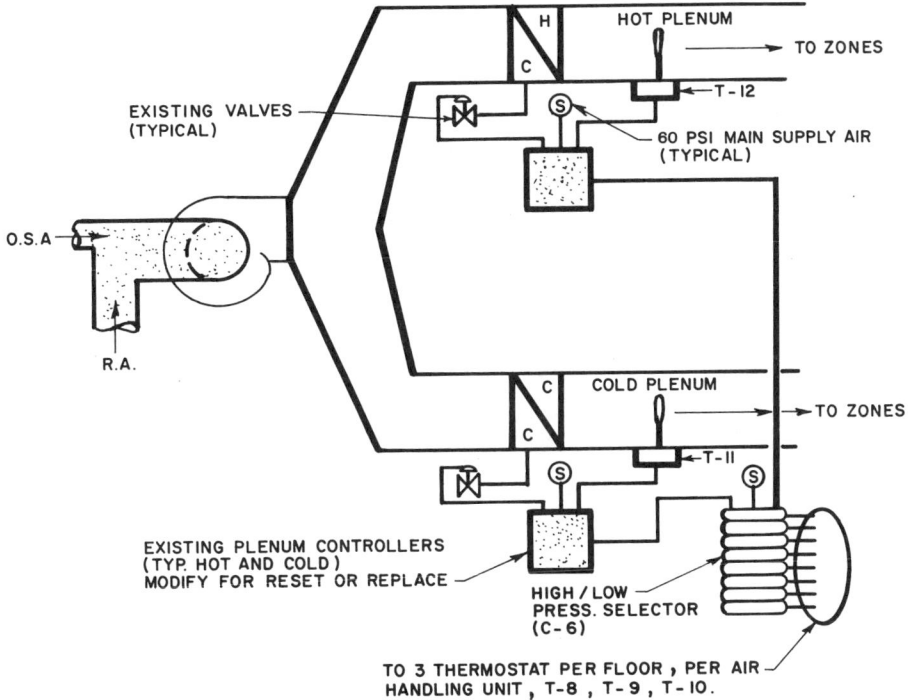

Figure 10-4. ECM Alternate #5.

the startup of the four main air handling system fans to the latest possible time to reduce the amount of fan energy consumed to the minimum necessary for comfort. This measure took into account the combined effect of the characteristic building thermal envelope and the ceiling induction units utilizing light cavity heat.

Building construction, orientation, design outside air quantities, and all other information necessary to determine building loads were taken from copies of as-built drawings of the building in question. (See Figure 10-17 for outside air temperature vs. system static pressure.) Actual equipment nameplate data, such as horsepower, air quantities, fuel oil inputs, etc., were used in all computer simulations of the air side and equipment systems.

In order to accommodate the two principal HVAC system types, multi-zone and high-velocity, dual duct, the building was divided into nine separate zones:

ZONE NO.	
1	Interior floors 1–18
2	North exposure floors 1–18
3	East exposure floors 1–18
4	South exposure floors 1–18
5	West exposure floors 1–18
6	Interior concourse and 19th floor
7	North exposure concourse and 19th floor
8	East exposure concourse and 19th floor
9	South exposure concourse and 19th floor

Although the west exposure of the concourse floor is not served from any of the HVAC systems, the westerly portion of the 19th floor principally comprises a mechanical room, which serves also as a plenum for the high-velocity, duct system serving floors 10 through

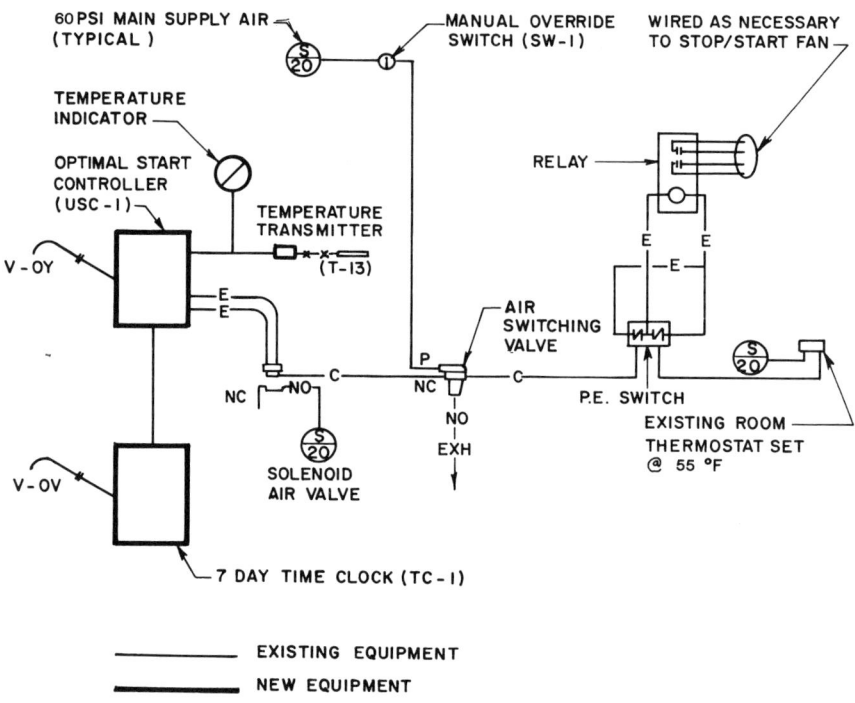

Figure 10-5. Optimal start/stop of high velocity air handlers: ECM Alternate #6.

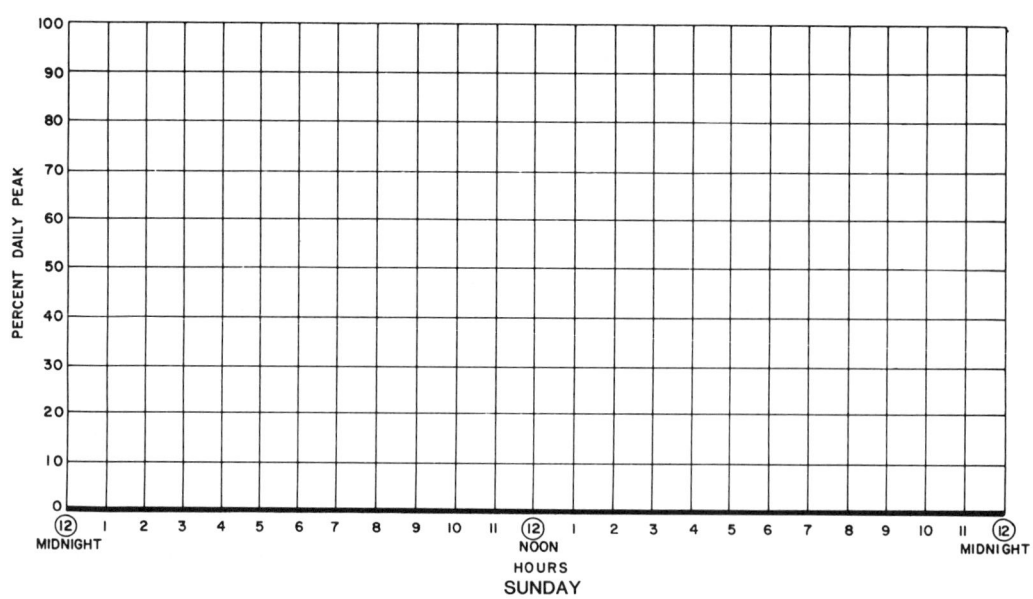

Figure 10-6. Occupants, elevators, interior lights, and HVAC equipment.

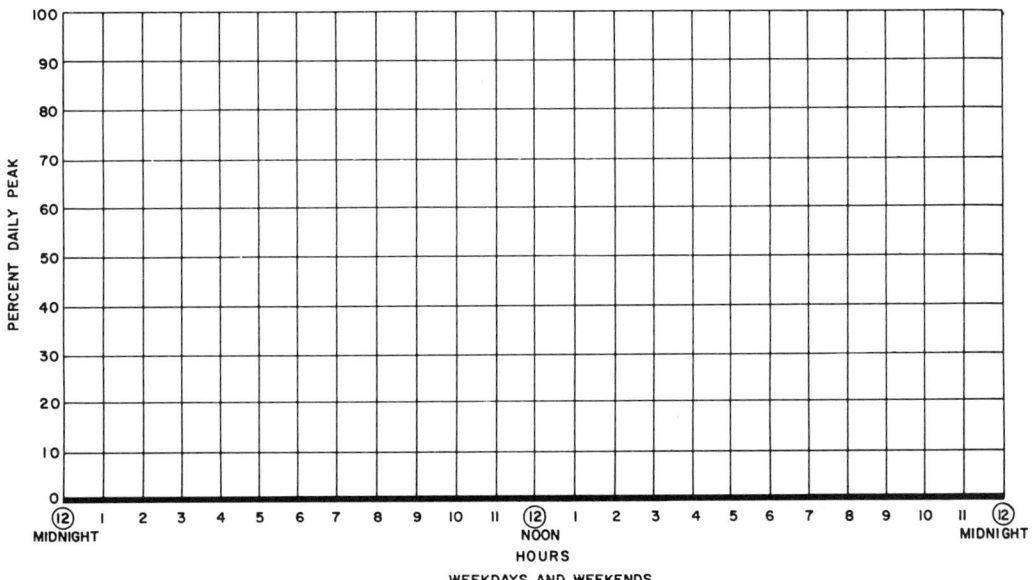

Figure 10-7. Exhaust fans.

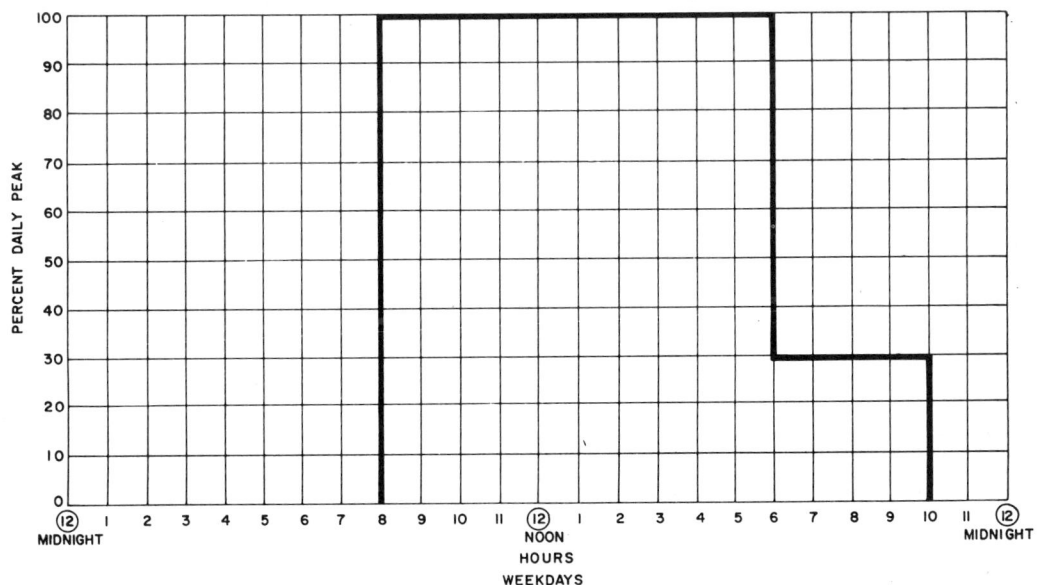

Figure 10-8. Interior lights.

197

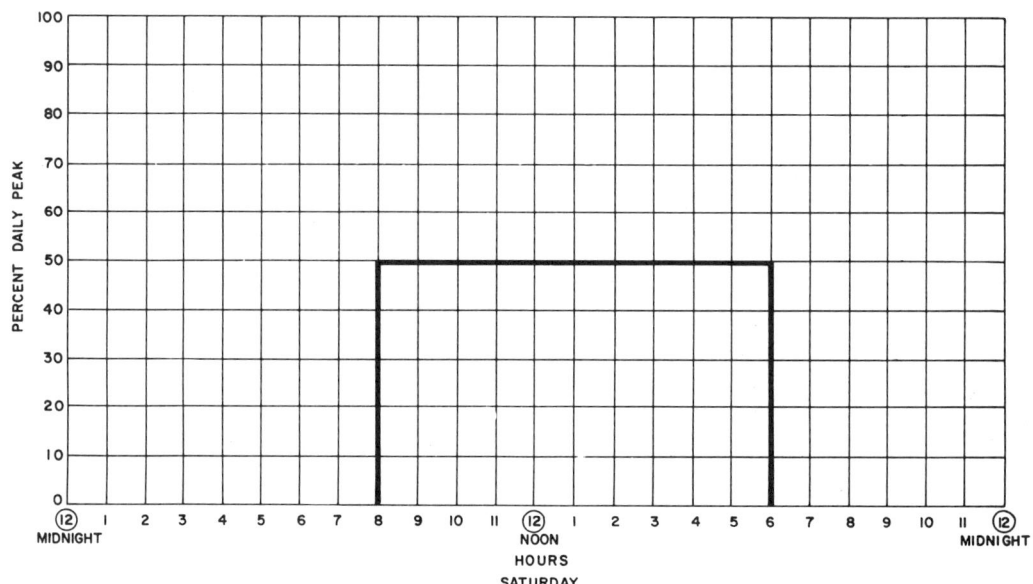

Figure 10-9. Interior lights.

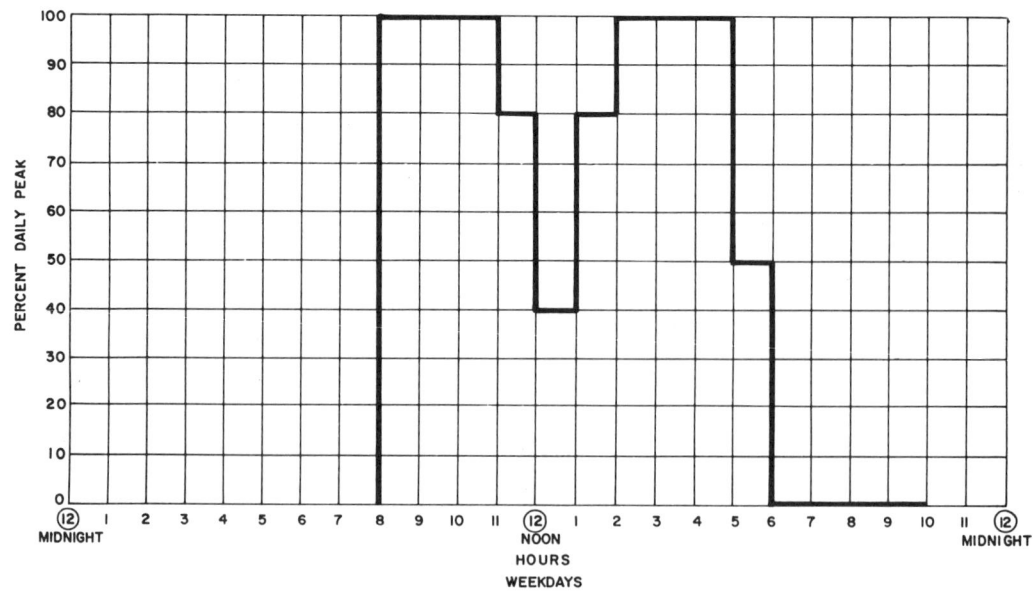

Figure 10-10. Occupancy.

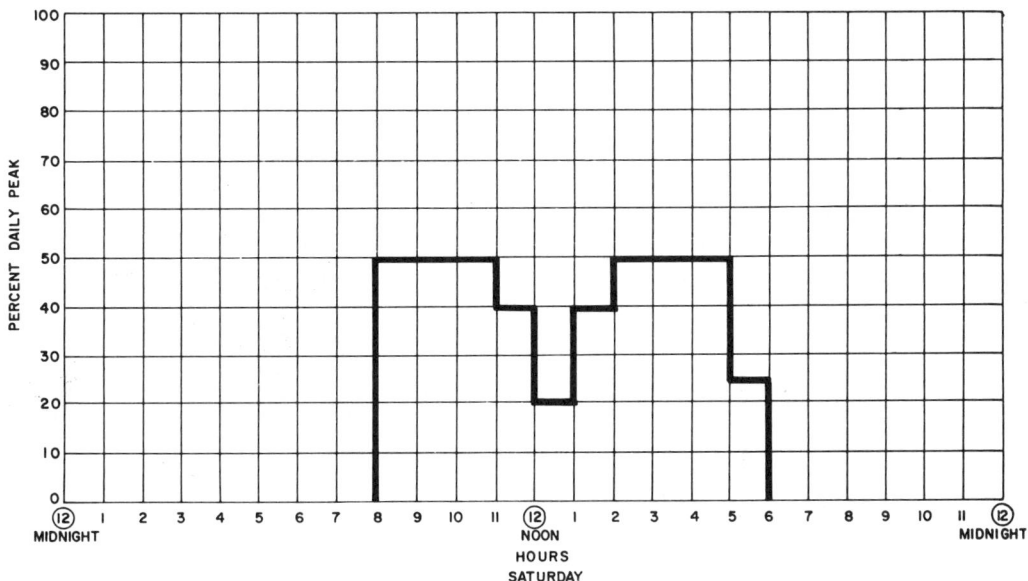

Figure 10-11. Occupancy.

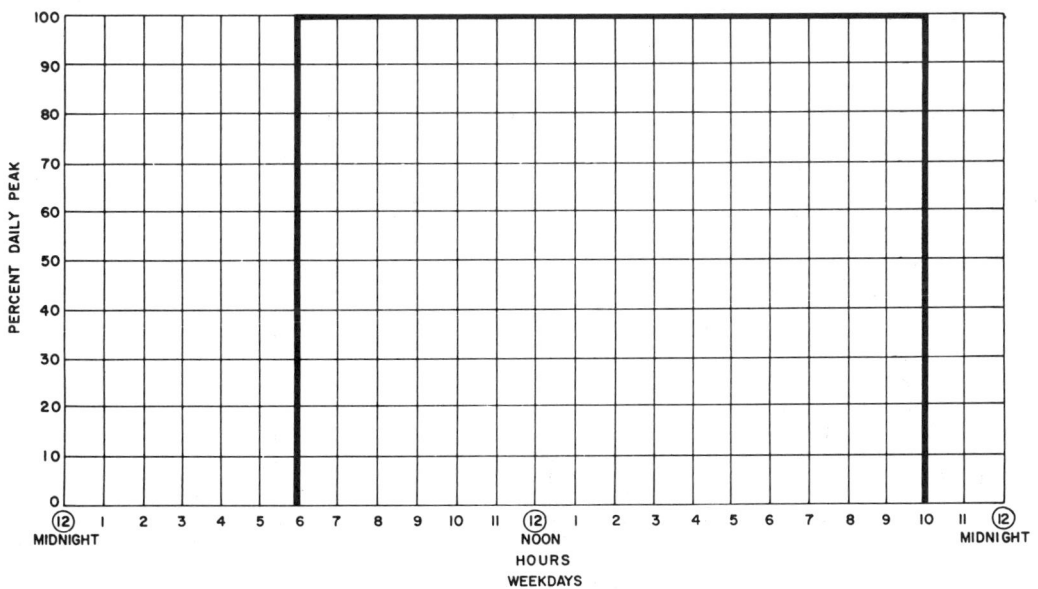

Figure 10-12. HVAC equipment.

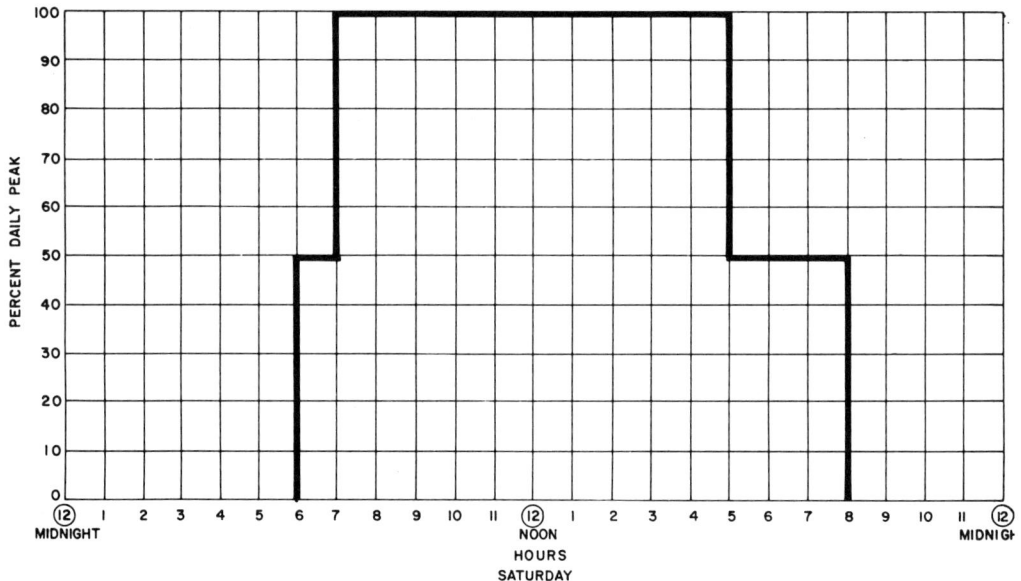

Figure 10-13. HVAC equipment.

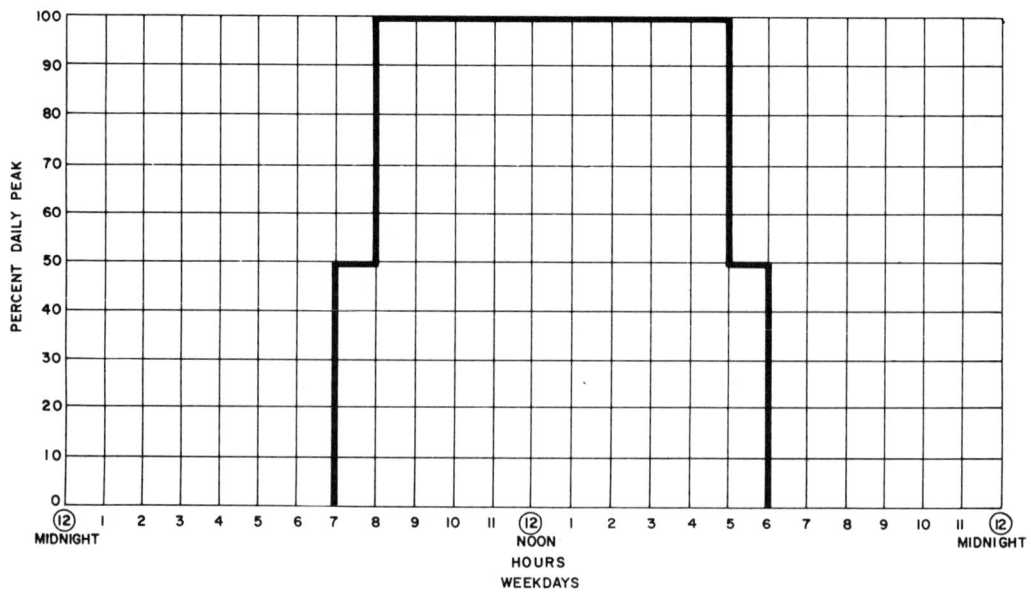

Figure 10-14. Elevators.

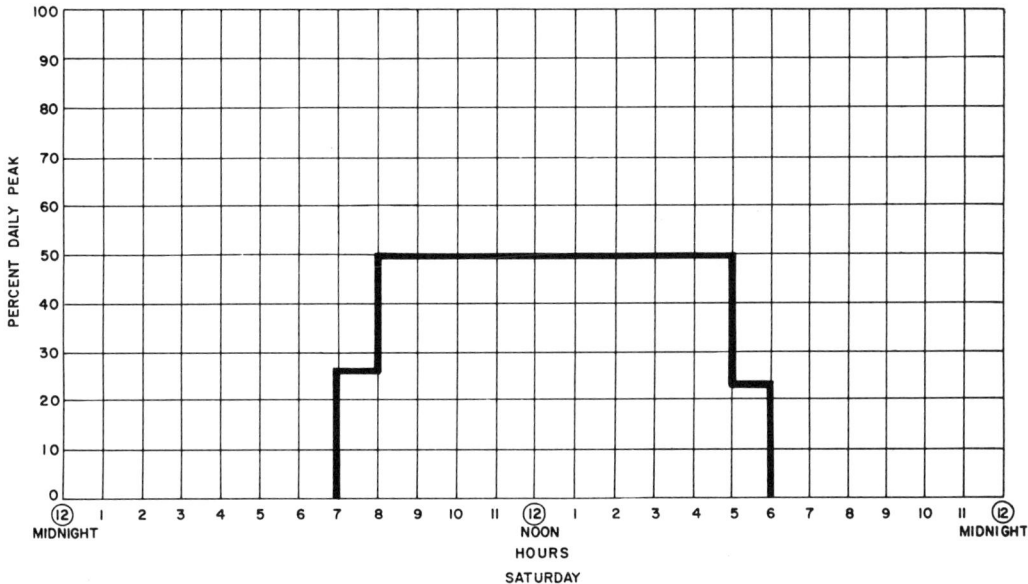

Figure 10-15. Elevators.

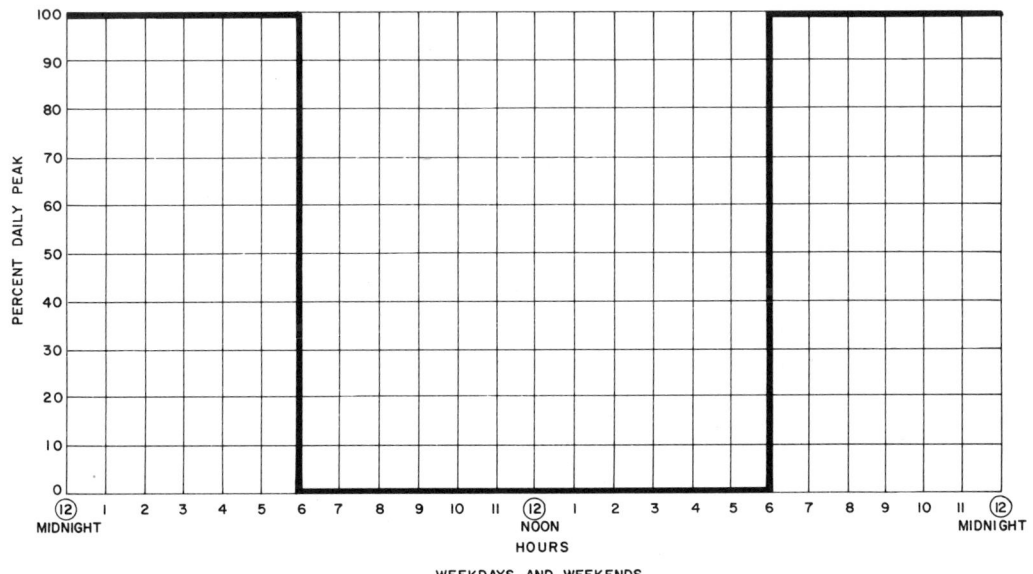

Figure 10-16. Exterior lights.

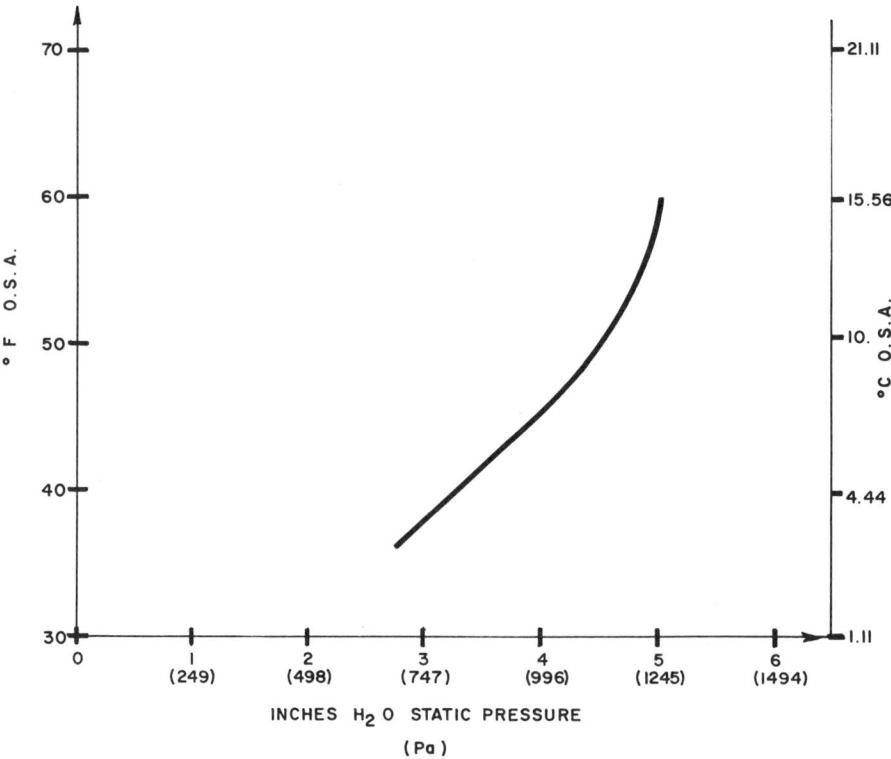

Figure 10-17. Outside air temperature vs. system static pressure.

18. The zones are served by the air side systems as follows:

ZONE NO.	AIR SIDE SYSTEM
1 through 5	High-velocity, double duct
6 through 9	Multi-zone system

Both air side systems were to be served from one equipment system located in the central power plant.

Two Johnson Controls Company computer programs were utilized in our study:

Johnson Controls Company Hot/Cold Deck Reset Computer Program. The function of the optimization program is to maximize the simultaneous heating and cooling potentials that occur under part load conditions for dual duct systems. The strategy is to monitor space temperatures and reset the hot and cold supply up and down to produce the smallest differential possible for space comfort.

For dual duct systems, both the heating and cooling functions are represented. Building areas are monitored, and both hot and cold decks are reset until one of the heating zones takes only the hot deck flow, and the cooling zones take only the cooling deck flows. Surveying all of the hours of the days provides the annual energy requirements for both heating and cooling systems in the cooling or heating modes. Program outputs are scheduled in the program input data, which utilize zone load profiles, a summary of annual energy use for specific hourly periods during the day, estimates of average allowable reset, and maximum reset to generate a summary of savings.

Johnson Controls Company Optimal Start/ Stop Computer Program. The function of the optimal start/stop program is to reduce the number of operating hours of each designated building HVAC system to the minimum necessary for achieving satisfactory occupant comfort based on ambient temperatures and the programmed envelope (i.e., heat transmission characteristics following initial system startup). The computer program surveys the weather data for each day of the year at a given locality. From programmed data on the building envelope and other input load factors, it computes the latest time that the HVAC systems can be started and stopped. The program then compares it with the normal HVAC systems daily start/stop time to compute the resulting energy savings. System stop is scheduled independently by a programmable time clock.

III. COMPUTER RUN BREAKDOWNS

Separate computer runs for each of the ECM alternates were conducted. Here is a brief description of each run:

Base Case (Computer Run #1). The building was simulated as presently operated. The various building HVAC systems and building operating schedules, as determined from field inspection and observation, were entered into the program, along with information from as-built drawings and available operating records. This run was necessary to establish a basis for comparison; there were no historical results of actual metered electrical and gas consumption levels. It was apparent from the results of our computer simulations that good agreement exists between the various equipment capacities for major equipment items and the indicated design loads shown on as-built construction drawings.

ECM Alternate #1 (Computer Run #2). In this computer simulation run, modifications were programmed in the operation of existing HVAC systems serving floors 1 through 18. Included was a simulation of the existing dual duct system with its mixed air temperature-controlled economizer cycle. The latter cycle assumes that outdoor air is used to maintain the design supply air dry bulb temperature during available weather periods. For this run, the multi-zone HVAC systems serving the basement and 19th floor areas were assumed unchanged (as in Base Case Run #1 described above).

ECM Alternate #2 (Computer Run #3). In this computer simulation run, each of the high-velocity, double duct systems was assumed to be modified to operate as a variable volume/double duct system, but without a mixed air temperature-controlled economizer cycle. The remaining multi-zone systems were also assumed unchanged.

ECM Alternate #3 (Computer Run #4). This computer simulation run is similar to Computer Run #3, except that it incorporates a variable volume dual duct system with an enthalpy-controlled economizer cycle (see Computer Run #2). Again, the other building multi-zone systems were assumed to remain unchanged.

ECM Alternate #5 (Computer Run #5). The Johnson Controls Company Hot/Cold Reset computer program (described elsewhere) was used to estimate the additional energy savings resulting from automatic resetting of hot and cold deck temperatures modified for all four main HVAC systems, as described under EMC Alternate #2. Selected representative thermostats control interior and exterior zones so that the temperature level requirements for simultaneous heating and cooling are reduced to the lowest (on heating) or highest (on cooling) levels consistent with maintaining satisfactory comfort conditions for building occupants.

ECM Alternate #6 (Computer Run #6). The Johnson Controls Company Optimal Start/Stop computer program was used to estimate the additional energy savings resulting from variable starting and programmed stopping of all four main HVAC systems, modified as described under ECM Alternate #2. This alternate was based on continuous monitoring of prevailing outdoor temperatures to reduce each daily HVAC operating time to the minimum necessary for maintaining satisfactory comfort conditions for building occupants.

Refer to Table 10-3 for a summary of annual energy savings broken down by heating, cooling, and air supply fan operations for two of the above-described computer runs.

It should be pointed out that, in both Computer Runs #2 and 4, the programmed economizer cycle was assumed to modulate automatically to lesser outside air quantities whenever the mixed air dry bulb temperatures fell below the desired supply air dry bulb temperature.

Construction cost estimates were prepared for each of the four promising energy conservation measures, ECM Alternates #2, 4, 5, and 6. Use of a separate general or mechanical contractor did not appear warranted because of the nature of the proposed work, which involved principally revisions or additions to the existing building automatic control system originally furnished by Johnson Controls Company.

11. Centralized versus Distributed Fan Systems in High-Rise Buildings

Gideon Shavit, Ph.D.

*Honeywell, Inc.
Arlington Heights, Illinois*

	Page
I. Introduction	205
II. Definition of the Problem	206
III. The Analysis	207
A. Description of the Building and Systems	207
B. The Simulation Program	208
IV. Results	208
A. Fan System Architecture	208
B. Retrofitting the Fan System	211
C. Retrofitting the Energy Control Strategies	212
V. Conclusions	212
References	213

I. INTRODUCTION

Traditionally, in commercial buildings conditioning of air is done by a fan system centrally located in the building. This approach usually has the least initial cost. The fan systems in a high-rise building are very often located in a mechanical room. This approach is used to minimize noise as well as to facilitate maintenance.

The development of the Life Safety Code for high-rise buildings in the early 1970s led to the creation of safety havens in buildings. This requirement is important because, in the event of a fire, the occupants can't all leave the building in just a few minutes. In order to allow the occupants to get to a safety haven, it is necessary to contain the propagation of smoke in the building. This requirement is implemented by pressurization of the floors above and below the fire floor and in stairwells. The centralized fan system now has an additional function of providing life safety. The pressurization process is achieved through dampening.

An alternative to pressurization by a central fan system is the compartmentization approach, in which each floor has an independent fan system. This approach provides a minimum number of openings, such as elevator shafts and stairwells, between floors. The shafts have fans dedicated to pressurization. The main difficulty in this approach is that the floor above and below the fire can't be pressurized if smoke starts to enter the floor.

Energy consumption has drawn increased attention ever since the energy crisis of 1973, and has become as important a factor in design of buildings as has life safety. Some designers continue with their traditional approach of designing centralized fan systems, resorting to a perimeter system that has heating and cooling, such as the four-pipe system. Redundant heating and cooling are eliminated through improved zoning and implementation of energy control strategies. For example, a fan system for each of the perimeter zones and one for the interior zone may provide a very good alternative.

The concept of a fan per floor looks attractive, since the system requires a relatively small amount of energy to transport the air around the floor [1]. However, in this approach, it is difficult to justify the HVAC system along the perimeter. With a constant volume system, the fan has to satisfy the load of all the zones at all exposures, which maintains a continuous high demand on the system. The tempering of the air requires some level of redundant heating and cooling. This problem is reduced by the installation of a variable air volume system (VAV) per floor and reheat along the perimeter. The system now matches the load of the zones at a lower penalty. However, during the heating mode in a given zone, redundant cooling and heating take place.

The purpose of this chapter is to study building performance when several fan system architectures (centralized, semicentralized, and distributed) and types of fan systems (reheat, double duct, and VAV) are subject to various energy conservation strategies. The results are used to determine promising configurations of fan systems that will minimize energy consumption.

The same problem of selecting optimum configuration of fan systems in new buildings exists with a somewhat different emphasis in the retrofitting of existing HVAC systems. Most of the centralized HVAC systems designed prior to the energy crisis in 1973 were reheat and double duct fan systems. These systems are now energy wasteful, and it becomes necessary to upgrade their performance. Presently there is activity in retrofitting double duct fan systems to VAV systems [2, 3]. However, there is no activity in retrofitting reheat fan systems to VAV systems. The improved performance of the few systems that have implemented integrated zero energy band control (enthalpy control, reset from the zone of highest demand, and dead band thermostat) shows that this strategy can be a viable alternative for retrofit. The purpose of this chapter also is to provide information on fan system retrofitting alternatives and their corresponding energy savings.

II. DEFINITION OF THE PROBLEM

The problem in designing a mechanical system for the life cycle of a building is to provide comfort with the least amount of energy consumption. The designer has to analyze the optimum configuration of the system and its operation, using the following parameters:

1. Centralized fan system versus distributed fan system
2. Type of mechanical system
3. Energy conservation strategies

The same parameters are involved when the designer retrofits a building. The magnitude of retrofit effort may include the following: establish new zones, converting from one mechanical system to another, and determining the energy conservation strategies that are important.

The issue of a centralized fan system versus a distributed fan system provides the designer with three major potential possibilities:

1. One fan system for the entire building
2. One fan system per floor
3. One fan system per exposure for the perimeter zones with or without fan coil systems and one for the interior zone (i.e., five fan systems)

Other possibilities in this category are rejected as not being advantageous. For example, it is possible to have a self-contained (heating and cooling) fan coil system in each office in the perimeter zone. This falls into the category of a fragmented approach, which creates great difficulty from a maintenance point of view.

Every fan system architecture encompasses the following possibilities:

1. A reheat fan system
2. A reheat fan system in the interior and a four-pipe fan coil in the perimeter zones

3. Double duct fan system
4. Variable air volume with reheat in the perimeter
5. Variable air volume with a four-pipe fan coil in the perimeter zones

Each of the above fan systems can be operated in different modes of control for energy conservation. The concept of zero energy band [4] with its requirement of a wide comfort range (68–78°F, 20–26°C) is very promising. The following strategies are considered for control:

1. Dry bulb economizer only
2. Enthalpy control and reset of the discharge temperature from the fan system on the zone of highest demand
3. The integrated concept of zero energy band (enthalpy, load reset, and deadband thermostat) [Strategy #3 includes a wide comfort range of 68–78°F (20–26°C) in the zone.]

The designer faces the following problem in designing a new building as well as in retrofitting an existing one: What is the best configuration to minimize energy consumption?

III. THE ANALYSIS

The purpose of the analysis is to determine the configuration that will provide minimum energy consumption during the life cycle of a building.

A. Description of the Building and Systems

A typical, but fictitious, office building in Washington, D.C. was selected for this analysis. It has 19 stories, and the northern exposure is directed to the north (no offset). The cross-section of the building is 150 × 100 ft (46 × 305 m) with a total area of 285,000 ft^2 (26,477 m^2). The core of the building, 40 × 75 ft (12.2 m × 23 m), is considered to be unconditioned space. The net area for conditioning is 228,000 ft^2 (21,181 m^2). The exterior wall structure is made of 6-inch (15-cm) concrete with 35% window area. The windows have two panes of standard glass with ¼ inch (0.64 cm) air space. White venetian blinds are installed on the windows (shading coefficient of 0.55). The floor thickness is 6 inches (15 cm).

Internal loads in the building are mainly artificial lighting and people. The lighting level maintained during daily operation is 2 watts/ft^2 (22 watts/m^2). One person occupies 100 ft^2 (9 m^2) of the net floor area. The mechanical system provides 0.2 CFM/ft^2 (.001 m^3/s/m^2) for ventilation all year long. The building's HVAC system and lighting are turned on at 7 A.M., and occupancy time is 8 A.M.

The support systems of the building are turned off at 6 P.M. A night setback temperature of 60°F (16°C) is maintained during the heating season. In summer, no air conditioning is provided during the night. The setting of the HVAC control system and the gain of the coils are given in Table 11-1. The space temperature is always held at +1.5°F (2.7°C) throttling range. In the conventional mode of oper-

Table 11-1. System Gain, Setpoint, and Throttling Ranges, °F (°C).

	REHEAT	DOUBLE DUCT	VAV	FAN COIL
Mixed air set point	54 (12.2)	54 (12.2)	54 (12.2)	—
Mixed air throttling range	10 (5.6)	10 (5.6)	10 (5.6)	—
Cooling coil setpoint	57 (13.9)	57 (13.9)	57 (13.9)	—
Cooling coil throttling range	10 (5.6)	10 (5.6)	10 (5.6)	—
Cooling coil gain	30 (16.7)	30 (16.7)	30 (16.7)	30 (16.7)
Reheat coil gain: exterior	40 (22.2)	40 (22.2)	30 (22.2)	35 (19.4)
interior	20 (11.1)	—	—	—
Preheat coil gain	—	—	30 (16.7)	—
Economizer damper cutoff	70 (21.1)	70 (21.1)	70 (21.1)	—
Night setback	60 (15.6)	60 (15.6)	60 (15.6)	—

ation, the space set point is maintained at 75 + 1.5°F (23.9 + 2.7°C), whereas with deadband it is maintained at 68 + 1.5°F (20 + 2.7°C) and 78 − 1.5°F (25.5 − 2.7°C).

The central plant consists of a boiler that has a capacity of 6.87×10^6 Btuh (7.29×10^6 kJ) and a single centrifugal chiller that has a capacity of 928 tons (3.26 MW). The chiller has a low limit cutoff at 100 tons (0.35 MW). The fan power requirement was determined for each individual configuration. An inlet vortex damper regulates the air flow in the VAV fan system.

B. The Simulation Program

Analysis of this type of problem is complex. Yet ability to analyze the performance of the system is necessary, since control strategies are an important alternative. The full interaction of the HVAC systems with the building subject to the control strategies has to be analyzed in the same manner as it is implemented in a real building. Honeywell's BLDSIM program simulates closed-loop operation of the HVAC system. An energy balance, made every minute, calculates the operating temperatures throughout the building for the mechanical and control system, as well as for the building structure. The operation of the HVAC components is then determined loop by loop, subject to the individual loop setting and control strategies coordinating the operation of the individual loops. A functional description of the program is illustrated in Figure 11-1. The effect of the central system on operation of the fan system is very important. For example, the temperature of the air leaving a cooling coil for two different settings of throttling range is illustrated in Figures 11-2 and 11-3. References 4, 5, and 6 give additional information regarding the capabilities of the BLDSIM program. The program has been used extensively for analysis of energy consumption in existing and new buildings.

A total of 45 individual cases were analyzed. In every case, the base included the type of fan system for a given fan system architecture and dry bulb economizer control for the intake of

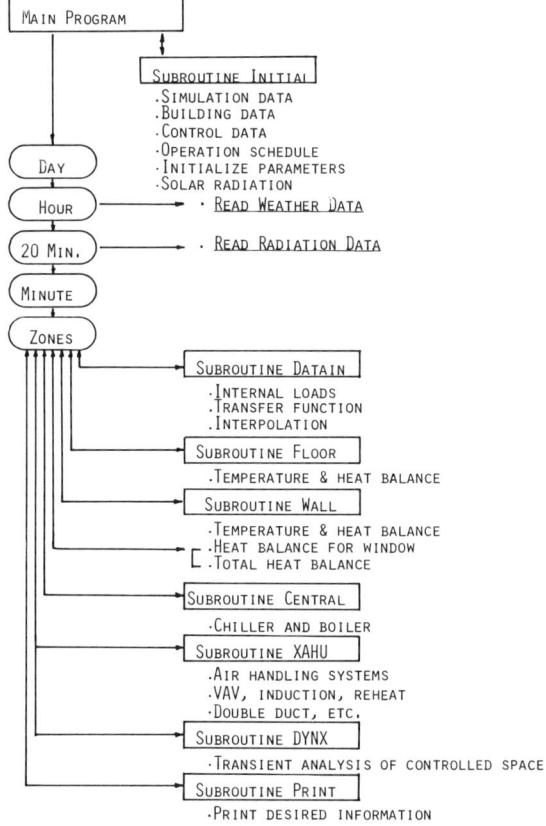

Figure 11-1. Honeywell's BLDSIM program.

the outdoor air. In each case, only one parameter was changed.

IV. RESULTS

The simulation program calculated annual building energy consumption for each of the following categories: artificial lighting, fans, chiller, total electricity, and natural gas (all heating is done with natural gas). The results were then analyzed to determine the optimum fan system architecture, the energy conservation due to the control strategies, the optimum type of fan system, and retrofit considerations.

A. Fan System Architecture

1. Reheat Fan System. Data in Table 11-2 show energy consumption for a reheat fan system with fan coils along the perimeter, for the

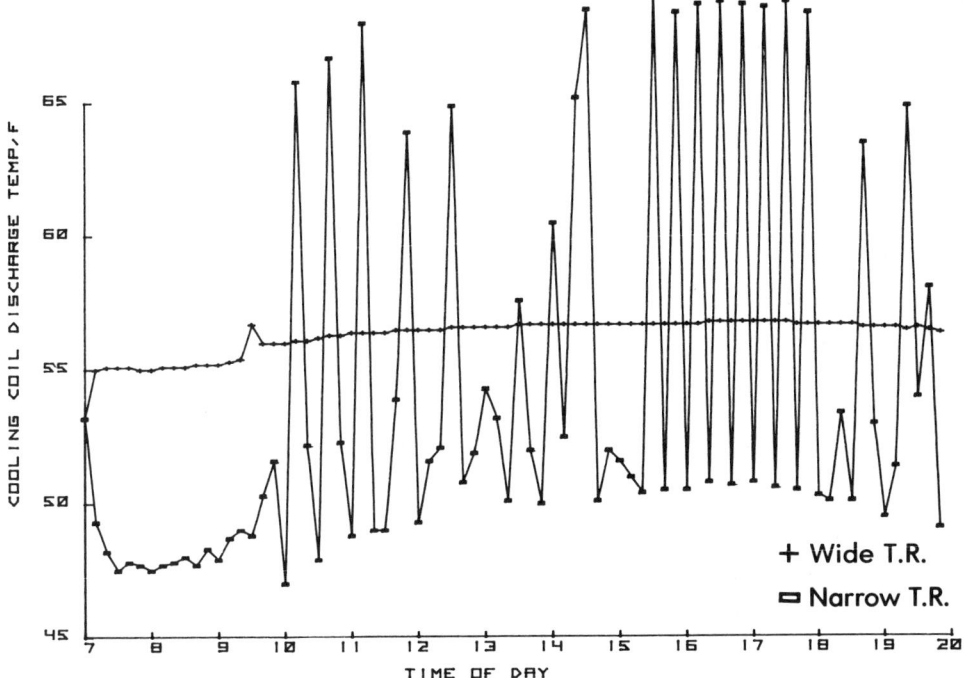

Figure 11-2. The effect of different throttling range settings of cooling coil. (Summer—north-facing exterior zone.)

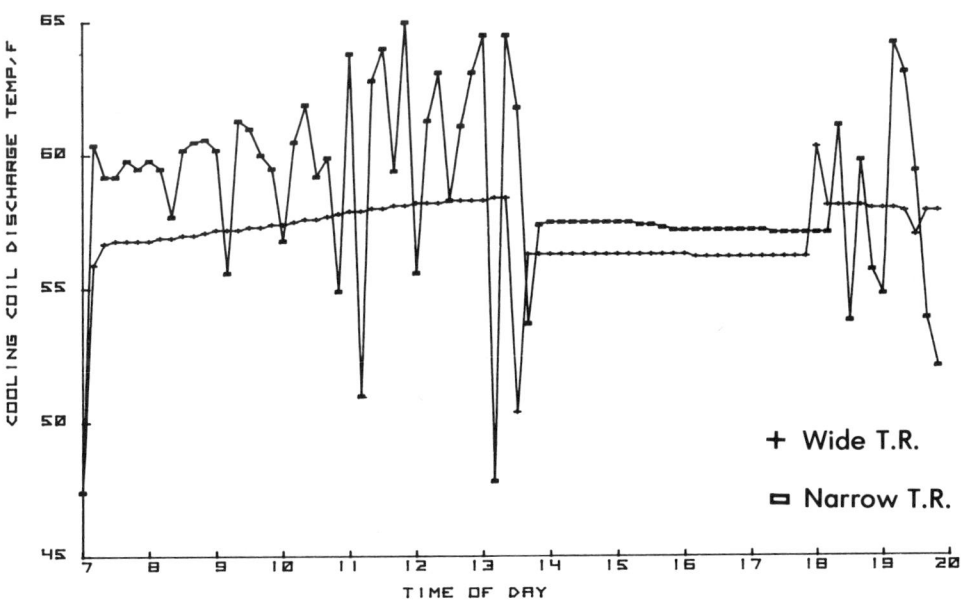

Figure 11-3. The effect of throttling range on discharged temperature. (Winter—south-facing exterior zone.)

Table 11-2. Reheat Fan System with Various Fan Systems Architecture and Energy Control Strategies.

	ECONOMIZER ONLY			INTEGRATED ZEB STRATEGY		
	FAN/BLDG REHEAT & FC*	FAN/FLOOR REHEAT ONLY	FAN/EXP REHEAT & FC	FAN/BLDG REHEAT & FC	FAN/FLOOR REHEAT ONLY	FAN/EXP REHEAT & FC
Lighting (kWh/sq ft/yr)[1]	(6.9) 74.5	(6.9) 74.5	(6.9) 74.5	(6.9) 74.5	(6.9) 74.5	(6.9) 74.5
Fan (kWh/sq ft/yr)[1]	(3.5) 37.8	(2.0) 21.6	(2.8) 30.2	(3.5) 37.8	(2.0) 21.6	(2.8) 30.2
Chiller (kWh/sq ft/yr)[1]	(5.5) 59.4	(7.6) 82.1	(5.5) 59.4	(4.1) 44.3	(4.6) 49.7	(3.8) 41.0
Total elec. (kWh/sq ft/yr)[1]	(15.9) 171.7	(16.5) 178.2	(15.2) 164.2	(14.5) 156.6	(13.5) 145.8	(13.5) 145.8
Boiler (cf/sq ft/yr)[2]	(48.5) 14.8	(111.6) 34.0	(47.5) 14.5	(17.7) 5.4	(63.6) 19.4	(13.8) 4.2

*FC: fan coil.
1. kWh/m^2/yr.
2. kl/m^2/yr.

fan/building and fan/exposure configurations. The fan/floor in Table 11-2 is for a reheat fan system only, since it is not cost-effective to have a fan system per floor, along with the fan coils around the perimeter. The data are given for two energy control strategies: conventional dry bulb economizer and integrated zero energy band. The systems with economizer control are discussed first. Fan/floor is the most economical for transporting the air. This is expected, since the power for transportation is proportional to the (CFM)^3 (m^3/s). The energy requirements for cooling and heating in the fan/floor are much higher than in the other two configuration, fan/building and fan/exposure. The main reason for this is the ability of the system with the fan coil to match the load of the zones with the least possible redundant heating and cooling, whereas the fan/floor performs as a conventional reheat system in which there is considerable redundant heating and cooling. The fan system per floor requires 38% more energy for cooling and 135% more energy for heating than the fan/exposure. The reduced energy for transportation and increased energy for cooling make the fan/floor less efficient by 8.5% than the fan/exposure when one is considering the total electrical energy requirement.

The situation is somewhat different when the integrated zero energy band strategy is implemented. The advantage of the fan/floor in energy required for transportation is offset by the disadvantage in increased cooling requirements. The total electrical energy requirement is the same for fan/floor and fan/exposure. It is expected that the fan/building would not be as efficient because of the higher level of energy required for transportation. The fan/exposure is much more efficient in heating requirements, since it provides a complete separation between heating and cooling in the fan coils, whereas the fan/floor has to provide colder air continuously to satisfy all the building's exposures during the day, and the off peak zones still require reheat to maintain comfort. Therefore, the fan system/exposure is most advantageous from the energy requirement point of view. In addition, to implement the integrated control strategy, it is much more cost-effective to have a five-fan system than to repeat it in 19 individual fan systems.

2. Variable Air Volume. Data in Table 11-3 represent energy consumption for a variable air volume fan system with fan coils along the perimeter for fan/building and fan/exposure. The data for the fan/floor in Table 11-3 are for a VAV with reheat, since other configurations are not cost-effective. Data are given for two energy control strategies: conventional dry bulb economizer and integrated zero energy band. The system with economizer control only is discussed first. The energy required to transport the air in all three fan systems is just about the same. Deviations in the data do not appear, because of the rounding off of numbers. The fan/floor requires greater amounts of energy for heating and cooling. The main reason for the difference in cooling is the nature of zoning in the system. In the fan/floor, a single fan supplies the entire floor for all exposures. The VAV terminal box does not shut off completely in order to maintain circulation

Table 11-3. VAV Fan System with Various Fan Systems Architecture and Energy Control Strategies.

	ECONOMIZER ONLY			INTEGRATED ZEB STRATEGY		
	FAN/BLDG VAV* & FC**	FAN/FLOOR VAV & RH***	FAN/EXP VAV & FC	FAN/BLDG VAV & FC	FAN/FLOOR VAV & RH	FAN/EXP VAV & FC
Lighting (kWh/sq ft/yr)[1]	(6.9) 14.5	(6.9) 74.5	(6.9) 74.5	(6.9) 74.5	(6.9) 74.5	(6.9) 74.5
Fan (kWh/sq ft/yr)[1]	(1.4) 15.1	(1.4) 15.1	(1.4) 15.1	(1.4) 15.1	(1.4) 15.1	(1.4) 15.1
Chiller (kWh/sq ft/yr)[1]	(4.4) 47.5	(5.3) 57.2	(4.3) 46.4	(3.8) 41.0	(3.6) 38.9	(3.6) 38.9
Total elec (kWh/sq ft/yr)[1]	(12.7) 137.2	(13.6) 146.9	(12.6) 136.1	(12.1) 130.7	(11.9) 128.5	(11.9) 128.5
Boiler (cf/sq ft/yr)[2]	(13.3) 4.1	(36.1) 11.0	(13.3) 4.1	(5.5) 1.7	(12.7) 3.9	(4.1) 1.2

*VAV: variable air volume.
**FC: fan coil.
*** RH: reheat (along the perimeter).
1. $kW/m^2/yr$.
2. $kl/m^2/yr$.

in the zone and to supply air for ventilation. In the configuration with fan coils along the perimeter, the control system can turn off valves, and energy will not be consumed. The higher heating load on the fan/floor is due to the reheating needed to maintain comfort in the perimeter zones. Therefore, fan/exposure or fan/building is more economical than fan/floor.

This situation is somewhat different when the integrated zero energy band strategy is implemented. There is no change in the energy requirement for transporting air. The energy required for cooling is now the same for fan/floor and fan/exposure and is somewhat lower than for fan/building. Therefore, total electrical energy requirements are the same in both cases and slightly higher in the fan/building. However, reheat in the perimeter zone of the VAV fan/floor causes higher energy requirements for heating than in the configurations of the fan/building and fan/exposure. It can be concluded that in the VAV fan system there is a considerable reduction in the energy requirement. There is no advantage in the fan/floor over the fan/exposure when electrical energy is considered, and when it is operated with the integrated zero energy band strategy. However, the fan/exposure and fan/building are considerably more economical than the fan/floor in regard to the energy requirements for heating.

B. Retrofitting the Fan System

At the present time, in the United States, there is a trend to retrofit double duct fan systems into variable air volume systems. The main motive for the retrofit is to improve fan system performance and to conserve energy. The double duct fan system is the primary target, since the mixing box can be retrofitted to VAV boxes with relative ease.

Annual energy consumption for the building in question using a single fan system with a reheat or a double duct fan system subjected to the dry bulb economizer and integrated zero energy band control strategies is presented in Table 11-4. The energy requirement for heating and cooling is lower in the double duct fan system than in the reheat fan system with the economizer control strategy. Therefore, the question is asked: Why is there no activity in the country to retrofit a reheat fan system to a variable air volume? The double duct fan system consumes 34% less heating energy and 11% less cooling energy than the reheat fan system. Additionally, there are more reheat fan systems than double duct fan systems in high-rise buildings.

The addition of integrated zero energy band conservation strategy to the reheat and the double duct fan systems reduces energy consumption in both fan systems. The impact of integrated zero energy band is greater on reheat fan systems than on double duct systems. The net effect is that the double duct fan system is now more efficient than the reheat by 7% and 2% for cooling and heating, respectively (Table 11-4).

Conversion of the reheat fan system to a VAV with reheat along the perimeter zone reduces energy consumption by a significant amount. Table 11-4 also gives annual energy

Table 11-4. Single Fan System/Building (Lighting Load of (74.52 kW/m^2/year) (6.9 kWh/sq ft/year)

TYPE FAN SYSTEM		ECONOMIZER ENERGY		INTEGRATED ZEB ENERGY		% SAVING
Reheat	Fans (kWh/ft^2/yr)[1]	(6.4)	69.1	(6.4)	69.1	0
	Chiller (kWh/ft^2/yr)[1]	(7.6)	82.1	(4.6)	49.7	39
	Total elect. (kWh/ft^2/yr)[1]	(20.9)	225.7	(17.9)	193.3	14
	Total gas (cf/ft^2/yr)[2]	(111.6)	34.	(63.6)	19.4	43
Double Duct	Fans (kWh/ft^2/yr)[1]	(6.4)	69.1	(6.4)	69.1	0
	Chiller (kWh/ft^2/yr)[1]	(6.8)	73.4	(4.3)	46.4	37
	Total elect. (kWh/ft^2/yr)[1]	(20.1)	217.1	(17.6)	190.1	12
	Total gas (cf/ft^2/yr)[2]	(73.9)	22.5	(62.7)	190.1	15
VAV & Reheat	Fans (kWh/ft^2/yr)[1]	(3.4)	36.7	(3.4)	36.7	0
	Chiller (kWh/ft^2/yr)[1]	(5.3)	57.2	(3.6)	38.9	32
	Total elect. (kWh/ft^2/yr)[1]	(15.6)	168.5	(13.9)	150.1	11
	Total gas (cf/ft^2/yr)[2]	(36.1)	11.	(12.7)	3.9	65

1. kWh/m^2/yr.
2. kl/m^2/yr.

consumption for a VAV with reheat. Energy savings are manifested now in all three categories: fan energy, energy for cooling, and energy for heating. The conversion of a reheat fan system with only economizer control to a VAV with reheat reduces energy consumption by 47, 30, and 67% for the fan, cooling, and heating energy, respectively. This is a considerable savings and, therefore, an attractive alternative for retrofit.

The VAV and reheat fan system also provides very attractive options for retrofit when integrated zero energy band strategy is implemented. Table 11-4 illustrates the potential of this alternative. The retrofit of a reheat fan system with zero energy band to VAV with reheat and zero energy band provides savings of 47, 22, and 80% on the energy for transportation, cooling, and heating respectively.

The maximum possible savings are obtained when a reheat fan system with economizer control only is retrofitted to a VAV fan system with reheat subject to the integrated zero energy band strategy.

C. Retrofitting the Energy Control Strategies

Tables 11-2, 11-3, and 11-4 contain data regarding the retrofitting of the control system to provide energy-efficient operation. The retrofitting of a mechanical system is rather costly and extensive. Conversion of fan system control from the simple dry bulb economizer to the sophisticated integrated zero energy band is less costly and results in significant savings. The integrated zero energy band strategy includes the following: enthalpy control, reset of the discharge temperature from the zone of highest demand, and a wider comfort range with a dead band thermostat.

Table 11-4 demonstrates that energy consumption in the reheat fan system with the zero energy band strategy is 39% and 43% less for cooling and heating, respectively, than it is with economizer control. The primary reasons for this are the minimization of redundant heating and cooling and the wider comfort range. The savings due to enthalpy control are secondary. Table 11-4 also demonstrates the attractiveness of this methodology in the double duct and VAV fan systems. For example, the reductions in energy consumption in the VAV fan system with zero energy band over the one without it are 32% and 65% for cooling and heating energy, respectively.

V. CONCLUSIONS

1. The architecture of a fan system/exposure is the most efficient, and the fan/building is more efficient than the fan/floor when the con-

ditioning is done with a reheat or VAV fan system and fan coils along the perimeter, subjected to the dry bulb economizer.

2. The architecture of a fan system/exposure is more efficient for heating consideration when the conditioning is done with a reheat or VAV fan system and fan coils along the perimeter, subjected to the integrated zero energy band strategy. There is no difference between the fan/floor and the fan/exposure in terms of total electrical energy usage.

3. The architecture of a fan/building is the least efficient for a reheat or VAV fan system and fan coils along the perimeter when one is considering only total electrical energy subject to the integrated zero energy band. It is more efficient than the fan/floor when considering heating energy when subjected to the integrated zero energy band.

4. It is not sufficient to consider only the energy savings due to transportation when comparing fan/floor and fan/exposure architecture. The energy conservation strategies for each architecture should also be considered.

5. Retrofitting of a reheat fan system per building to a VAV fan system plus reheat along the perimeter provides 67% and 30% reduction in the requirement for heating and cooling, respectively. Compared to an economizer control system, the integrated zero energy band provides a reduction of 47, 80, and 22% in the energy requirement for transportation, heating, and cooling, respectively.

6. A double duct fan system is more efficient than a reheat fan system. Significant effort should be directed to retrofitting reheat fan systems to VAV fan systems.

7. The architecture of a VAV fan system per exposure and fan coils along the perimeter is the most efficient configuration.

8. Implementation of integrated zero energy band control provides significant energy savings without retrofitting of the fan systems.

9. The results presented here are true only for the given structure, fan systems arrangement, control setting, dynamics of the given control and HVAC system, and the given location. The results should not be extrapolated, since they may result in erroneous data.

REFERENCES

1. T. C. Gilles, "The Modular Approach to Large Building Air Conditioning," *Heating/Piping/Air Conditioning,* September, 1978.
2. Henry Obler, "VAV System Eliminates Overcooling," *Heating/Piping/Air Conditioning,* September, 1979.
3. C. E. Brown, "Retrofitting Dual Duct Systems with VAV Components," *Building System Design,* June 7, 1977.
4. Gideon Shavit, "Energy Conservation and Fan Systems: Computer Conduct with Floating Space Temperature," *ASHRAE Journal,* October, 1977.
5. Gideon Shavit and Don Spethmann, "A Dynamic Simulation Analysis of Alternative Control and Operating Strategies for a Typical Office Building," 2nd Symposium on HVAC Systems, Purdue University, April, 1976.
6. Gideon Shavit, "Total System Consideration for Design of Building Envelopes," ASME Winter Annual Meeting, November 27, 1977.

Part 5

Institutional Retrofitting and Case Studies

12. Institutional Rehabilitation: An Opportunity for Significant Energy Savings

John J. Halas, P.E.
Edward A. Sears Associates
Consulting Engineers

	Page
I. Introduction	217
II. General Characteristics and Overall Results of the Survey	218
III. Central Plant Type Retrofits	218
A. Boiler Maintenance and Operation Improvements	220
B. Oil Burner Replacement	222
C. Waterside Scale Buildup	222
D. Fireside Soot Buildup	223
E. Heat Loss Due to Domestic Hot Water Leak	223
F. Energy Losses Due to Malfunctioning Steam Traps	224
G. Preheating Combustion Air	225
H. Hot Water Storage Tank Insulation	226
I. Insulating Hot Pipes	227
IV. HVAC Retrofits	228
A. Chiller Retrofits	228
B. Energy Transfer Retrofits	232
C. Control and Regulatory Retrofits	233
V. Lighting Retrofits	234
Acknowledgment	236
References	236

I. INTRODUCTION

The aim in institutional energy conservation is twofold: (1) to establish and prepare a present energy consumption profile, including all categories of energy, and convert all forms of energy to one common base (Btu or calorie) for comparison and evaluation purposes; and (2) to generate actual energy conservation measures, and information about their costs, benefits, and paybacks, resulting in the reduction of energy consumption in the particular institution.

To meet these goals we generally undertake the following tasks:

1. Establish and study data relating to energy consumption. In this respect, it is generally helpful to analyze sharp "peaks" in electrical or steam demand curves. Further, establish a priority system based on consumption of energy and conduct an institutional survey from the largest energy user to the smallest in descending order, as time and costs permit. With this in mind, the largest and most beneficial energy savings are identified first, before large sums are spent for relatively small and unimportant projects.

2. Collect and analyze weather data, energy

costs, and projected energy cost increases.
3. Establish the energy conservation measures and list them in order of cost savings, energy savings and payback periods.

The following sections describe the techniques for identifying energy conservation measures that grew out of the author's work on behalf of the New York State Energy Office and various other clients of the firm of Pope, Evans & Robbins, Inc. In response to plans of the United States Department of Energy, the New York State Energy Office launched major energy conservation studies directed at several classes of facilities, health care institutions, schools and municipal buildings.

This chapter deals with the results of the health care study. Clearly, one cannot hope to enumerate all energy conservation measures in such a short format, nor can one cite the numerous individual unique examples that apply to a single institution. These latter are usually the result of a thorough survey conducted by a skilled professional engineer. Rather, the general discussions here can provide an important base required for typical energy conservation reports.

II. GENERAL CHARACTERISTICS AND OVERALL RESULTS OF THE SURVEY

It is well known from past energy audits that, compared to commercial, educational and speculative type structures, hospitals are major users of steam and/or hot water as well as large users of heating and, in most cases, year-round air conditioning. Further, to comply with a unique set of rules and regulations embodied in the Federal Hill-Burton guides, they are users of large amounts of outside air in "once-thru" systems and have relatively minor diurnal load variations.

A recent joint United States Department of Housing and Urban Development USDOE study concluded that the mean annual energy consumption of heating, cooling and lighting for hospitals is almost three times that of office buildings or universities.

During the fall of 1978 and the spring of 1979, teams of engineers visited 14 hospitals and 10 nursing homes (selected randomly by computer analysis) throughout New York State to establish a data base for common energy conservation measures as well as energy reduction by the institutions themselves. The hospitals surveyed were identified by number and analyzed according to their energy consumption per unit area.

In performing such surveys, care must be taken to remember that no ECM, no matter how beneficial, can interfere with the critical nature of the care provided by the facility. Audit efforts should always be aimed at institutional, HVAC, non-process-energy-consuming systems. Most ECM's identified during the survey resulted in a five year payback or less. It is generally not cost-effective for most hospitals to consider longer-payback ECM's except in exceptional circumstances, such as with solar projects. For the ECM's discussed here, energy savings were based on state-of-the-art information available from weather charts and manufacturers. Generally, no fuel escalation costs were considered, and no life cycle studies were performed. In most cases, it was found that a 30% or better reduction in energy consumption can be achieved by considering only the largest and most obvious measures. See Figure 12-1 for details.

The study indicated that an investment of $47,000,000 (1978 dollars) applied to the ECM's for the 234 voluntary nonprofit hospitals in New York State would result in a 6×10^{12} Btu/yr (6.33×10^{12} kJ/yr) savings, or the equivalent of about 43 million gallons (162,970 m^3) of oil per year.

Results for the nursing homes are shown in Figure 12-2. The savings are somewhat smaller; for a $5,900,000 (1978 dollars) investment, a 0.9×10^{12} Btu/yr (9.495×10^{11} kJ/yr), or 6.5 million gallons (24,635 m^3) of oil per year energy reduction is anticipated.

III. CENTRAL PLANT TYPE RETROFITS

The following central plant type retrofits have been found in most institutions:

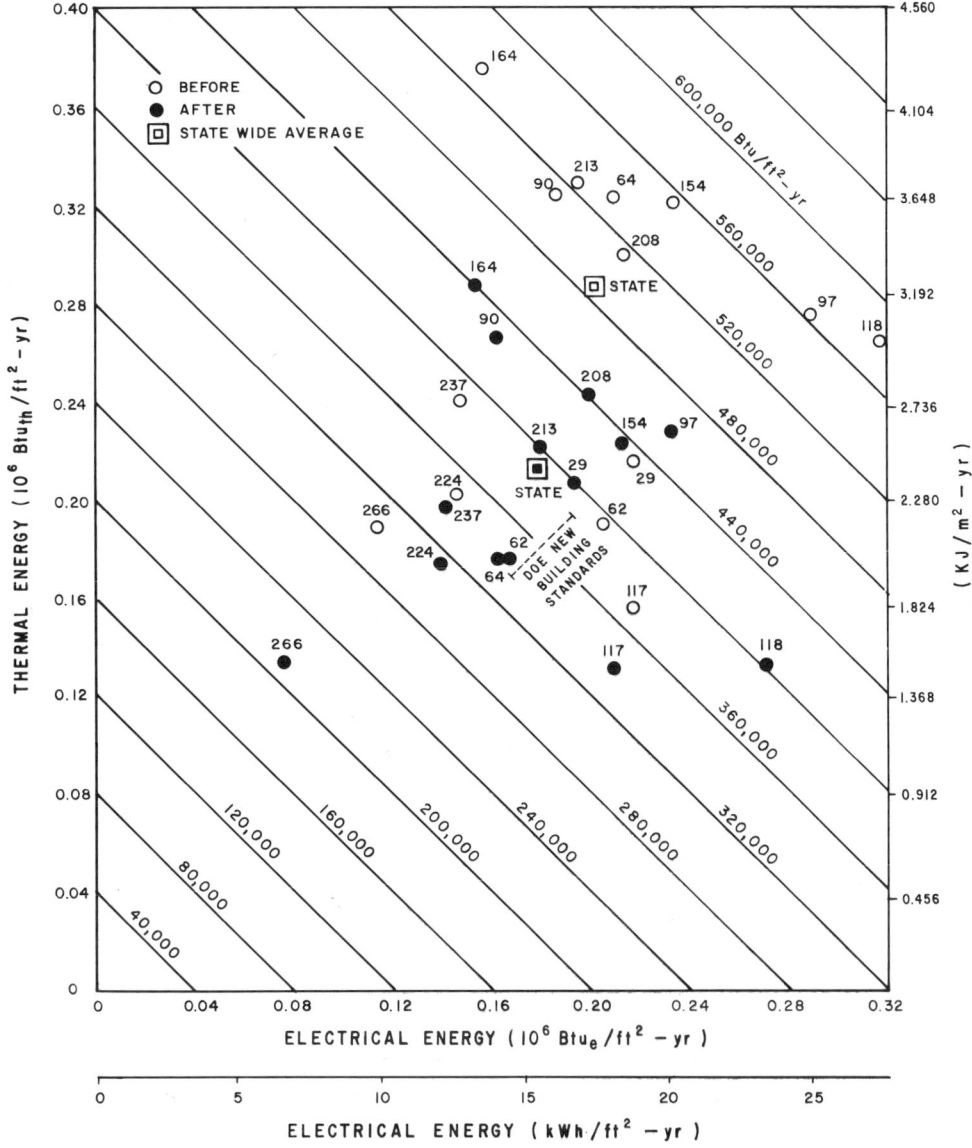

Figure 12-1. Hospitals' annual energy consumption per unit area. (Reproduced by permission of Pope, Evans and Robbins, Inc.)

A. Boiler combustion efficiency increase
B. Oil burner replacement
C. Waterside—Scale buildup removal
D. Fireside—Soot buildup removal
E. Reduction of energy losses due to domestic hot water leaks
F. Reduction of energy losses due to malfunctioning steam traps
G. Preheating combustion air for higher efficiency
H. Hot water storage tank insulation for reduced heat loss
I. Insulating hot pipes for reduction of heat loss

The following examples provide somewhat simplified illustrations of these ECM's. It is hoped that the reader may enlarge and further develop these examples to fit particular applications.

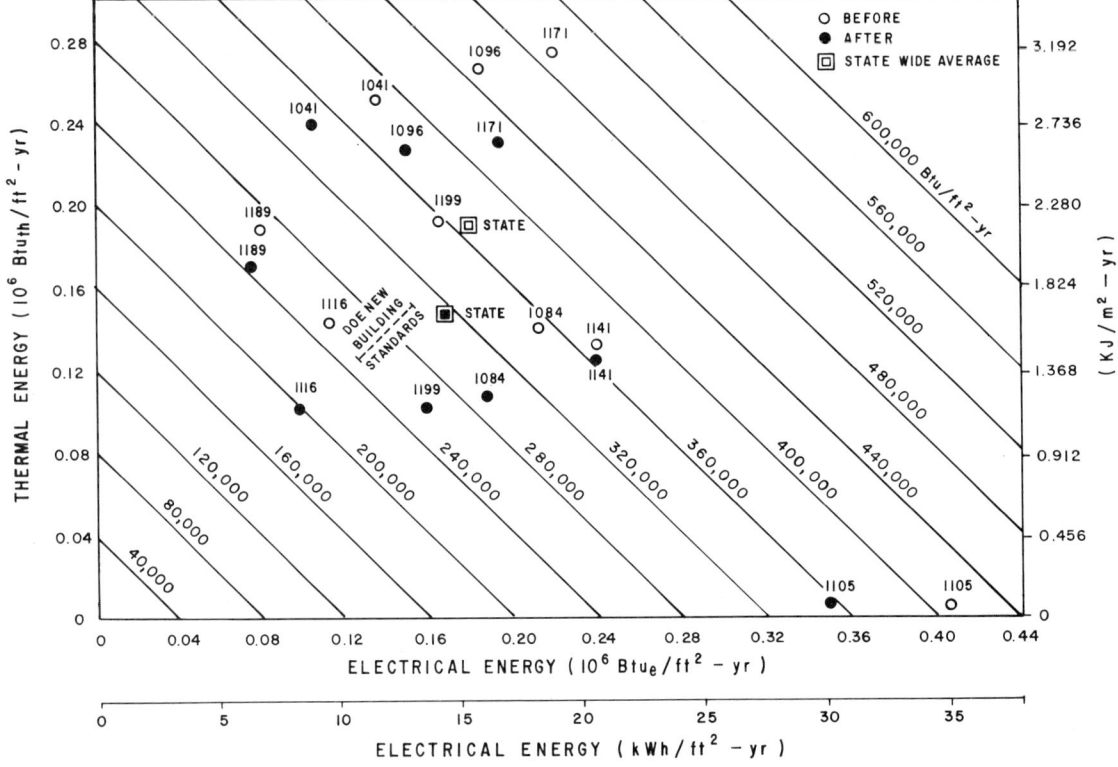

Figure 12-2. Nursing homes' annual energy consumption per unit area. (Reproduced by permission of Pope, Evans and Robbins, Inc.)

A. Boiler Maintenance and Operation Improvements

It is important to remember that general boiler plant retrofits usually result in large energy savings. The boiler room or plant, as the primary consumer of thermal energy in most hospitals, deserves special consideration for energy conservation. It is imperative that all boilers operate at maximum efficiency to derive maximum energy from the fuel. For smaller boilers, of less than 30 million Btuh input (8.79×10^6W), it is recommended that:

1. A permanent stack gas thermometer should be located in the breeching. This provides the operator with a positive indication of stack temperature. An increase in stack temperature is an indication of a decrease in boiler efficiency.
2. Stack gases should be tested weekly for temperature, draft, CO_2, O_2 and smoke, and logged to assure that the boiler is operating at the recommended efficiency. Burner settings, dirty gas, water side conditions, or improper draft should be corrected to maximize efficiency.
3. The fireside of all boilers should be cleaned at least annually.
4. Water treatment programs should be made adequate, to prevent scaling of the waterside. If scaling is present, the boiler waterside should be cleaned. A daily log of water treatment tests should be maintained.
5. Makeup water should be metered to monitor possible leakage or excessive blowdown on steam boilers.

Boilers greater than 30 million Btuh input (8.79×10^6W) should be investigated for possible application of on-line O_2 control and/or economizers to preheat feedwater. Steam boiler pressure should be reduced, if possible, since steam leakage and waste are reduced at lower pressure, and efficiency is increased. It is often possible to increase boiler gross efficiency 2 to 5% by careful operation and maintenance.

Example #1. Boiler Combustion Efficiency Increase. To determine the combustion efficiency increase and related annual savings resulting from adjusting the secondary air damper to reduce excess air, assume that:

1. Preadjustment CO_2 is 7% at 500°F (260°C).
2. Postadjustment CO_2 is 11% at 350°F (176.67°C).
3. Annual heating oil fuel cost is $40,000.

Procedure:

1. Determine the pre-adjustment combustion efficiency and post-adjustment combustion efficiency as follows: Enter the graph in Figure 12-3 at 7% CO_2 and proceed vertically to intersect the 500°F (260°C) curve; then proceed horizontally to determine efficiency, 74%. Proceed similarly for post-adjustment efficiency, 84%.
2. Determine increase in efficiency:

   $$(84\% - 74\%)/74\% = 13.5\%$$

3. Calculate annual $ savings:

   Annual $ savings = 13.5% $\times$ $40,000 = $5,400

4. Note that for natural gas the corresponding increase in efficiency is the same, while the savings are somewhat smaller owing to the lower price of the fuel. See Figure 12-4.

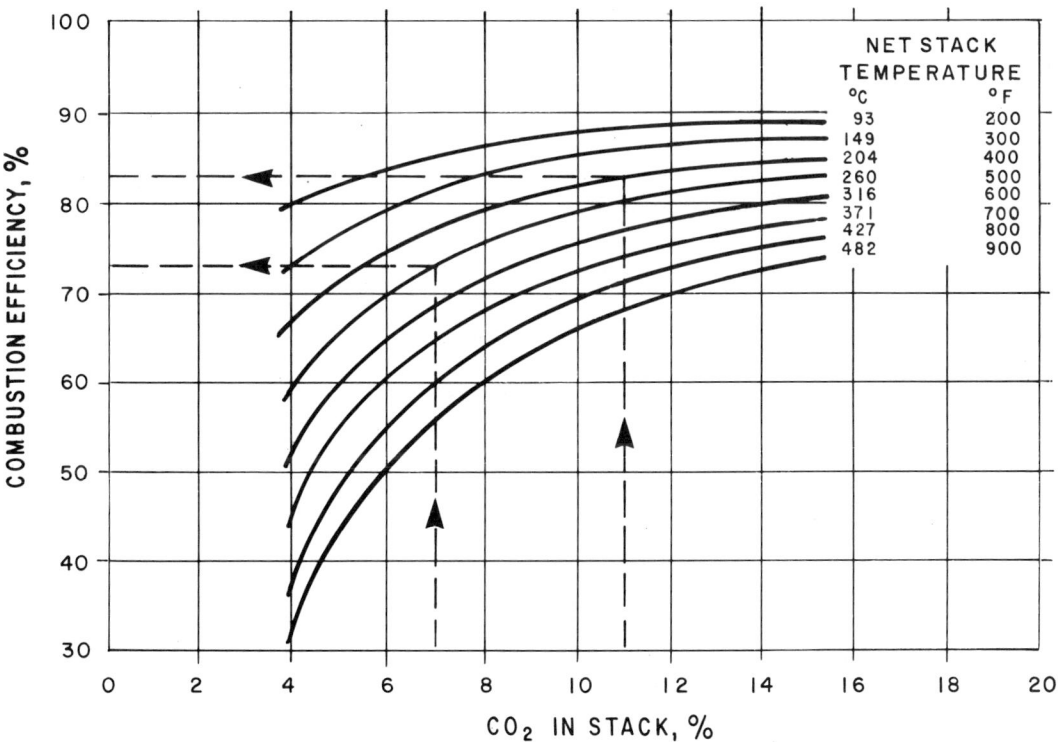

Figure 12-3. Boiler combustion efficiency increase. (Reproduced by permission of Pope, Evans and Robbins, Inc.)

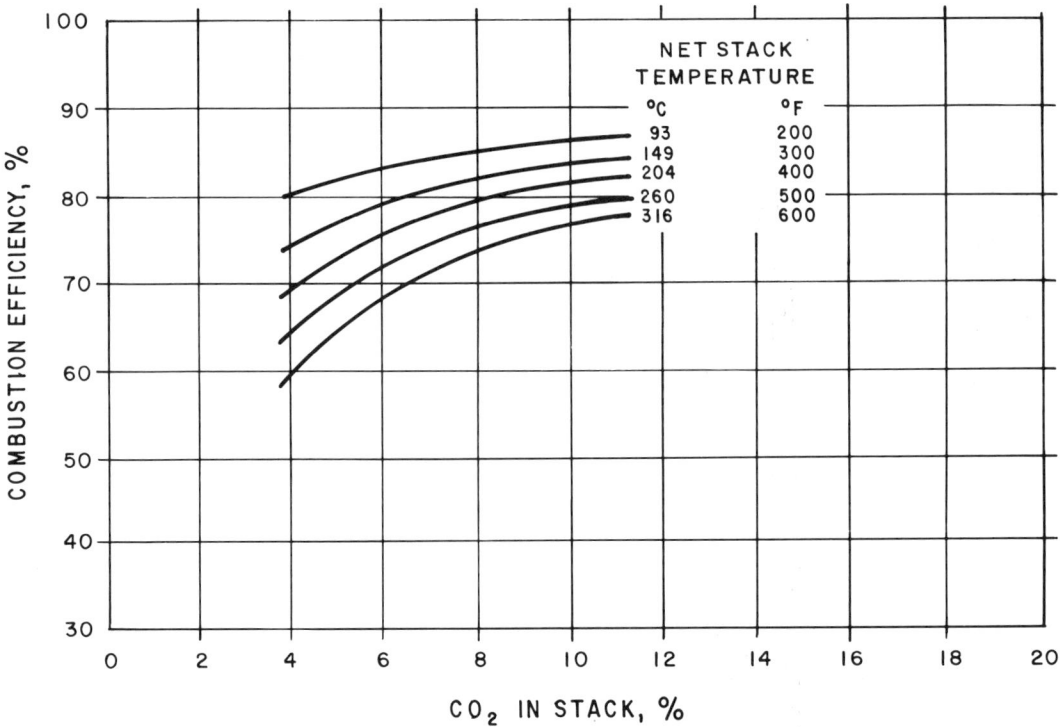

Figure 12-4. Natural gas combustion efficiency. (Reproduced by permission of Pope, Evans and Robbins, Inc.)

B. Oil Burner Replacement

Example #2. An old, inefficient burner with 60% efficiency is replaced by a 75% efficient new burner. If the burner consumes 25,000 gallons (94.750 l) of oil per year, determine the annual energy saved by this replacement.

Procedure:

1. To find the new burner's improvement in percent, subtract 60% from 75% and divide by 60%.

   Percent improvement
   $= (75\% - 60\%)/60\% = 25\%$

2. Multiply 25% by 25,000 gallons to determine the annual fuel savings:

   Annual fuel savings
   $= 25\% \times 25,000$ (94,750 l)
   $= 6,250$ gallons (23,690 l) of oil

3. Determine the annual $ savings when fuel cost $.80 per gallon ($0.211 l): 6,250 × 0.80/gallon (23,690 × 0.211/l) = $5000/yr.

C. Waterside Scale Buildup

Example #3. Determine the energy loss associated with a scale buildup of 1/32 inch (0.79 mm) on the waterside of a boiler. Annual fuel cost is $40,000.

Procedure:

1. Enter the Figure 12-5 graph at 1/32 (0.79 mm) and proceed vertically to intersect the "normal scale" curve; then proceed horizontally to determine gross energy loss: 2%.

2. Calculate annual $ savings by multiplying energy lost by annual fuel cost:

   Annual $ savings = 2%
   × $40,000 = $800

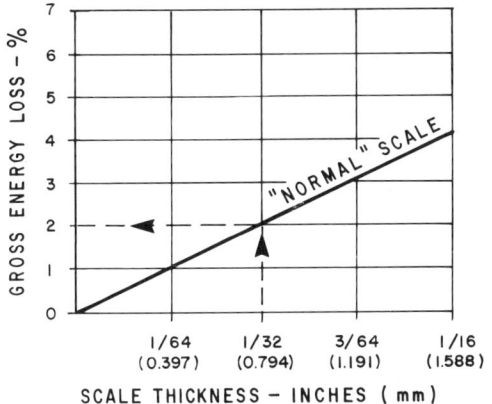

Figure 12-5. Waterside scale buildup. (Reproduced by permission of Pope, Evans and Robbins, Inc.)

D. Fireside Soot Buildup

Example #4. Determine the associated fuel losses resulting from 1/32 inch (0.79 mm) buildup on the fireside of an oil-fired boiler. The normal measured net stack temperature, when the boiler is clean, is 350°F (176.67°C), and CO_2 is 10%. Annual fuel cost is $40,000.

Procedure:

1. Enter the Figure 12-6 graph at 1/32 inch (0.79 mm); proceed vertically to intersect the curve; and then proceed horizontally to determine the increase in flue gas temperature, 80°F (26.67°C).
2. Determine the increase in flue gas temperature by adding the clean temperature, 350°F (176.67°C), to the increase in temperature due to the 1/32 inch (0.79 mm) soot buildup, 80°F (26.67°C). Flue gas temperature = 430°F (211.11°C).
3. Determine the combustion efficiency for the dirty and clean boilers by referring to Figure 13-3. Enter the graph at 10% CO_2 and proceed vertically until intersection with the 430°F (211.11°C) curve (dirty boiler); then proceed horizontally to determine the combustion efficiency, 81.5%. Perform the same procedure for the clean boiler except use the 350°F (176.67°C) curve, 83.5%.

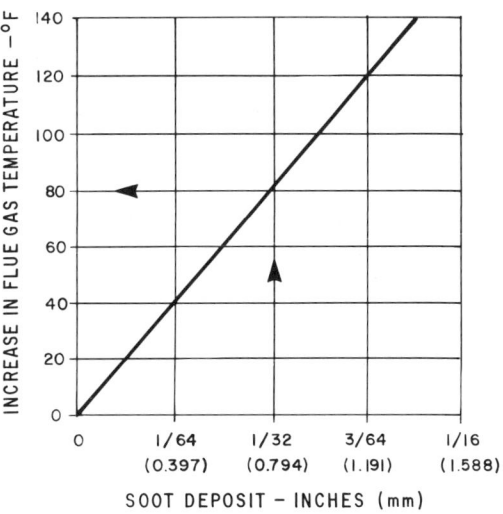

Figure 12-6. Fireside soot buildup. (Reproduced by permission of Pope, Evans and Robbins, Inc.)

4. Determine the percent energy loss in efficiency by subtracting:

    % loss = Clean efficiency
    − Dirty efficiency
    % loss = 83.5% − 81.5% = 2%

5. Calculate annual $ savings by operating a clean boiler:

    Annual $ savings
    = 2% × $40,000 = $800

E. Heat Loss Due to Domestic Hot Water Leak

Example #5. A domestic hot water system is leaking hot water at the rate of 5 gal/hr (5.25 ml/s). If #6 oil is used as fuel for the hot water system, determine the annual energy savings that will be achieved from stopping this leakage.

Procedure:

1. Enter the Figure 12-7 graph at 5 gal/hr (5.25 ml/s) of hot water and proceed vertically upward to intersect the curve;

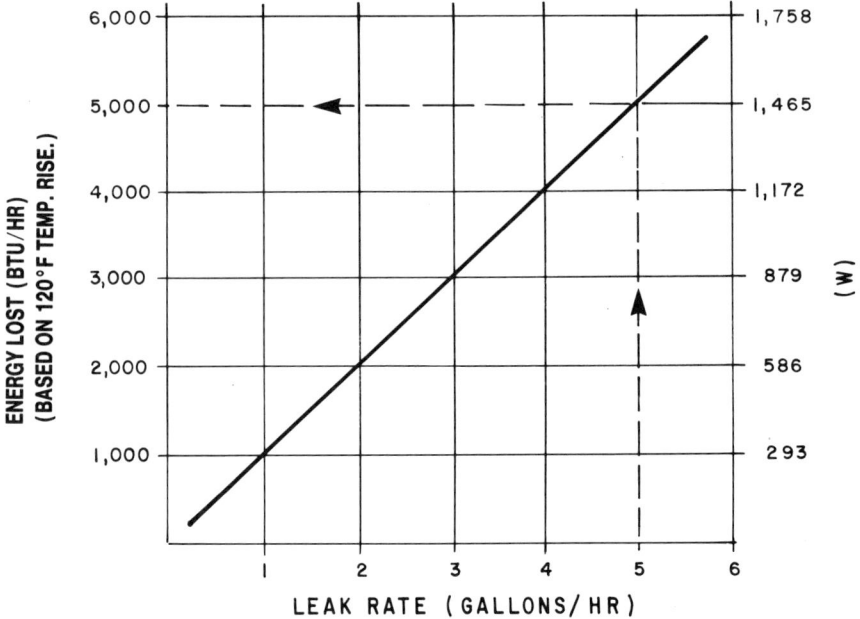

Figure 12-7. Heat loss due to domestic hot water leak. Hot water temperature rise 100°F (38°C); system efficiency 83%. (Reproduced by permission of Pope, Evans and Robbins, Inc.)

then follow horizontally to energy lost, 5,000 Btu/hr (1,465 W).

2. Multiply 5,000 Btu/hr (1,465 W) by 365 days and by 24 hours to get annual energy savings:

Annual energy savings = 5,000 (1,465 W) × 364 × 24
= 44 million Btu (46.42 × 10⁶kJ)

3. Fuel conversion: Divide 44 million Btu (46.42 × 10⁶kJ) by 149,700 (41,766.3 kJ/l) to convert Btu to gallons of oil.

$$\text{Annual Fuel Savings} = \frac{44 \text{ million Btu } (46.42 \times 10^6 \text{kJ})}{149,700 \ (41,766.3 \text{ kJ/l})}$$
× .75 (Boiler efficiency)
= 392 gallons (1485.68 l) of #6 oil.

4. Annual $ savings when #6 oil costs $.80 per gallon ($0.21/l):

Annual $ savings = 392 × $.80 (1485.68 × 0.21)
= $314

F. Energy Losses Due to Malfunctioning Steam Traps

Example #6. Determine the total annual energy saved by repairing a steam trap in a building with an orifice size of 0.20 inch (49.8 Pa) for a 15-psi (10,335 kPa) steam system. The heating fuel used is #6 fuel oil, and steam is generated 30% of the total plant operating time of 4,800 hours.

Procedure:

1. Enter the Figure 12-8 graph at 0.20 steam trap orifice size and proceed horizontally to intersect the 15-psig (103.35 k Pa) curve; then proceed vertically to the steam loss, 35 lb/hr (4.41 g/gs).
2. To find annual energy savings, multiply 35 lb/hr (4.41 g/s) by 30% and the number of plant operating hours, 4,800, and

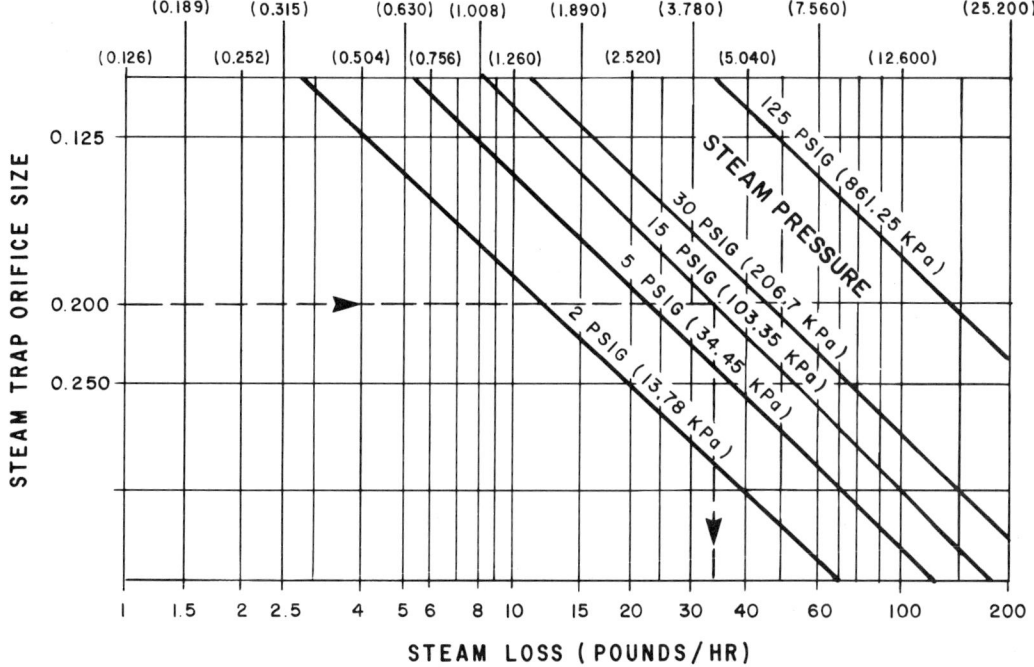

Figure 12-8. Energy losses due to malfunctioning steam traps. (Reproduced by permission of Pope, Evans and Robbins, Inc.)

by the steam enthalpy of 1,000 Btu/lb (2330 kJ/kg):

Annual energy savings = 35 × 0.30 × 4,800 × 1,000
(4.41 g/s × 0.3 × 3,600 × 4,800 × 2,330 J/g)
= 50.4 × 10^6 Btu (53.172 × 10^6 kJ)

3. To find annual fuel savings, divide the result of step 2 by the Btu content of #6 oil (used in this example), which is 149,700 Btu/gal (41,766.3 kJ/l), and the system efficiency, estimated to be 75%. Annual fuel savings:

50.4 × 10^6 Btu/yr (53.172 × 10^6 kJ/yr)/[149,700 Btu/gal (41,766.3 kJ/l) × 0.75] = 449 gal/yr (1,701.71 l/yr)

4. Determine the annual $ savings with #6 oil costing $.70 per gallon ($0.18/l):

Annual $ savings = 449 gallons (1701.71 l) × $.70 per gallon ($0.1845/l)
= $314

G. Preheating Combustion Air

Example #7. The combustion air temperature is raised from 80°F (26.67°C) to 300°F (148.89°C) by using an air preheater (economizer) prior to reaching the #2 fuel oil burner. If the annual boiler fuel consumption is 25,000 gallons (94,750 l), determine the achieved savings by raising the combustion air temperature.

Procedure:

1. Refer to the Figure 12-9 graph and determine the fuel savings for the 300°F (148.89°C) air. Enter the graph at

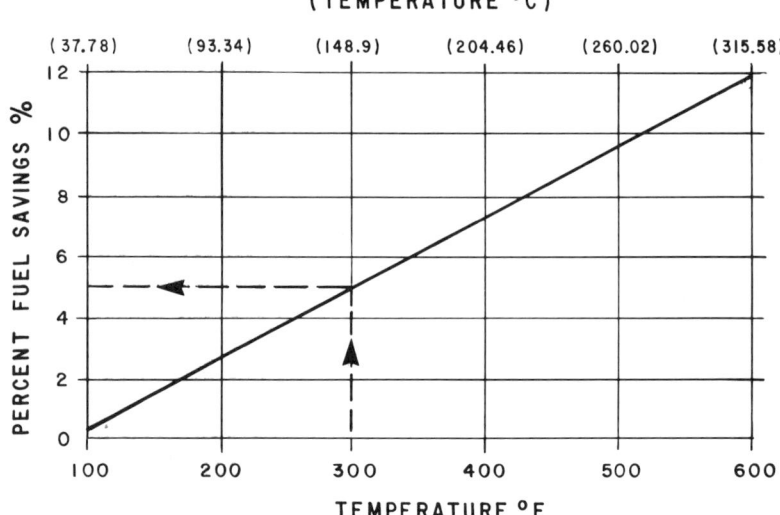

Figure 12-9. Preheating combustion air. (Reproduced by permission of Pope, Evans and Robbins, Inc.)

300°F (148.89°C) and proceed vertically until the curve is intersected; then proceed horizontally to determine percent of fuel saved (5%).

2. The base-line temperature for the above graph is 80°F (26.67°C) air, which corresponds to 0%.
3. Multiply the percent savings by the amount of annual fuel consumption (25,000 gallons) (94,750 l) to determine the annual fuel savings:

Annual   = 5% × 25,000 (94,750 l)
fuel savings = 1,250 gallons (4,737.5 l)
For #6 oil, = 1,250 (4,737.5 l)
Btu saved   × 149,700 (41,766.3 kJ/l)
               × .75 (System efficiency)
             = 140,344,000 Btu
               (148,062,920 kJ)

4. Determine the annual $ savings when fuel cost $.80 per gallon:

Annual $ = 1,250 × $.80 (4737.5
savings   × $0.211)
           = $1,000

H. Hot Water Storage Tank Insulation

Example #8. Determine the savings incurred by insulating a bare hot water storage tank, 6 ft (1.828 m) long × 3 ft (0.914 m) diameter with 3 inches (76.2 mm) of insulation. The water temperature is 140°F (60°C), and the average surrounding ambient air temperature is 80°F (26.67°C). Fuel is #2 oil.

Procedure:

1. Find the temperature difference:

$$140°F - 80° (60°C - 26.67°C)$$
$$= 60°F (33.33°C).$$

2. Find, from the Figure 12-10 graph, the heat loss for the 3-inch insulated tank and the uninsulated tank by entering the graph at 60°F (33.33°C) and proceeding vertically until the appropriate curves are intersected; then proceed horizontally to determine values.

Insulated: 40,000 Btu/yr/ft² (456,000 kJ/m²)

Uninsulated: 400,000 Btu/yr/ft² (4,560,000 kJ/m²)

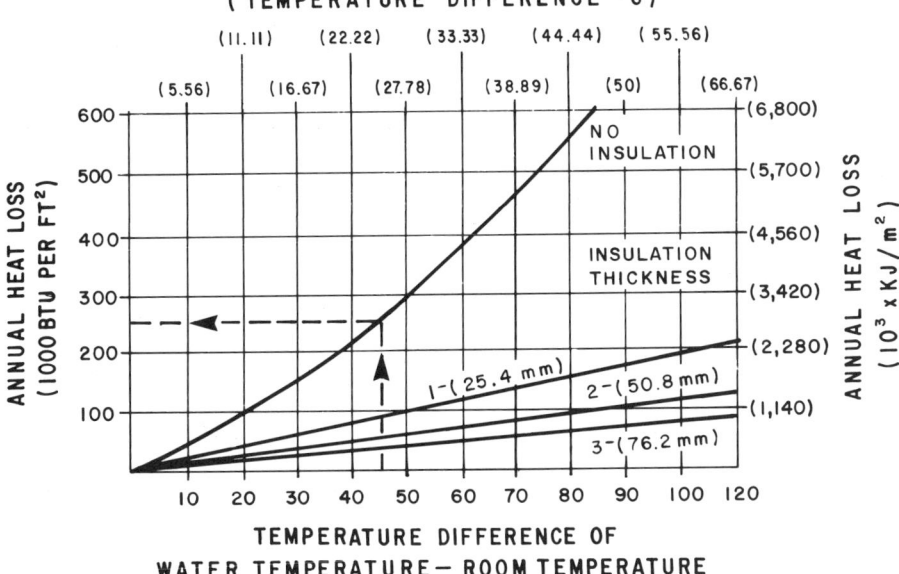

Figure 12-10. Hot water storage tank insulation. (Reproduced by permission of Pope, Evans and Robbins, Inc.)

3. Find the difference in heat loss:
400,000 − 40,000 = 360,000 Btu/yr/ft^2
(4,560,000 − 456,000) = (4,104,000 kJ/m^2)

4. Determine the exposed area of the hot water storage tank. For cylindrical tanks:

Area = (3.14 × Dia) × (Length × Dia/2)
= [3.14 × 3 ft (0.914 m)]
× [6 ft (1.83 m)]
× 3 ft/2 (0.914 m)
= 84.78 sq ft (2.402 m^2)

5. Calculate the annual energy savings:

Annual energy savings = 84.78 sq ft × 360,000 Btu (2.402 m^2 × 379,800 kJ)
= 30.5 × 10^6 Btu (912,279.6 kJ)

6. Determine annual fuel savings by dividing step 3 by 138,700 Btu/gal (38,697.3 kJ/l) (fuel conversion number for #2 oil), and by the estimated system efficiency of 75%:

Annual fuel savings = 30.5 × 10^6 Btu/yr (32.178 × 10^6 kJ/yr)/(138,700 Btu/gal × 0.75) (38,697.3 kJ/l × 0.75)
= 293.2 gallons (1,111.2 l)

7. Determine the annual $ savings when fuel oil costs $.85 per gallon ($0.22/l):

Annual $ savings = 293.2 gal (1111.2 l) × $.85/gal ($0.22/l)
= $249

I. Insulating Hot Pipes

Example #9. Determine the annual fuel savings achieved by insulating a 100-ft (30.48 m) long, bare hot water pipe, 1½ inches (38.1 mm) in diameter with ½ inch (12.7 mm) of insulation. Hot water temperature is 180°F (82.22°C), and the building heating fuel is natural gas.

Procedure:

1. Enter the Figure 12-11 graph at 1½-inch (38.1-mm) pipe size and follow horizon-

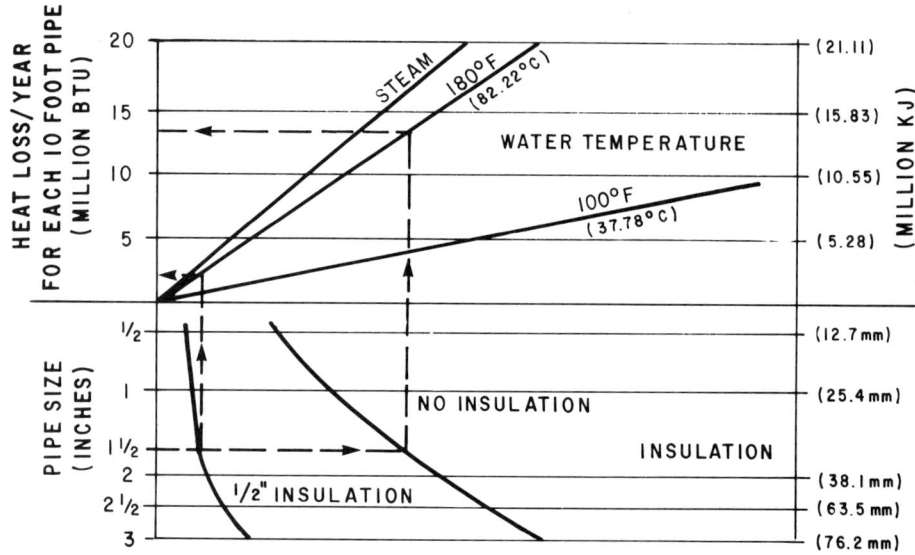

Figure 12-11. Insulating hot pipes. (Reproduced by permission of Pope, Evans and Robbins, Inc.)

tally to the right to intersect the no-insulation curve; then proceed vertically upward until the 180°F (82.22°C) curve is intersected; then proceed horizontally to the left to determine the heat loss for each 10 feet (3.048 m) of pipe (13 × 10⁶ Btu/yr) (13.715 × 10⁶ kJ/yr).

2. Enter the graph at 1½-inch (38.1-mm) pipe size and follow to the right to intersect the ½-inch (12.7-mm) insulation curve; then proceed vertically upward until the 180°F (82.22°C) curve is intersected; then proceed horizontally to the left to determine the heat loss for each 10 feet (3.048 m) of pipe (2.1 × 10⁶ Btu/yr (2. 216 × 10⁶ kJ/yr).

3. Determine the annual energy savings by subtracting Step 2 from Step 1 and *multiplying by 10,* the number of lengths of 10 feet that equal 100 feet:

Annual energy savings = [13 million (13.715 × 10⁶ kJ) − 2.1 million (2.216 × 10⁶ kJ)] × 10
= 109 million Btu (114.990 × 10⁶ kJ)

4. Determine the annual fuel savings by dividing Step 3 by the fuel conversion for natural gas, 1,030 Btu/cu ft (38.419 kJ/m³):

Annual fuel savings = (114.99 × 10⁶ kJ/38,419 kJ/m³ × 0.75 (System Efficiency)
= 144,900 cu ft (4100.67 m³) natural gas

5. Determine annual $ savings when natural gas costs $.35 per therm ($12.36 per 1,000 m³):

Annual $ Savings = [144,900 (4,100.67 m³)/1,000] × .35 ($12.36)
= $51 per 100 feet (2.83 m)

IV. HVAC RETROFITS

HVAC retrofits (as opposed to central plant type retrofits) are best discussed with few specific examples, since circumstances at each installation usually vary. The best tool for dealing with HVAC retrofits is a thorough understanding of the principles of heat transfer as well as the limits and capacities of the hardware used.

A. Chiller Retrofits

The four most commonly found chillers in terms of their part load energy efficiencies are:

1. Steam absorption units
2. Reciprocating chillers
3. Centrifugal chillers
4. Screw-type chillers

Most chillers generally run at less than 100% loading most of the time. If we compare at part loading the amount of energy input versus output (in terms of Btu's supplied and removed) the least efficient is the steam absorber, while the most efficient is usually the screw-type chiller. Reciprocating units are generally small in size and somewhat difficult to control in small steps for part load operation. In addition, they present no energy savings if equipped with hot-gas bypass-type capacity controllers.

For new water-cooled units, the trend is to use "double bundle" condensers where one bundle may be used for domestic hot water generation, terminal air, reheat, or other purposes. Older machines may benefit from the addition of a heat exchanger in the condenser water circuit to generate warm water from the condenser waste heat. One must be careful, however, not to increase the condensing temperatures of existing units, since such an increase often consumes more energy than the amount gained in the subsequent heat exchanger.

Because the variations of additions and alterations are almost limitless, only three chiller examples with wide potential application are presented here.

Retrofitting an Older Absorption Chiller.

Older absorption chillers, often in service for 20 years or more, should always be retrofitted to increase their efficiency and cut their steam consumption per ton of refrigeration produced. The retrofit generally consists of one or more of the following types of measures:

1. Improve heat transfer between coolant and refrigerant by utilizing "on-line" cleaning systems that keep the heat exchanger surfaces clean.
2. Provide necessary controls to permit operation with lower condenser as well as higher chilled water temperatures. As these liquids approach each other's temperature, the efficiency of the operation is improved.
3. Improvement of the operation by providing more accurate temperature and load sensing devices, thereby reducing excess energy and waste.

The most commonly applied measures are:

1. *Full range economizer valve.* The valve controls the flow of dilute (weak) solution to the concentrator (generator) by means of a solenoid or three-way by-pass valve. If less weak solution is permitted in the concentrator, less steam is required; hence part load efficiency improves.
2. *Uncontrolled condenser water operation.* Older machines usually need their condenser water temperature regulated to around 86°F (30°C) to 90°F (32.22°C) regardless of the availability of cooler condenser water from the cooling tower. The affinity of the brine solution for water is higher if the temperature is lower. Older machines often crystallized if the condenser water was too "cold" and the concentration high. In uncontrolled condenser water operation [down to 55°F (12.78°C) condenser water], a signal is usually generated by the condenser water temperature. The signal is then sent to the steam supply valve to control its opening, and simultaneously the solution in the absorber is diluted to prevent crystallization. The diluted and colder brine solution still retains its affinity for water; therefore chiller operation is not affected. For each 1°F (0.56°C) temperature drop of the condenser water, 1.5 to 2% savings in steam consumption are possible.
3. *Load anticipation and chilled water modulation.* By sensing outside air or return water temperature or a combination of both, the chilled water temperature is allowed to "float" upward until a set temperature difference, say 5°F (2.78°C) to 10°F (5.56°C) is established between supply and return water temperatures.

While such a variation of the leaving chilled water temperature effectively eliminates humidity control for most cooling coils, the benefits of energy savings generally outweigh the negative aspects of close humidity control for units serving noncritical hospital areas. Increasing the leaving chilled water temperature of the chiller 1°F (0.56°C) generally results in a 1.5 to 2% decrease in steam consumption or overall efficiency. An added side benefit for most applications is the additional savings of not having to reheat the overcooled supply air.

4. *On-line brush cleaning of condenser and absorber tubes for better heat transfer.* Overall heat-transfer reduction due to layers of materials deposited on tube-wall surfaces makes all chillers become less and less efficient until the steam rate (lb/ton) (kg/kW) becomes high enough to warrant chemical and mechanical cleaning. Because of fouling, secondary effects such as water velocity changes, feed systems irregularity, algae treatment failure, or microscopic particle deposition may speed up the naturally occuring time-dependent fouling. To overcome fouling and to reduce energy and maintenance costs, a small plastic brush is fitted into the tubes and is made to travel the length of the tubes by the flow of the water. A four-way valve is used to reverse the water flow according to a predetermined schedule. The traveling brushes remove the slowly accumulating scale and dirt and restore the badly fouled tubes' heat transfer to the original "clean" condition.

5. *Control modifications for operational changes.* It was found that depending on actual job conditions the addition of a few control components can make overall operation more reliable. Mention is made of the following items for reference only:

   (a) *Steam flow control for optimum efficiency.* By limiting the open position of the steam valve at transient conditions, steam is saved, since the wide-open steam valve wastes steam.
   (b) *Chiller insufficiency compensation.* This prevents chiller operation if any transient conditions due to mechanical malfunctions exist.

A somewhat shortened example of typical calculations and anticipated results is included below.

Upgrading for Uncontrolled Condenser and Chilled Water. Economic analysis for a 426 nominal-ton (1499.52-kW) machine to be upgraded is based on the following actual conditions found during a field survey:

1. Average load on machine of 60%.
2. Condenser water regulated to be about 85°F (29.44°C) during the cooling season.
3. Usage about 3,600 hr/yr.
4. Steam costs of $7.20/1,000 lb ($15.86/1,000 kg).
5. Usage factor taken from manufacturer's catalog.
6. Present steam rate: 19.8 lb/ton (255 kg/kW). (Actual, measured at full load).

Estimated present use:

$$19.8 \text{ lbs/ton} \times 3,600 \text{ hr/yr} \times 0.6 \times 426 \text{ tons/hr} = 18.2 \times 10^6 \text{ lb/yr}$$
$$(2.55 \text{ kg/kW} \times 3,600 \times 0.6 \times 1499.52 \text{ kW/hr}) = 8.26 \times 10^6 \text{ kg/yr}$$
$$\text{Cost: } (18.2 \times 10^6 \text{ lb/yr}) \times (\$7.2 \times 10^{-3}/\text{lb}) = \$131,040/\text{yr}$$
$$(8.26 \times 10^6 \text{ kg/yr} \times \$15.86 \times 10^{-3}/\text{kg} = \$131,004/\text{yr})$$

Estimated future use:
Average condenser water temp. (future uncontrolled) estimated:

72°F down 13°F from 85°F to 72°F

22.22°C down 2.22°C from 29.44°C to 22.22°C

Average chilled water temp. (in future allowed to rise if load is reduced):

46°F, a rise of 4°F from 42°F to 46°F
7.78°C, a rise of 2.2°C from 5.56°C to 7.78°C

Total combined:

4°F + 13°F = 17°F @ 1.5% of
(2.2°C) + 7.22°C = 9.42°C @ 1.5% of

Savings are:

25.5% reduction of steam consumption (estimated)

Usage factor: 0.6 × 0.745 = 0.447

Reduced consumption:

19.8 × 3,600 × 0.447 × 426
= 13.57 × 10⁶ lb/yr
2.55 × 3,600 × 0.447 × 1,499.52
= 6.16 × 10⁶ kg/yr

Savings:

(18.2 − 13.57)
= 4.63 × 10⁶ lb/yr, or at $7.20/1,000 lb
8.26 × 10⁶ − 6.16 × 10⁶
= 2.10 × 10⁶ lb/yr, or at $15.86/1,000 kg

The cost savings is $33,336/yr.
On-line Brush Cleaning Savings:
Estimated average fouling (from field tests): $f = 0.002$.

Increase in steam consumption per ton due to fouling as per manufacturer's charts is approximately 22%.

Rated consumption: 19.8 lb/ton (2.55 kg/kW).

Increase in steam consumption per ton due to normal fouling is 3%.

Present increase due to excessive fouling: 22% − 3% = 19%.

Consumption at present fouling:

19.8 lb/ton × 1.22 = 24.156 lb/ton
2.55 kg/kW × 1.22 = 3.11 kg/kW

Annual present energy cost: $131,040/yr.

Estimated annual savings due to brush cleaning:

Energy cost/(1 + % increase in lb/ton)
= 131,040 − [131,040/(1 + 0.19)]
= $20,922/yr

At steam costs of $7.20/1,000 lb ($15.86/1,000 kg), this represents 2.9 × 10⁶ lb (1.32 × 10⁶ kg) of steam saved per year.

While the above measures are considered individually, the combined results of both would cut steam consumption drastically. It is obvious that if a machine is retrofitted for uncontrolled condenser, chilled water, and other modifications, the brush cleaning will not result in additional savings of 2.9 × 10⁶ lb/yr (1.32 × 10⁶ kg/yr). Savings will be somewhat less, since, owing to other measures, less steam is used to begin with. In such situations, one should start with the already reduced steam consideration a smaller than calculated further drop of steam consumption. It is also important to remember that many old machines will seldom have their steam rates reduced 20 to 25% or consume less than the original "new" rate regardless of the retrofit. This is mainly due to internal design as well as external conditions.

Heat Exchanger Installation. Our third example is based on a hermetic reciprocating chiller operating at present with a C.O.P. of 2.99. The unit provides 69 tons (242.88 kW) of cooling (nominal), with a power input of 80.9 kW [chilled water at 42°F (5.56°C)] and recorded condenser water temperatures of 85°F (29.44°C) water entering the 102.3°F (39.05°C) water exiting from the machine. Water temperatures are regulated at the tower, since this older machine cannot safely operate with much colder condenser water. Saturated discharge temperature is set at 120°F (48.89°C), and 127.6 gpm (8.051 l/s) of condenser water is circulated.

Assume a heat exchanger (double-walled for safety) is installed in the condenser water circuit to provide tempering for the incoming domestic hot water. The heat exchanger selected would cool 127.6 gpm (8.051 l/s) of water from 102.3°F (39.06°C) to 87.3°F (30.72°C), a 15°F (8.34°C) drop, by preheating the incoming 40°F (4.44°C) domestic hot water to 70°F (21.11°C), a rise of 30°F (16.67°C). Although 63.8 gpm (4.025 l/s) of domestic water could theoretically be heated, because of inherent inefficiencies, only 60 gpm (3.786 l/s) is considered. Also note that the approach temperature of 87.3°F (30.72°C) − 70°F (21.11°C), or 17.3°F (9.61°C), is rather high to reduce the size and first cost of the heat exchanger. It is very likely that water warmer than 70°F (21.11°C) would be available to the domestic hot water heater; this consideration makes these calculations conservative:

The energy gained is:

60 gpm × 8.33 lb/gal
  × 60 min/hr × 30°F
  = 899,640 Btu/hr
0.2274 m³pm × 999.6 kg/m³ × 16.67°C
  = 263,595.0 W

For 3,600 full-load hr/yr, this is equivalent to $3,238 \times 10^6$ Btu ($3,416.09 \times 10^6$ kJ). Or, if oil is used to generate the steam used in the existing domestic hot water heater, the oil saved is:

($3,238 \times 10^6$ Btu/yr)/(140,000 Btu/gal
  × 0.8 × 0.9) = 32,123 gal/yr
(3.416×10^9 kJ/yr)/(39,060 kJ/l
  × 0.8 × 0.9) = (121,466 l/yr)

If boiler efficiency is 0.8 and estimated hot water generation efficiency is 0.9, at a rate of $0.75/gal ($0.198/l) the cost savings are $24,092 ($24,050) per year.

It is important to note that the above examples are only approximations, generally on the conservative side. For example, no credit has been given to savings brought about by reduced cooling tower operation. The reader is therefore urged to evaluate each situation independently for more accurate results.

B. Energy Transfer Retrofits

Energy transfer retrofits generally consist of devices that transfer energy from one stream of fluid such as air (exhaust) to another, usually incoming outside air. They are best used when the incoming outside air stream is heated and/or cooled without the addition of return air. Such "once-thru" systems are often found in hospitals, serving operating rooms or other critical areas where the addition of return air may result in cross contamination. They may also be applicable for kitchens, where the exhaust air steam is usually too hot, fouled, or both, for reuse. The best application is for systems that use both heating and cooling. In such systems, heat is transferred from the exhaust stream to preheat the incoming outside air in the winter, while in the summer, the heat is transferred from the hot outside air to the cooler exhaust air. Depending on the relative locations of the intake and exhaust ducts there are generally three types of heat transfer systems available for a simple retrofit:

1. Run-around coils
2. Heat pipes
3. Heat wheels

Run-around coils consist of two air-to-water heat exchangers placed in the appropriate air streams. The coils are interconnected by a pipe loop incorporating a pump and a three-way valve for freeze protection. A glycol/water mixture is usually circulated through both coils to transfer energy from one to another. During winter, especially in places where the climate is severe, 30 to 50% of the heat can be easily recovered. The run-around coils can only transfer sensible heat; this limits their usefulness in the summer. The major advantage is that the air streams need not be near each other. It is generally true that long distance between the air streams reduce efficiency, because of the heat loss in transfer by the piping system, but large quantities of air at large temperature gradients make even distant coil systems cost effective.

If the two air streams (exhaust and supply) are close enough, a minor alteration of ductwork may make it possible to use either heat pipes or a heat wheel. Both devices require parallel and opposite flows of air at different temperatures. A heat-pipe device is usually a coil consisting of Freon elements. The elements are concentric tubes containing a liquid (usually refrigerant) that is allowed to evaporate on the warmer side and migrate to the cooler half of the coil. At the cool side, the refrigerant condenses and flows by gravity back to the warmer side of the coil. Summer/winter switchover is relatively simple; only the tilt of the coil needs to be changed, since the relative temperatures of the air streams are reversed. The heat pipe is inserted in the ductwork, one half in the intake the other in the exhaust. The edges are sealed, and the system transfers up to 70 to 80% of the energy. Energy transfer is still sensible only, and the air streams must be near each other.

Both the run-around coil and a properly sealed heat pipe prevent cross contamination, since no part of the system moves from the contaminated air stream into the fresh air stream.

The heat wheel uses a wheel-type heat transfer unit inserted into the parallel and near air streams, in a configuration somewhat similar to the heat pipe. The wheel, however, rotates, and cross contamination may become a problem if organisms attach themselves to the heat transfer material. The great advantage of the heat wheel is that latent heat transfer is possible, thereby increasing the overall summer/winter efficiency, especially in areas where the summers are long and humid. While selection of the optimum energy-transfer method should be left to the qualified energy auditor, bear in mind that a once-thru system with large temperature gradient, running 24 hours a day, will result in the shortest payback. Most manufacturers make computer-based economical analysis available, eliminating the necessity for complicated and time-consuming manual calculations.

C. Control and Regulatory Retrofits. Many control and regulatory retrofits may be identified in the course of hospital energy audits. While there are too many to mention, some stand out as applicable to most facilities. Brief descriptions of the three most important ones are presented below in descending order of usage:

Thermostatic Control Valves on Hot Water or Steam Heating Elements: Lack of temperature control is a frequent cause of overheating and excessive energy consumption in low pressure steam and hot water heating systems. Installation of thermostatic control valves allows room-by-room control of temperature, matching the thermal energy consumption to the heat losses in the space. These valves consist of self-contained, nonelectric thermostatic operators and control valve bodies suitable for modulating steam or hot water flow to radiators, convectors, fan coils, etc. Set temperatures may be locked to minimize tampering. Thermostatic temperature dial and sensor may be mounted remotely, easing installation and operation.

Energy savings is usually a function of existing temperature and controlled future temperature. Any ongoing installation of new thermopane windows or added insulation will reduce heat loss in occupied spaces, increasing the tendency toward overheating. Installation of these valves will compensate for this tendency and provide maximum savings. Savings of 10 to 15% are typical in complete hospital installations where overheating due to insufficient control is a problem.

Cycle of Shutdown of Heating Ventilation and Air Conditioning Equipment: Many areas in the hospital are either unoccupied, used only in emergencies, or lightly staffed during the off-hours of 7 P.M. to 7 A.M. Significant energy and dollar savings can accrue if heating, ventilating, and air conditioning (HVAC) equipment serving these areas is duty-cycled or shut down during those periods. Cycling and shutdown should be provided, however, for off-hour or emergency operation. Most HVAC equipment is designed to handle peak loads. Since such loads are infrequent, some equipment may be cycled during the day. Overdesign or equipment functional changes may also allow equipment to be cycled within code or other requirements. Two methods are available for such control:

1. Local electromechanical timers can be installed on each piece of equipment. These timers typically have a seven-day cycle and can turn loads on and off several times a day. They are relatively inexpensive—about $300 per control point installed. However, they do not have the versatility to cycle equipment frequently and do not perform predictive demand control.
2. A central microprocessor based load and demand controller can be installed to turn equipment on and off. These units are programmable; they provide versatility in cycling, shutdown and demand limiting. Installed costs run about $600 to $800 per control point. This type of unit is connected to each load with low voltage wiring and can monitor, control and override the operation and shutdown of equipment from a central location. A significant improvement in operation, resulting in cost reductions as well, is the central microprocessor generating high frequency signals that utilize the hospital's existing wiring for transmission. The units to be controlled are equipped with relays only that are activated by the signal generated. Since no additional wiring is required, the installed first cost is reduced from the range of $600 to $800 to the range of $400 to $600 per control point.

Installation of Enthalpy-Economizer Controls. Many areas of a hospital are ventilated and heated or cooled using a mixture of outside air and recycled return air. When the outside air conditions are similar to the conditions required for delivery from the air handling unit, it is desirable to have outside air introduced into the building. If the outside air conditions do not closely match the inside requirements, it is desirable to introduce the minimum fresh air allowed by applicable codes. An enthalpy economizer cycle allows outside air and return (inside) air to be mixed in the proper proportions so the least amount of energy need be expended to heat or cool the supply air to the required condition. This controller automatically controls the dampers in the air ducts (based on the enthalpy wet bulb temperature) of the air streams.

V. Lighting Retrofits

The present trend in lighting ECM's is to reduce the overall general illumination and provide levels of light consistent with requirements in task areas; this is known as task lighting. The U.S. Department of Energy has recommended:

Fifty foot-candles (538 lumen/m^2) should be maintained in office or work space task areas, 30 foot-candles (323.0 lumen/m^2) in office non-task areas, and 10 foot-candles (108.0 lumen/m^2) in other nonwork areas such as storage areas, stairs, and corridors. Special work areas are to be treated according to their specific requirements. Where necessary, ceiling-mounted fixtures should be supplemented with desk lamps in office spaces or with portable lamps in assembly spaces.

Each area should be separately analyzed for task lighting. Two rooms identical in size, containing the same number of lighting fixtures and the same number of desks, may have different recommendations. The furniture arrangement must be considered. Ideally, desks should be located together in a task area; file cabinets, tables, book shelves, etc., should be grouped in non-task areas.

Foot-candle readings should be taken at task or work areas at the surrounding non-task areas and in open unallocated spaces. If there are more than 65 foot-candles (700 lumen/m^2) or 70 foot-candles (753 lumen/m^2) in the task areas, or more than 40 foot-candles (431 lumen/m^2) or 50 foot-candles (538 lumen/m^2) in the nontask areas, lamps should be removed and the foot-candle (lumen/m^2) levels lowered to approximately 50 and 30 (538 and 323 lumen/m^2), respectively.

ECM's can be established for lighting by

two basic means: (1) reduction in wattage per given area, or (2) reduction in the hours of usage.

Wattage reduction can be achieved by "de-lamping" of fixtures; (1) replacing fluorescent lamps with lower-wattage energy saving lamps; and (2) replacing incandescent lamps with lower-wattage lamps or with self-ballasted screw-in fluorescent lamps where feasible.

Reduction in the hours of usage can be achieved by: (1) manually switching off lights when they are not needed, (2) using manual timers to control lights in small offices, and (3) using photocells to switch lights off when sufficient daylight is available for requirements of the area involved.

Noncritical areas should have their levels of illumination reduced in accordance with the DOE recommendations. Of course in a hospital, many areas require exacting visual acuity. The lighting in these areas should be treated according to the specific requirements applicable.

The following example demonstrates one of the most common energy retrofit methods. It also indicates that most lighting retrofits offer short term paybacks.

Example #10. Determine the amount of energy saved by replacing 3,000 watts of incandescent lighting with fluorescent lighting to achieve an equivalent lighting level. The lamps are lit 12 hours per day.

Procedure:

1. Enter the Figure 12-12 graph at 3,000 incandescent lamp wattage and proceed horizontally to the right to intersect the curve; then proceed vertically downward to fluorescent lamp wattage (600 W).
2. Compute the difference of wattage for the two lamps of equivalent light level:

$$\text{Incandescent lamp} = 3{,}000 \text{ W}$$
$$\text{Fluorescent lamp} = 600 \text{ W}$$
$$\text{Difference} = 2{,}400 \text{ W}$$
$$= 2.4 \text{ kW}$$

3. Calculate the annual energy savings by multiplying the difference found above by the number of hours illuminated (12) and days per year.

$$\text{Annual energy savings} = 2.4 \text{ kW} \times 12 \times 365$$
$$= 10{,}512 \text{ kWh}$$

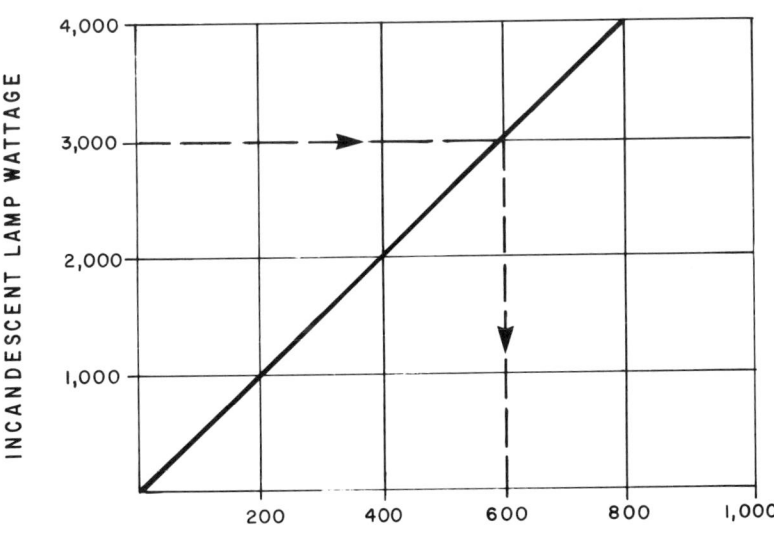

Figure 12-12. Replace incandescent with fluorescent lighting. (Reproduced by permission of Pope, Evans and Robbins, Inc.)

4. Determine the annual $ savings when electricity costs $.08/kWh:

$$\text{Annual \$ savings} = 10{,}512 \text{ kWh} \times \$0.08/\text{kWh} = \$841$$

ACKNOWLEDGMENT

All illustrations in this chapter are courtesy of Pope, Evans and Robbins Inc., New York, New York.

REFERENCES

"Hospitals Can Save Energy," New York State Energy Office, Albany, N.Y., undated.

"A Study of Energy Conservation in the Non-Profit Hospitals and Nursing Homes of New York State," Pope, Evans and Robbins, Inc., New York.

Stephen L. Baron, ed., *Manual of Energy Saving in Existing Buildings and Plants*, Vol. 1, *Facility Operation and Maintenance*, Prentice-Hall, Inc., Englewood Cliffs, N.J. 1978.

Stephen L. Baron, ed., *Manual of Energy Saving in Existing Buildings and Plants*, Vol. 2, *Facility Modifications*, Prentice-Hall, Inc., Englewood Cliffs, N.J. 1978.

Colorado Energy Conservation and Alternatives Center for Commerce and Industry, *Energy You Can Bank On: A Practical Energy Conservation Guide for Business and Industry*, Energy Conservation and Alternatives Center, Denver, Colo., 1978.

Colorado Energy Conservation and Alternatives Center for Commerce and Industry, *Making Cents of Your Energy Dollar*, Vols. 1 and 2, U.S. Department of Energy. Washington, D.C.

Department of Health, Education and Welfare, *Total Energy Management for Hospitals*, DHEW Publication No. (HRA) 77-613, Washington, D.C.

New York State Energy Office, *Energy Study: Hospitals and Nursing Homes in New York State*, New York State Energy Office, Albany, N.Y., 1979.

Albert Thumann, *Plant Engineers and Managers Guide to Energy Conservation*, Van Nostrand Reinhold Company, New York, 1977.

U.S. Department of Energy, *Energy Audit Workbook for Hospitals*, National Technical Information Service, United States Department of Commerce, Va., 1978.

Part 6

Retrofitting for Industry

13. Industrial Energy Observations and Opportunities

Robert C. LeMay

R. C. LeMay Associates, Inc.
Lafayette Hill, Pennsylvania

		Page
I.	Introduction	239
II.	Steam Boilers	239
	A. Boiler Condition or Control Arrangement	240
	B. Boiler Location	240
	C. Winter Comfort/Process Steam Boilers	240
	D. Economizers and Air Preheaters	240
	E. Waste Heat Boilers	240
III.	Steam Distribution and Use	240
IV.	Metal Melting	241
V.	Hot Forming Metals	242
VI.	Metal Heat Treating	242
	A. High Temperature Insulating Wool	243
	B. Flue Gas Analyses and Flame Adjustment	243
VII.	Vitreous Enameling	243
VIII.	Glass Plants	244
IX.	Rotary Kilns	244
X.	Bakery Ovens	244
XI.	Drying Processes Generally	245
XII.	Electrical Power Factor	246
XIII.	Electrical Equipment and Demand	246
XIV.	Illumination	246
XV.	Building Structures	246

I. INTRODUCTION

The observations in this chapter are the result of hundreds of industrial plant energy surveys in widely diverse industries. Additional insights were gained from presenting 34 industrial management seminars. The findings herein have been augmented by the observations of ten other energy specialists, and it is hoped that the numerous recommendations we have included here will prove to be timely and useful in industrial application.

II. STEAM BOILERS

Despite the fact that approximately half our industrial fuel is burned under steam boilers, and the fact that many well-operated steam plants exist, we are convinced that *the majority of all industrial boiler operators neither know their boilers' efficiencies nor have on hand the relatively simple instruments required to determine them*. In other words, they are still "flying blind," hoping that those who service their boilers at infrequent intervals have properly adjusted boiler combustion, and that favorable adjustments have survived. Because fuel compositions and combustion conditions continually change, however, we suggest that boiler operators purchase and use these devices if they are not already employed.

A. Boiler Condition or Control Arrangement

Boiler condition or control arrangement is at times less than satisfactory. In a brass mill surveyed, our portable instruments showed much excess air at the stack, although the burner adjustment indicated no such condition. How could this be with a tight metal surrounding enclosure? A member of our survey team was resourceful enough to go down *beneath* the boiler, and found that the boiler floor was perforated and sucking in much air beyond the burner wall. This unwanted air chilled the tubes and severely reduced overall heating efficiency.

A brewery had boilers whose fuel–air ratio had always been manually adjusted. After new boilers were installed with automatic fuel–air ratio control, the chief engineer reported that he cut his fuel usage by one-half.

B. Boiler Location

Boiler location is often too far from the places where steam is used. In two large plants surveyed, we found that steam line and condensate return losses were so large that steam costs at the points of use were twice as high as at the boilers. In one old Connecticut plant, it was determined that $200,000 could be saved annually by abandoning the remote, overaged boilers and locating smaller, packaged boilers near the steam-using equipment.

C. Winter Comfort/Process Steam Boilers

Boilers providing steam for both winter comfort heating and smaller process steam uses are often too large for their comparatively small summer loads. With extensive line losses, they are consequently inefficient for summer use. In such cases we recommend either that smaller boilers be employed at points of steam use, or that the process be direct-heated. Either plan permits the large boiler(s) to be shut down during the inefficient summer use period.

D. Economizers and Air Preheaters

Economizers which transfer heat from departing flue gases to incoming boiler feedwater, and preheaters which transfer the heat to incoming combustion air, are all too frequently missing from many of the larger industrial water tube boilers where today's fuel prices dictate that they be employed. Serious thought should now be given to their addition, and an economic survey should be conducted.

E. Waste Heat Boilers

Waste heat boilers are not yet very numerous in the United States. In the past, this situation resulted partially from the relative unavailability of standard packaged units; most of those installed were custom-designed. Today, however, there are many standard catalog units available, and several organizations are prepared to size and sell them. These boilers come with or without auxiliary firing burners to take over when waste heat is not sufficiently available. Wherever high temperature gases are discharged and steam is needed, their use should be considered.

In one plant making heavy forgings, steam for hammers and presses came from a remote boiler plant and suffered serious line losses en route. An on-site study showed that flue gases from the forge furnaces could be run through waste heat boilers to generate *all* the steam needed by the forming equipment. Consequently, plant management concluded that its large central steam plant should be abandoned in favor of forge shop waste heat boilers. This change reduced manpower, maintenance, and fuel consumption.

III. STEAM DISTRIBUTION AND USE

All too frequently, steam systems fail to return condensate, or to have fully insulated steam lines, return lines, and condensate receivers. Generally, most plant steam lines have been insulated once. In one textile plant, we found 3,000 ft of high pressure steam main that had never been covered. Hastily made extensions and repairs are often left uninsulated, and to find condensate return lines insulated at all is rather unusual. At one plant, slightly opened valves at the ends of steam lines substituted for both traps and return system!

The next most common shortcoming is the failure to check and repair steam traps with regularity. It is more common to wait until obvious trouble develops before traps are repaired or replaced. Unfortunately, many trap failures are not obvious. Recently, a large midwestern grain processing plant hired a new plant engineer who found that three-fourths of all steam traps were defective.

There are many uninsulated steam and hot solution vessels in industrial plants, most frequently in metal cleaning rooms, plating rooms, and dairies. Using the excuse that inspectors fear product contamination from loose insulation, most operators leave pasteurizers, evaporators, cheese vats, spray dryers, and heated storage vessels entirely uninsulated in dairies. This causes much heat loss, resulting in uncomfortable working conditions. We believe that all of these vessels can be safely insulated behind carefully fabricated stainless steel jackets, but we suspect that operators have been hesitant to make the necessary investment. Tire molds and other rubber molding presses are seldom insulated today, resulting both in high energy consumption and in less than comfortable working conditions.

An entirely unexpected situation was found in the plant of a well-known company that manufactures steam turbine generators at other locations. High pressure steam *with superheat* was, for some reason, reduced at the boiler room wall to 15 psig (103 kN/m^2) for distribution to other buildings, through a pressure regulating valve. This pressure reduction loses much energy that could be harnessed by pressure reduction through one of the company's own steam turbine generators! This is somewhat akin to the shoemaker's children going barefoot.

IV. METAL MELTING

Much energy is wasted in metal melting, and this is increased by the fact that melting temperatures are generally high. Ferrous scrap preheating before melting can now be accomplished by drawing hot combustion gases from a burnered refractory cover down through a scrap-containing bucket into a heat fan. This is much preferred to firing burners against scrap piled beneath sheet metal covers. Yet, too many instances of the latter still exist.

Ladle preheating wastes much fuel in many foundries, because of the all too common practice of simply firing down into the ladles with open burners. In one Pennsylvania steel foundry, we observed this practice with ladles 8 ft (2.4 m) in diameter by 8 ft deep. Flame tails rose 10 ft (3 m) into the air above the ladles. Since our first visit, this foundry has correctly installed a burnered refractory cover on one of the ladles, and will presumably do this to confine the heat in the other two as well.

Most reverberatory melters are less thermally efficient than cupolas or several other improved furnace designs now available. A few years ago during the natural gas shortage, we were shown five reverberatory furnace-fed production lines in a large midwestern aluminum extrusion plant. The manager wanted to add a sixth line but could not then purchase more gas. After seeing that high temperature flue gases were being discharged from furnaces directly to the outside, we expressed the belief that metal flue recuperators would recover sufficient heat in combustion air to release enough fuel for that sixth production line. Since our visit, we have been informed that the recommended change has been carried out and the desired result achieved.

Too many melting furnaces have recently been observed with burners firing through open holes in their walls; this arrangement permits very little control of fuel–air ratios. These burners should be replaced with sealed-in units to bring fuel–air ratio control into an acceptable range.

V. HOT FORMING METALS

Steel mills customarily have full-time combustion and energy management engineers. Yet while visiting one of the larger Eastern mill's reheat furnace department, we recently observed large 2,000°F+ (1,100°C+) furnaces with doors open 2 ft above their sills because operators couldn't be bothered to close them between one slab and the next. This, of course, reveals a labor situation that should be corrected.

High production forge shops such as those found in automotive plants are usually well set up; it is normally the job forging establishment whose furnaces and hot bending machines are large energy wasters. We were once called from Philadelphia to Ohio, where an automatic hot bending machine was failing to maintain rated production. Simply by climbing a ladder and removing a dirty air filter, we were quickly able to restore full production. Maintenance of heating equipment is always worthwhile.

Other thermal inefficiencies are more difficult to overcome. The job forging industry is largely energy-wasteful. Unnecessary furnace openings radiating at about 2,400°F (1,316°C), broken refractories, uninsulated linings, ill-fitting doors, crumbling walls, too few burners, maladjusted combustion, and poorly arranged operating schedules are some of the problems observed. We even saw a well-known Pittsburgh-area plant with a forge furnace too short for its car hearth. Some enterprising person had knocked out the rear of the furnace and then piled loose bricks at the car's end in an attempt to reduce heat losses at that point. The resulting condition was both unsafe and a major source of heat/energy loss. We reached the conclusion that approximately half of the plant's $3 million annual natural gas purchases could be eliminated.

In another plant, long rolldown furnaces had been designed so that burners at the low ends would send hot gases up the sloping hearths, over the work in a counterflow manner, and out the flues at the high ends where the work was fed in. Along each side of the furnace, however, was a row of holes presumably closed with rolling cast iron covers. (Holes were used by operators who inserted bars to nudge along the large work pieces.) Careful inspection showed that the covers were sufficiently loose that upper covers discharged flue gases while lower covers admitted sufficient unwanted air to lower furnace temperature and to oxidize the ingots excessively. Our final report stated that one properly designed and insulated furnace with tight doors could do the work of both furnaces present, while using one half the fuel.

Not many forge shops yet employ heat recovery from those hot flue gases leaving at temperatures up to about 2,500°F (1,371°C). Where steam presses and hammers are used, waste heat boilers may perhaps be the best investment. In other cases, high temperature recuperators for combustion air preheating should be considered. And steel forge furnaces operate in a temperature range where oxygen additions to combustion air may possibly save both fuel and money: this is so because added oxygen reduces the necessity of raising nitrogen in the air to high temperatures needlessly.

VI. METAL HEAT TREATING

It is their use on metal heat treating furnaces for which flue recuperators have most widely been publicized—for the preheating of combustion air. Most frequently, they are attached to the ends of radiant tubes. They are helpful and should be encouraged, but most metal heat-treating furnaces today still continue to waste heat in flue products.

One useful and economical device too seldom employed today is thermal cascading, where hot flue products from high temperature processes are transferred to heat lower-temperature processes nearby, thus saving fuel at the lower-temperature equipment. In a Wisconsin plant with a long heat treating furnace aisle, steel hardening furnaces stood on the left and lower temperature "draw" furnaces on the right. Fortunately, management was preparing to cascade heat between one pair of furnaces across the aisle. If this worked satisfactorily, management would then duplicate the system all along the aisle.

Thermal cascading can convey heat to "beneficiaries" other than furnaces. In a Connecticut plant, we observed gas-fired comfort heaters being installed in a room adjacent to a heat treating room. In that instance, furnace flue heat should have been cascaded through the wall, either directly or via heat exchangers. Cascading requires only a heat fan, insulated ductwork, a bit of engineering, and sometimes a heat exchanger where flue atmosphere from the higher temperature unit is unacceptable in the downstream location. This hasn't been advertised widely, probably because expensive equipment is usually not required.

A. High Temperature Insulating Wool

Recently, a new family of high temperature insulating materials has been developed in the form of kaolin or alumina wool blankets or blocks, which can and often should replace hot face refractories in linings of furnaces where there is neither abrasion nor exposure to molten materials or liquids. Proper substitution of these wool materials has saved users significant amounts of energy and heat-up time. Such lining conversions should definitely be encouraged after proper installation has been assured. Make sure that any materials against the higher-temperature face of these wools will not overheat as a result of their addition and suffer resulting deterioration.

B. Flue Gas Analyses and Flame Adjustment

Gas analyses are taken on boiler flues with some regularity to assure proper combustion; so why not also at heat treating furnaces which usually operate at higher temperatures? Where multiple-burnered furnaces require individual burner ratio adjustments, this is usually more complex than adjusting single-burnered boiler furnaces. Yet this is no reason why the effort should not be made, since excess air or gas at high temperature units causes losses greater than in lower-temperature ones. Furthermore, excess air oxidizes away steel at higher temperatures. Once flue analyses are made, furnace fuel–air ratios should be adjusted accordingly.

Excess flame should not be permitted to protrude from furnaces, because fuel that burns outside heating units does no good inside them.

VII. VITREOUS ENAMELING

Traditionally, most vitreous enameling furnaces have been indirectly fuel-fired or electrically heated, probably because early attempts with direct fuel firing did foul some of the enamels with incomplete combustion products. Yet we have in recent years conducted sufficient tests with good combustion equipment to be confident that a good line of fused colors can usually be produced in direct-fired furnaces. This has been proved now both in this country and in Europe. The only apparent limitation is that a few delicate colors cannot be exactly matched against those earlier fired out of contact with combustion products.

Because much vitreous enameling can now be direct-gas-fired, radiant tube (indirect) furnaces for this purpose usually consume much more gas than is needed. This is so because radiant tubes are less efficient (and more costly to purchase and maintain) than direct firing.

One midwestern manufacturer relied upon radiant tube firing to fuse vitreous enamel on the interior walls of so-called glass-lined water heater tanks, but complained that he had to start heating his furnace at 5 P.M. the previous evening in order to reach the required 1,600°F (871°C) operating temperature the following morning. Yet a competitor not far away was performing the same operation in a direct-fired furnace that heated so quickly that operators shut it down during the lunch hour. To overcome this long heat-up problem, we referred the complaining operator to the fast-heating-equipment's manufacturer.

The most promising opportunity for thermal cascading we have seen recently was in a Wisconsin plant whose parallel production lines perform vitreous enameling and organic enameling plus hot solution washing and drying, with all units separately fired and flued through the roof. There appeared no good reason why exhaust gases from the higher-temperature radiant-tube-fired vitreous enameling furnace could not be employed to heat the lower-temperature organic enameling oven. Likewise, its exhaust gases, or those from the immersion-tube-fired solution tank, could be used to heat the drying oven. The only reason given for failure to cascade this heat was that no one had proposed it.

VIII. GLASS PLANTS

Large glass melting tanks have long employed checker work brick-type regenerators to preheat combustion air, as do those steel mill open hearth furnaces that remain in service. Usually, that is all that is done to recover heat from the melter, yet temperatures measured in flue gases departing glass plant regenerators often show readings of 1,000°F (538°C). This is too much heat to throw away, and we are delighted to learn that waste heat boilers have been added downstream from the regenerators in a few plants, with successful results. Batch dust carryover probably requires incorporation of "soot blowers" or some other means for cleaning boiler tubes, however.

Glass "day tanks," unit melters, pot melters, and other small melting units are all too seldom equipped with appropriate heat recovery devices. However, one West Virginia plant recently reported saving 21% of its fuel by adding recuperators and insulation to day tanks.

Glass annealing and decorating lehrs are usually lined with insulating fire brick (IFB). We believe this brick should now be replaced with kaolin wool, with corresponding decreases in burner input ratings. Only inertia seems to have delayed this beneficial step.

Glass fire finishing or "fire polishing" is one of the least scientific steps in glass making, and the oft employed "flame bath" is usually highly wasteful of energy. In the light of what is now known about proper fire finishing, we believe most of the "lazy" flame baths in use should give way to efficient modern equipment with better-controlled flames.

IX. ROTARY KILNS

Too many inclined rotary kilns neither preheat incoming feed nor recover heat for combustion air from the departing product. In a midwestern paper mill, the operators reported inability to calcine desired quantities of lime. Because they employed heat recovery at neither end of the kiln, the reason was quite apparent. The engineering and revisions to recover that heat would certainly cost far less than the contemplated purchase of a second kiln, and would require very little additional floor space.

In the United States, with labor costs high and past fuel costs low, most Portland cement plants have chosen the long, inclined rotary kilns with comparatively high shell heat losses. In Europe, where fuel costs have long been a prime consideration, the more energy-economic vertical kilns are often encountered. Near the firing ends of inclined rotary kilns, external shell temperatures around 500°F (260°C) are often measured. Because they usually operate in the open, this design loses much heat to passing winds and rain.

X. BAKERY OVENS

Whereas there are today many fine ovens in use, there are nevertheless some rather sad examples in industrial food plants. A row of older ovens in a Michigan plant was losing so much hot air and combustion product from both ends that a study of internal gas flow was indicated, or, alternatively, a study of heat recovery overhead. Most serious of all, however, was the fact that several of the oven's lateral pipe burners were discharging gas–air mixtures without ignition. That, of course, was both wasteful and hazardous. Some of the ovens employing central gas–air premixing were operating well, but those with burners separately ratio-controlled were poorly adjusted. Our recommendation was to call in the oven builder's service engineer, and possibly convert all ovens to complete gas–air premixing.

Success in several countries with an electrostatic device for increasing oven output and reducing fuel usage appears to justify its increased use here. This so-called Acceletron device provides a simple and relatively inexpensive means for imposing an electrostatic field between metal baking pans and an overhead metal grid. This results in speed-up of moisture migration to the surface of the dough, where the usual oven heat transfer modes can take over. Bakers we visited were happy about the improvements, which they claimed also benefited product quality.

XI. DRYING PROCESSES GENERALLY

Production drying and curing processes require very large amounts of oil and natural gas. As a consequence, these processes challenge us to find ways to reduce fuel consumption. Applications can be divided between the group where water must be removed (such as for drying foods, ceramics, lumber, paper, textiles, and aqueous coatings) and the other group, where hydrocarbon solvents must be carried away (such as is required in baking enamels, foundry cores, and other materials that employ hydrocarbons).

Most older ovens curing organic coatings have built-in energy waste because they were designed to carry much excess air that would assure that oven atmospheres remained well below the lower explosive limit (L.E.L.) for the solvent being used. There is now a way to operate such ovens much more economically, by employing instrumentation that will constantly determine the condition of the oven atmosphere and hold it closer to, yet safely below, the L.E.L. By resorting to this procedure, an East Coast textile plant was able to double product output with no increase in fuel. We are inclined to believe that this control system can be readily adapted for curing organic finishes on all types of materials.

The previously discussed thermal cascading can also be used to good advantage in many drying processes. Whereas ceramic plants have long employed kiln exhaust gases to predry and preheat incoming greenware, it is now time to look around every production dryer for sources of higher-temperature waste heat. Of course such hot waste gases must be adequately clean and well distributed into the dryer, but the dryer's automatic temperature controls can then be expected to cut back heating energy inputs and effect the desired savings.

Should the waste heat be too hot for the dryer to use directly, automatic dampers can admit enough room air to moderate its temperature. Most washer/dryer sets should employ thermal cascading; manufacturers of new systems are, for the most part, incorporating it, particularly where solution heating employs immersion tube firing.

In discussing bakery ovens, we described the Acceletron process, which speeds moisture migration to the surface of the dough. There is good reason to believe that comparable accelerated moisture migration could take place in drying green brick, other food products, wallboard, lumber, and other unfired ceramic items. The manufacturer of the process believes that the drying of water-based coatings would also benefit.

While conducting an energy survey in a Michigan food products plant, we discovered one drying room that was so hot as to be quite uncomfortable. While taking flue gas analyses from overhead air heaters, we observed that a pressurized hot duct had been left uncapped, causing the room and occupants to dry more effectively than the food product. We arranged for the duct to be capped, after which the hot air was redirected to the dryer. We wondered, of course, how long that heat had been misdirected.

In a poorly maintained Minnesota brick factory, hot kiln exhaust products were conveyed to drying ovens that seemed strangely ineffective. The hot gas transfer duct overhead was covered with fiber glass insulation. By climbing up alongside the overhead duct, we learned that the fiber glass insulation was water-soaked from a roof leak, and was consequently serving as an evaporative cooler for the gases being transported. This incident emphasizes the fact that the condition of insulation and refractories needs periodic checking.

XII. ELECTRICAL POWER FACTOR

Because uncorrected electrical power factors in an industrial plant often increases power bills, in-plant correction should be made by adding either capacitors or large synchronous motors as circumstances permit.

XIII. ELECTRICAL EQUIPMENT AND DEMAND

Many industrial plants employ a few pieces of equipment that create heavy electrical loads during relatively short operating cycles. If these high loads are not staggered, total electrical demands will be higher than otherwise. Load control equipment can be added that will substantially reduce demand charges, lower the power capacity needs, and yet, in most instances, allow the same productive capacity. Substantial power cost savings will result. Demand control equipment need not always be computerized; simple clock switches can do the job in many cases.

Surveys also indicate that it is not uncommon to find electrical equipment turned on for extended periods when no work is being done. Unreliable manual operation can be eliminated by installing automatic switching equipment to control the situation.

Some plants continue to use obsolete equipment until it becomes impossible to purchase or fabricate replacement parts. While such a practice does postpone capital expenditure, in some cases it wastes as much as half of the electrical power. We surveyed one plant where obsolete equipment was generating electricity at costs approximating 15 cents per kilowatt-hour!

Transmitting power over a spread-out industrial complex can cause electrical energy waste. As the plant grows and electrical load increases, the voltage feeding the complex should be increased to offset distribution power losses.

XIV. ILLUMINATION

Many large commercial and industrial establishments allow their lights to operate much longer than necessary. While cleaning and maintenance are sometimes done outside of regular working hours, it is very wasteful to use the entire lighting system for these purposes when a fraction of it would provide sufficient illumination. Obsolete incandescent lighting systems, which emit relatively small amounts of illumination per watt, should be replaced by newer, more efficient fixtures that utilize halogens, mercury, or sodium vapors. At the least, incandescent bulbs should be replaced with fluorescent lamps.

XV. BUILDING STRUCTURES

This presentation would be incomplete without a discussion of the causes of excessive building heat losses. Factories have been surveyed, in northern states, with holes in their walls, broken windows, negative internal pressures, large amounts of single glass, unshielded loading docks, poorly fitting doors, and uninsulated sheet metal outer shells. Because of higher costs for energy, steps are now being taken to correct such defects or to erect better structures. Large single-glass roof monitors and some of the side-wall single glass type are being insulated, in the belief that more heat goes out than light comes in. Enclosures are being installed around loading docks, makeup air heaters are returning work areas to positive pressure, and sources of air leakage are being tightened.

A particularly helpful and inexpensive procedure used when factory heat collects near the tops of high bays is to install controllable-speed propeller fans to blow heat down to working levels. In new construction, convection heating is giving way to radiant overhead gas heaters that heat workers, machinery, and floors preferentially and with much lower fuel consumption.

Improvement of comfort heating in industrial plants generally requires the application of common sense, and it is often accompanied by recovery and reuse of process heat now wasted.

14. Energy Conservation in the Industrial Combustion Field

John H. Hirt

Hirt Combustion Engineers
Montebello, California

		Page
I.	Introduction	247
II.	Process Curing Oven	247
III.	Oven with Thermal Oxidizer	248
IV.	Resin Manufacturing Process	249
V.	Aluminum Scrap Melting Furnace	249
VI.	Gasoline Station Savings	251
VII.	Conclusion	252

I. INTRODUCTION

Everyone knows the price of fuel has skyrocketed, but that, in itself, is not a sufficiently dramatic statement to get the full attention of most plant management. In many plants, the energy bills exceed the net corporate profit, or the total plant labor costs. In some cases, energy costs are in excess of 50% of the total operating cost.

By far the most practical remedy for this is to perform a meticulous energy audit on each individual piece of energy-consuming machinery. This consists of analyzing every possible source of energy loss, and then deciding on the best way to solve the problems. In this way, it is possible to approach optimum operating conditions step by step.

Additionally, to increase the awareness of the individual departments, it is wise to charge each department with its own energy costs and allocate individual worker production bonuses on the basis of how well the department meets a predetermined energy goal.

The saving of energy is a hard concept to sell; arms often must be twisted to get the audit project under way, but there are immense savings (profits!) to be made on almost any energy-consuming device. I have selected four practical cases to demonstrate the potential savings.

II. PROCESS CURING OVEN

Our first case is a typical process curing oven (Figure 14-1). It will generally have the following sources of energy loss, which must be justified if not remedied:

1. The conveyor belt itself conveys heat out of the oven.
2. The conveyor belt carries solvent or fume out of the oven where it must be collected and incinerated.
3. The belt entrance and exit area may be excessive, allowing excess air to enter the oven, adding to the heat load.
4. The burners may not be operating with the proper fuel–air ratio; they are burning too lean or too rich.
5. The recirculation may be inadequate or

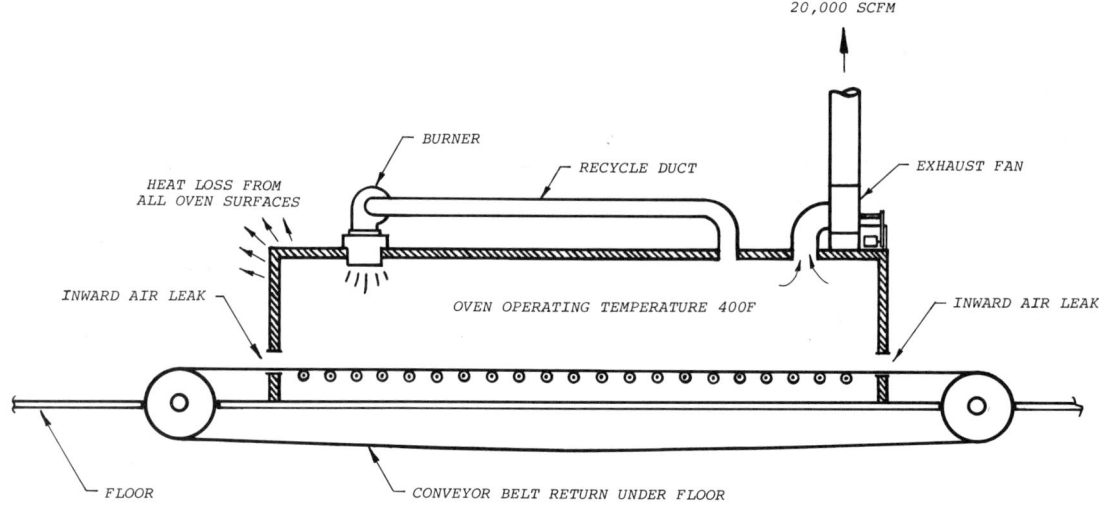

Figure 14-1. Typical conveyor belt oven.

inefficient, requiring excessive air movement or operating temperature.
6. The oven shell insulation may not be adequate.
7. The oven shell seams/joints may be leaking, causing heat loss.
8. The exhaust rate may be excessive.

An energy audit of this oven would reveal each problem and its magnitude. When the problems are broken down into little problems, the solutions become more apparent.

III. OVEN WITH THERMAL OXIDIZER

Our second case is the same type of oven, to which has been added a thermal oxidizer with heat exchange provisions (Figure 14-2). It serves to preheat the incoming fume and to add to the heat of the recycled air from the

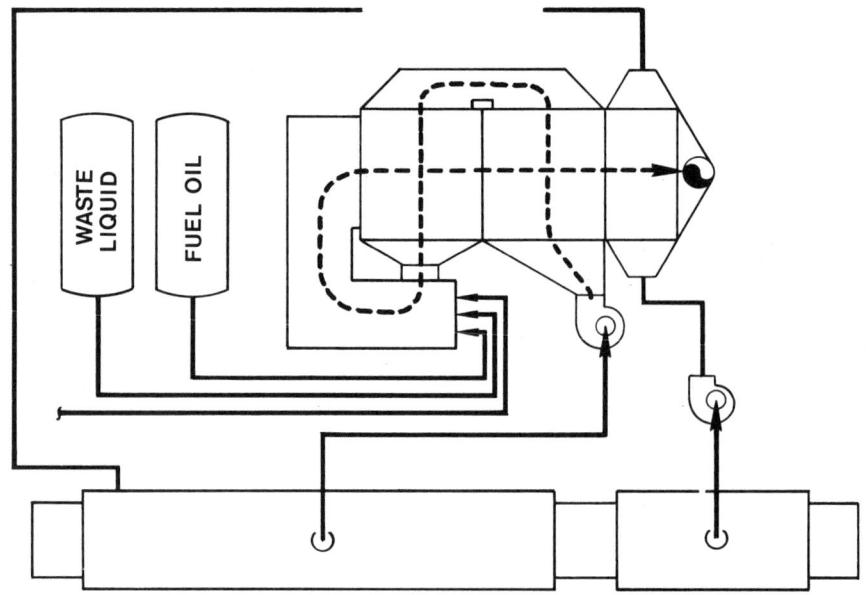

Figure 14-2. Conveyor belt oven and thermal oxidizer.

ENERGY CONSERVATION IN THE INDUSTRIAL COMBUSTION FIELD

Table 14.1. Conveyor Belt Oven

	BEFORE	AFTER
Operating temperature, °F	400 (204.4°C)	400 (204.4°C)
Exhaust rate, SCFM	20,000 (9,440 l/s)	10,000 (4,720 l/s)
Exhaust rate, MMBtu/hr	7.62 (2.23 × 10⁶W)	2.3 (0.67 × 10⁶W)
Lower explosive limit, %	3	6
Operating cost, $/yr	228,600	69,000
(Fuel $4.00/MMBTU—7,500 hr/yr)		
Operating cost using liquid waste	—	-0-
1 sq ft inward air leak, $/yr	2,407	3,292
Same, without heat recovery		9,487

cooling section to the curing section. Additionally, the thermal oxidizer is capable of burning natural gas, fuel oil, and liquid waste. The hydrocarbon/odor level in the exhaust has also been reduced to acceptable levels.

Instead of the thermal oxidizer's increasing the operating costs, it has actually reduced the total fuel consumption. On this particular oven, it was possible to reduce the area of the conveyor belt openings by 50%. The air from the cooling section, which could no longer be exhausted to atmosphere, was preheated in a heat exchanger and recycled to the curing section. Oven production was increased by 25%.

Table 14-1 shows the "before" and "after" comparison between these two oven arrangements. We have reduced the cost of operating a 1 sq ft (0.0929 m²) inward air leak to $/yr so that it is possible to point out to the operator how much the privilege of viewing the curing process is costing.

IV. RESIN MANUFACTURING PROCESS

Figure 14-3 is a resin manufacturing process. The cooking resin is blanketed with inert gas manufactured from natural gas. Inert gas produced this way is only 5¢/100 SCFM (47.2 l/s), as opposed to 30¢ to 50¢/100 SCFM (47.2 l/s) for nitrogen gas. Having this inert gas generator also allows the plant to blanket the hydrocarbon storage tanks, greatly reducing the fire hazard. Gases from the blend tank and storage tanks are diluted with air for safety reasons and then burned in the thermal oxidizer to eliminate noxious odors. The hydrocarbon content of these gases reduces fuel consumption. In this system, the noncondensable vapors are also used as a fuel source, and the water of reaction is oxidized. Since they contain sufficient organic material to contribute to the net heat requirement, there is a cost benefit in burning them. Additional savings are realized by the elimination of disposal costs. Disposal costs are running as high as $50/barrel ($.31/l), with responsibility for transport and noncontamination of the dump site remaining with the disposee. Net results are that the customer has no air pollution, no waste requiring disposal, and a bonus of well over $100,000 worth of steam per year produced in the waste heat boiler.

Table 14-2 shows comparative costs before and after adding the thermal oxidizer/waste heat boiler.

V. ALUMINUM SCRAP MELTING FURNACE

In Figure 14-4, we see a typical aluminum scrap melting furnace. The aluminum is melted in the well, often causing smoke problems. The furnace may or may not have proper hot gas circulation for proper heating of the metal. When firing natural gas, the burner should be capable of producing a 12% CO_2 Orsat with zero combustibles. As the gases ap-

Table 14.2. Resin Plant

	BEFORE	AFTER
Misc. organic material disposal	$12,000	-0-
Water of reaction disposal	$123,000	-0-
Air pollution odor complaints	Many	-0-
Steam production, lb/hr	-0-	3,450
Value at $6/M lb—7500 hr/yr	-0-	$155,000
Total savings/yr	—	$290,000

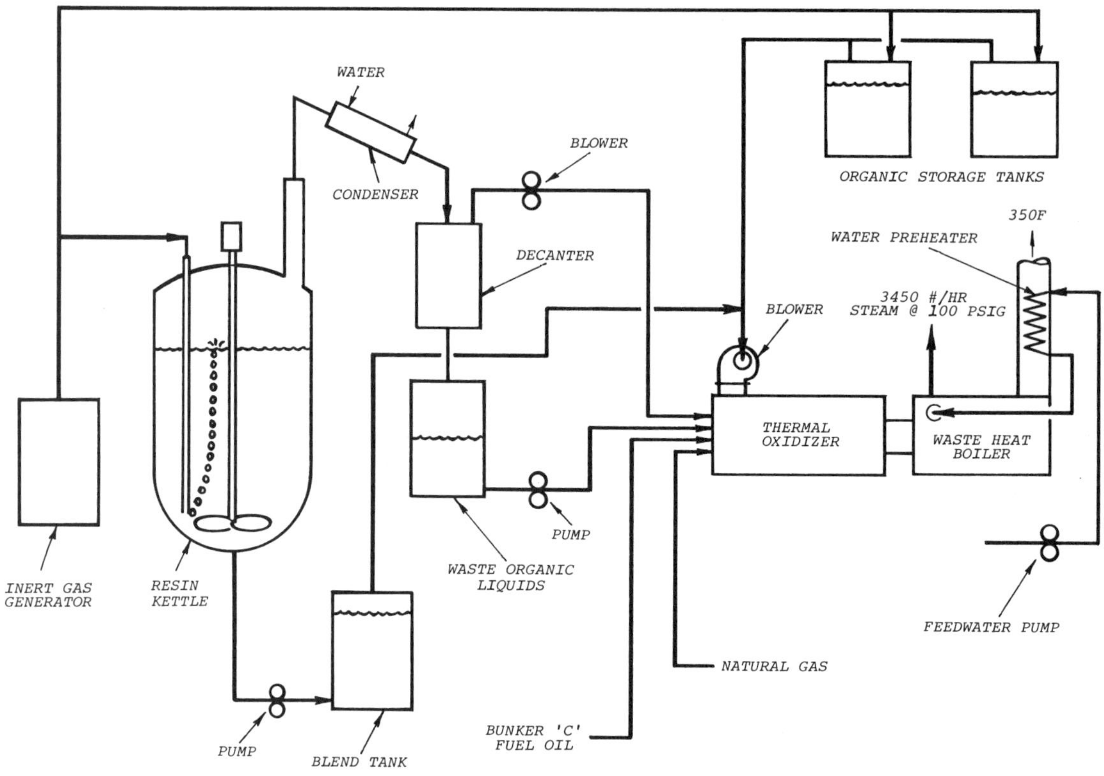

Figure 14-3. Resin plant process schematic.

Table 14.3. Aluminum Melting Furnace.

	BEFORE	AFTER
Operating hours per year	5,000	5,000
Firing rate, MMBtu/hr	22.5 (6.593 × 10^6)	16.9 (4.95 × 10^6)
CO_2, % of flue gas	10.5	12
O_2, same	2	0.2
Preheat, $/yr	450,000	337,000
Dross loss, %	3	1
Same, $/yr	1,688,000	563,000

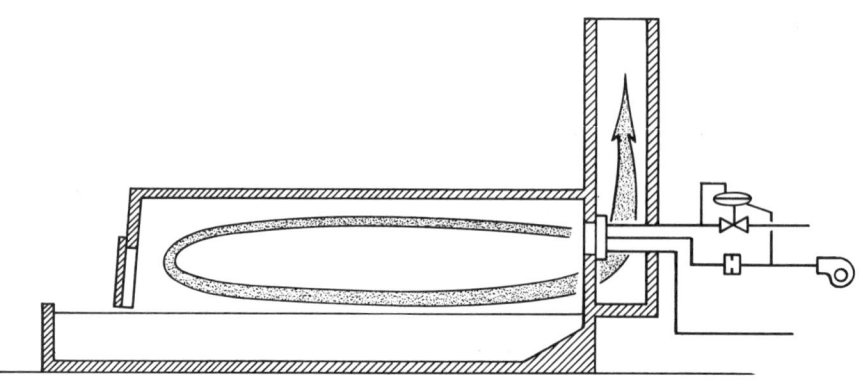

Figure 14-4. Aluminum melting furnace.

proach this 12%, the excess air is reduced, increasing the thermal efficiency. Additionally, the gases are hotter, so they transfer their heat faster, causing a lower stack temperature. This further increases the thermal efficiency. A heat exchanger can be added to the stack to preheat the combustion air, adding to the radiant effect of the flame and further increasing combustion efficiency.

Table 14-3 shows the savings that can be achieved by proper burner and heat exchanger application. Obviously, the burner must be capable of accepting the preheated air. A large side benefit is the reduction of dross loss achievable by proper combustion and the elimination of inward air leaks.

VI. GASOLINE STATION SAVINGS

The application of air pollution control equipment often results in economic benefits. Figure 15-5 represents a typical gasoline service station equipped with vapor control devices. The California Air Pollution Board has determined that the venting of fumes to the atmosphere during the filling of automobile gas tanks is unacceptable. By adding a vapor collection hose to the fill nozzle, the vapors are returned to the underground storage tanks. Similarly, as the tanker fills the storage tanks, the fumes being displaced are returned to the tanker.

A 200,000 gal/month (758 m^3/month) station achieves the following benefits from the addition of this system (known as the Hirt system):

1. By returning vapors to the underground storage tanks, the air–hydrocarbon equilibrium is maintained. This, in turn, prevents further evaporation of fuel in the tank.
2. In warm weather, the returned vapors actually condense out as additional gasoline.

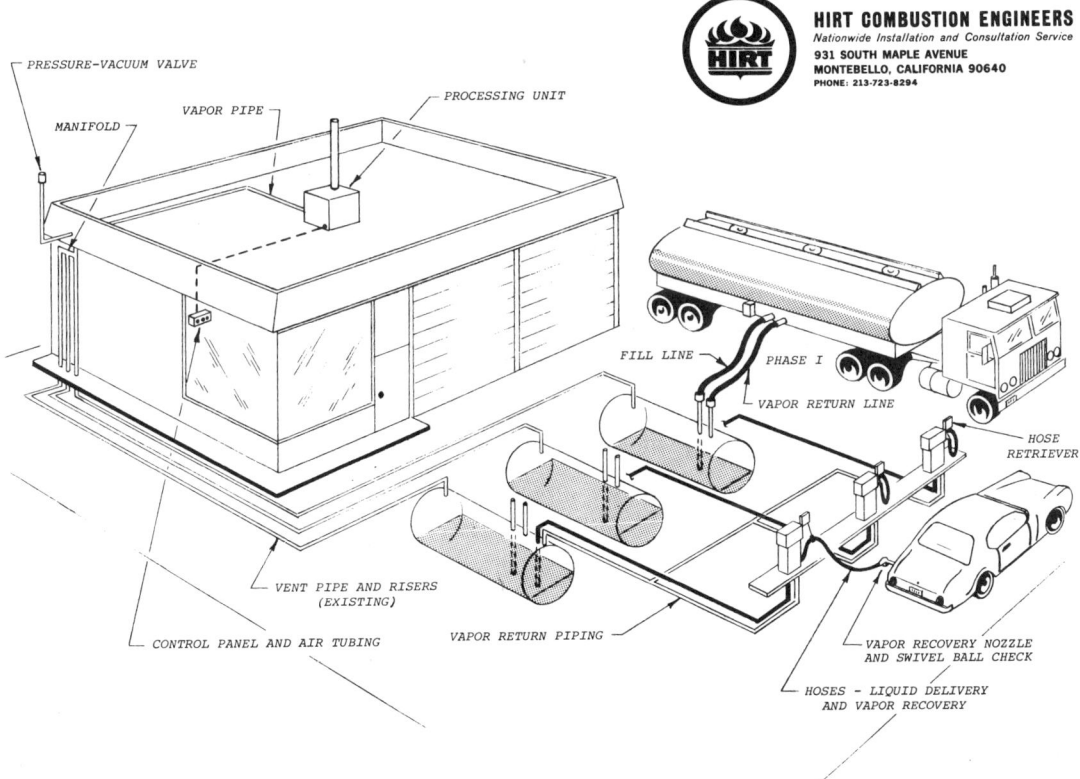

Figure 14-5. Gasoline station vapor control system.

Table 14.4. Gasoline Service Station Vapor Destructor.

	BEFORE	AFTER
Station pumping rate, gal/month	200,000	200,000
Evaporation loss, gal/month	200	-0-
Recovery from filling auto tanks	-0-	100
Net gain, gal/month	—	300
Yearly net gain at $1.10/gal		$3,960
Potential liability from customer breathing of vapors	No limit	-0-

3. The owner eliminates any future claims for disability from breathing the fumes.

Table 14-4 shows the savings possible with the system.

VII. CONCLUSION

It is not possible to cover in a brief article the entire spectrum of equipment to which the foregoing principles apply. I can only say that the energy audit approach is by far the best, and that the subsequent application of ingenuity and good engineering principles will reduce wasteful energy consumption to acceptable levels, helping all of us stay in business.

15. Retrofit Economics and the Utilities*

James M. Archibald

*Vice President, Smith Environmental Corporation
Duarte, California*

		Page
I.	Introduction	253
II.	The Plant Survey	253
III.	Tube-Type Heat Exchangers	254
IV.	Direct Furnace Gases	256
V.	Space Heat	257
VI.	Recovery Back to Process	259
VII.	Cool Gases with Heat Exchanger Rather Than Bleed-in Air	260
VIII.	Iron Foundry Cupola	260
IX.	Preheat Air Prior to Incinerator	261
X.	Waste Heat Boilers	262
XI.	Fuel Cost Savings for Heat Recovery on Open Flame Furnace	262

I. INTRODUCTION

With the advent of a shortage in fossil fuels and high costs for fuels and electrical power, it is incumbent upon plant operators to determine points of potential savings and to develop maximum fuel and power utilization for their facilities. The purpose of this chapter is to inform the reader about reductions in fuel and electrical energy usage that can be had by the full utilization of air-to-air, tube-type heat exchangers and the installation of equipment that can make use of heat now being wasted.

II. THE PLANT SURVEY

The first step in a fuel and power savings program is an obvious one: a plant survey. The person assigned to make or supervise the survey must be qualified and not too burdened with other responsibilities to devote a major portion of his or her time to doing a thorough job. If you do not have qualified personnel, there are many companies that can provide this service to you. Make certain, however, that the company you select has had experience in combustion, heat transfer, oven and furnace operation, air flow and duct design, fuels and fuel efficiencies, electrical energy savings potential, etc.

Any survey should start with a plant layout listing all of your energy- or fuel-consuming equipment. All data concerning each piece of equipment should be obtained and analyzed. When information is sketchy, meters should be installed. After you have examined combustion rates and correlated fuel use to production, you can determine the current equipment process efficiency and compare it to industry standards.

---

*This chapter, including all figures, is reprinted by permission of the Society of Manufacturing Engineers, Dearborn, Michigan, from SME Technical Paper EM80-282, 1980.

INFORMATION NEEDED FOR DESIGN OF HEAT EXCHANGER
FOR GIVEN APPLICATION

NAME_____ TITLE_____
COMPANY_____ ADDRESS_____
CITY_____ STATE_____ ZIP_____ PHONE_____

1. AMOUNT OF AIR FROM HEAT SOURCE _____ SCFM.
2. A. TEMPERATURE OF AIR _____ MAX. _____ MIN.
   B. DESIRED EXHAUST TEMPERATURE FROM EXCHANGER _____.
3. ROOM AVAILABLE FOR INSTALLATION.
4. PRESSURE AVAILABLE FROM HEAT SOURCE, " OF WATER COLUMN____.
5. AMOUNT OF AIR REQUIRED BY USING SYSTEM_____ SCFM.
6. RISE IN TEMPERATURE OF AIR REQUIRED BY USING SYSTEM_____ °F.
   TEMPERATURE IN _____ TEMPERATURE OUT_____.
7. TYPICAL CHEMICAL ANALYSIS OF FUME SOURCE. ESPECIALLY INTERESTED IN ANY FORM OF CHLORIDES, FLUORIDES, OR SULFIDES. ANY ALKALIES THAT MAY TEND TO BECOME STICKY OR GUMMY AT HIGH TEMPERATURE OR ANY OTHER FUME CONTAMINANT THAT COULD CREATE PROBLEMS IN THE EXCHANGER.
8. ANY ABRASIVE PARTICLES IN THE FUME SOURCE AND THE GRAIN LOADING AND SIZE IN MICRON.
9. HEAT RECOVERY DESIRED IN % RECOVERY_____.
10. OTHER POSSIBLE USES OF HEAT _____
    _____.
11. GENERAL DESCRIPTION OF PROPOSED SYSTEM _____
    _____.
12. OTHER COMMENTS RELATIVE TO THE PARTICULAR INSTALLATION ___
    _____.

Figure 15-1. Information needed for design of heat exchanger for given application.

An integral part of the fuel usage evaluation of each process is a study of whether substitute methods or energy recovery will reduce energy use. Ask yourself any conceivable questions regarding a better way to use fuel. If you cannot answer them about specific subjects, do not hesitate to call in outside help.

After all the data are gathered and evaluated and you have answered all the questions, you are prepared to outline an energy use program. One method your program may call for is the use of a heat recovery system on one or more of your processes. A questionnaire for making an analysis of the system and what you want it to accomplish is given in Figure 15-1.

III. TUBE-TYPE HEAT EXCHANGERS

If your evaluation leads to the conclusion that a tube-type heat exchanger can meet your needs, let us look at design considerations. The primary variables in the design of a tube-type heat exchanger are:

1. The square foot area of the tubes in the exchanger
2. The pressure drop in the exchanger of air over and through the tubes
3. The heat transfer coefficient in Btu/sq ft/°F/hr (W/m^2°C)

Computer runs with the same air flow and percentage of heat recovery efficiency indicate the effect of surface area per tube on pressure drop and the size of the unit (see Figure 15-2). Final design is a compromise. Efficient design will use the optimum combination of pressure drop, spacing and arrangement of tubes, tube area, and size of unit to fit the user's installation requirements.

Pressure drop is partially based on velocity and, therefore, in addition to area or footage of tube, can be adjusted by changing the diameter of tubes (pressure drop through the tubes) and the spacing of tubes (pressure drop over the tubes).

A good design will have a film coefficient of approximately 4 to 5 Btu/hr/sq ft/°F (22.72 to 28.4 W/m^2 · °C). Adjusting the variable factor to arrive at this film coefficient provides the most economical design. Heat exchangers are designed for percentage of recovery. The higher the percentage of recovery, the higher the Btu savings. If we compare a 10,000 SCFM (472 l/s), 1,000°F (537.78°C) rise, 80% efficient heat exchanger with the 50% efficient heat exchanger, the Btu savings are increased from 5,800,000 to 9,280,000 Btu (6,119,000 to 9,790,400 kJ). The tube will increase four times, so the cost of the tube in the unit will increase by four times, and the cost of the unit will increase three times from, say, $25,000 to $75,000.

A typical curve for tube usage versus heat exchanger efficiency is shown in Figure 15-3. In Figure 15-4, a curve for fuel savings versus percent efficiency is shown. The effect of electrical power cost versus pressure drop is also indicated.

In evaluating heat exchanger units, check

RETROFIT ECONOMICS AND THE UTILITIES

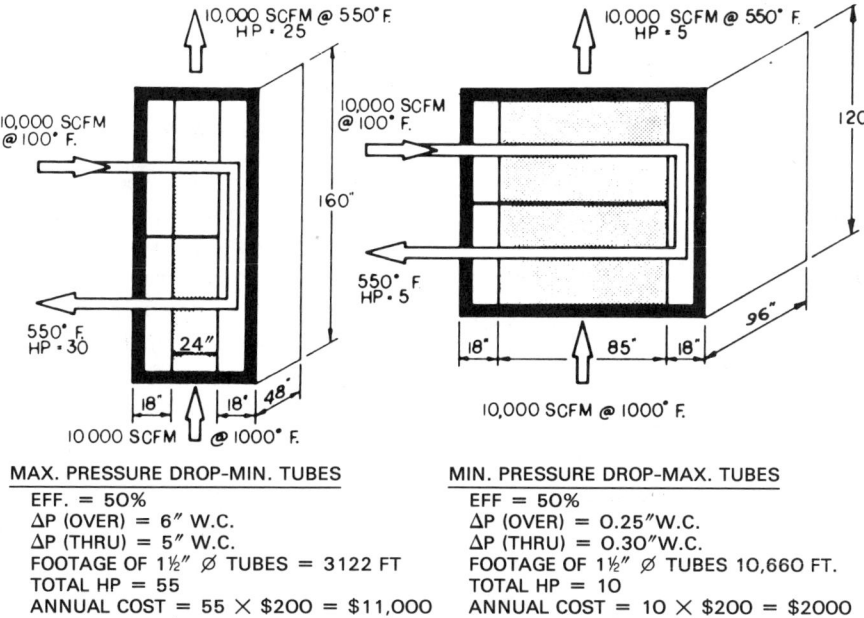

Figure 15-2. Effect of surface area per tube on pressure drop and size of unit.

pressure drop or horsepower when comparing cost. In other words, if your final design saves 50% of your fuel but increases horsepower 50%, you have made no real savings. The advantage of the typical tube-type heat exchanger is that the design lends itself to variable efficiency of heat recovery. By using various alloy steels, temperatures as high as 1,800°F (982.22°C) can be satisfactorily recovered. As a user, you are interested in long,

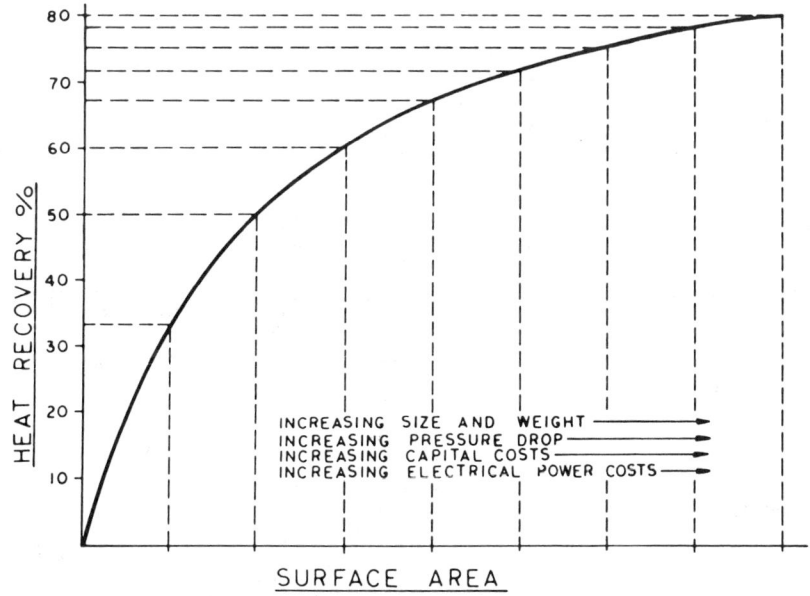

Figure 15-3. Tube usage vs. heat exchanger efficiency.

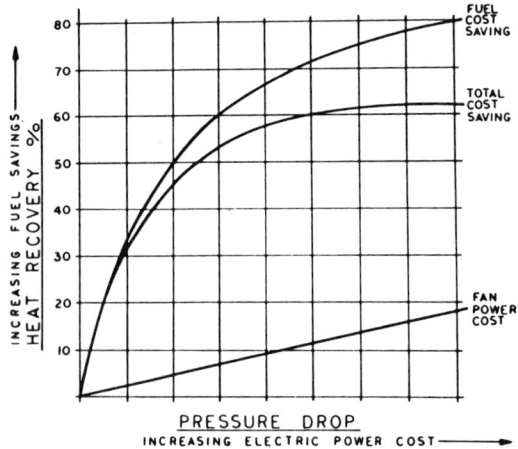

Figure 15-4. Fuel savings vs. percentage of efficiency, electrical power cost vs. pressure drop.

maintenance-free life for your heat exchanger installation. It must be designed correctly to withstand your process problems. Potential problems that will affect exchanger life, along with suggested solutions, are shown in Figure 15-5. Units can be custom-designed to meet your requirements. Next, let's examine some typical applications of tube-type heat exchangers. The examples shown will have theoretical air quantities, air temperatures, etc., to make the computations easier to follow.

IV. DIRECT FURNACE GASES

Examine the use of direct furnaces or oven gases to heat your ware. An example of this would be a paint spray operation using barrels, buckets, or parts prior to painting cans. The hot products of combustion from your oven or furnace dry the cans from the washer prior to painting. In most cases, the products of combustion will not contaminate your finished ware. Such a system can be an economical application of heat; all that is needed is ductwork and some form of minor temperature control.

Other applications of direct furnace gases might be the preheating of process materials such as steel scraps prior to dumping into an electric furnace, preheating ceramic ware prior to entering a firing kiln, or placing raw materials for cement manufacturing in a preheater design system.

If a heat exchanger is placed on a furnace that discharges to the atmosphere, it is desirable to keep pressure drop through the tubes to a minimum so that no fan is necessary on the

PROCESS AND DESIGN PROBLEMS

PROBLEM	SOLUTIONS
1. EXPANSION	1. BELLOWS-ALLOWANCE FOR MOVEMENT IN ALL DIRECTIONS. 2. INDIVIDUAL TUBE EXPANDERS. 3. FIX ONE END. LET OTHER END FLOAT.
2. ABRASION MATERIAL	1. RUN DIRTY GASES THROUGH THE TUBES. 2. USE STRAIGHT WALL TUBING. 3. USE SMITH PATENTED REPLACEABLE WEAR INSERTS. 4. REDUCE VELOCITY THROUGH THE TUBES. 5. INCREASE WALL THICKNESS OF TUBE.
3. HIGH TEMPERATURE	1. ALLOY STEEL TO MATCH TEMPERATURE REQUIREMENTS. 2. BLEED IN AIR TO COOL TO MEET TEMPERATURE OF TUBE MATERIAL. 3. BYPASS IN EVENT OF FURNACE UPSET CONDITION.
4. CARBIDE PRECIPITATION	1. RUN UNIT CONTINUOUS. REDUCE CYCLING OF FURNACE. 2. USE MATERIAL TO WITHSTAND CARBIDE PRECIPITATION. 3. INSTALL HIGHER TEMPERATURE ALLOY STEEL.
5. TUBE PLUGGING DUE TO STICKY MATERIALS OR CONDENSATION	1. KEEP TEMPERATURE TO TUBE WALL ABOVE DEW POINT. 2. INCREASE VELOCITY THROUGH TUBES. 3. RUN DIRTY GAS THROUGH PLAIN WALL TUBES. 4. PREHEAT FIRST PASS OVER TUBES TO PREVENT CONDENSATION. 5. COOL ALL GASES BELOW STICKY TEMPERATURE. 6. PROVIDE POSITIVE MECHANICAL CLEANING.
6. DETERIORATION OF TUBES CAUSED BY CONTAMINATION OF CORROSIVE MATERIAL	1. DETERMINE COMPOSITION OF GAS STREAM. 2. USE CORROSIVE RESISTANT MATERIALS. 3. USE PILOT TEST MODEL. 4. BYPASS TO NON-CORROSIVE OPERATING TEMPERATURE. 5. BUILD REPLACEABLE SECTIONS TO REDUCE MAINTENANCE. 6. USE THICKER WALL TUBES. 7. RUN DIRTY GASES THROUGH TUBES. 8. NEUTRALIZE CONTAMINATE WITH SPECIAL ADDITIONS.

Figure 15-5. Process and design problems.

RETROFIT ECONOMICS AND THE UTILITIES 257

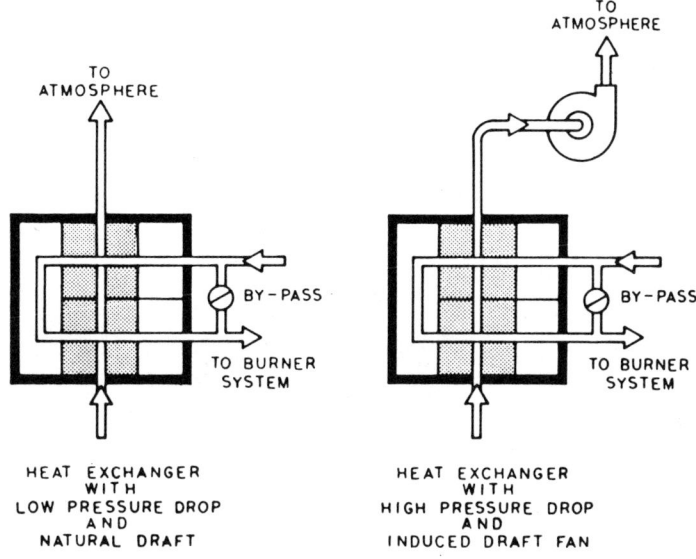

Figure 15-6. Heat exchangers on furnaces that discharge to atmosphere.

furnace discharge. This saves horsepower (kW) and keeps the cost of installation to a minimum. Extra footage of tube in the exchanger will be warranted to accomplish these savings. Figure 15-6 illustrates these savings. Your supplier can accomplish this design by reducing pressure drop through tubes to less than 0.25 inch (62.25 Pa) water column.

V. SPACE HEAT

The heat from your furnaces or ovens or other heat devices can be recovered by installing a heat exchanger on the exhaust gas stack. The exhaust gases will discharge through the tubes of the exchanger, and fresh outside air will be preheated by passing over the tubes prior to entering the plant. See Figure 15-7 for details.

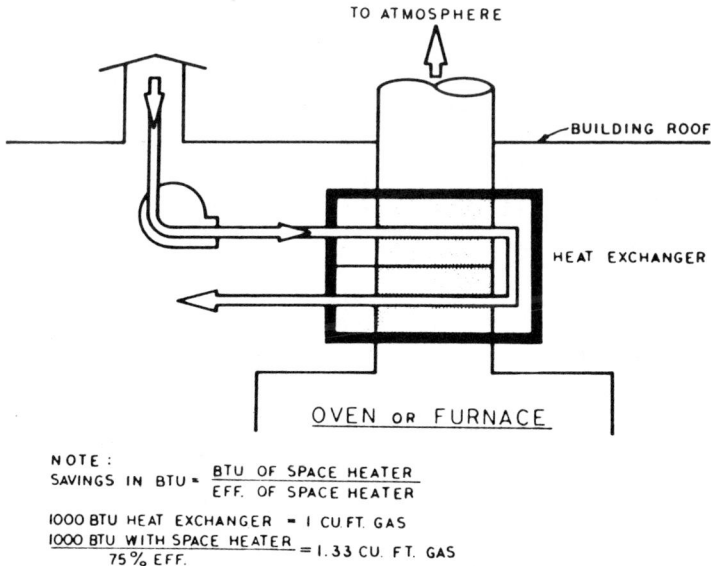

NOTE:
SAVINGS IN BTU = $\dfrac{\text{BTU OF SPACE HEATER}}{\text{EFF. OF SPACE HEATER}}$

1000 BTU HEAT EXCHANGER = 1 CU. FT. GAS
$\dfrac{1000 \text{ BTU WITH SPACE HEATER}}{75\% \text{ EFF.}}$ = 1.33 CU. FT. GAS

Figure 15-7. Recovering heat by use of a heat exchanger on the exhaust gas stack.

258 6/RETROFITTING FOR INDUSTRY

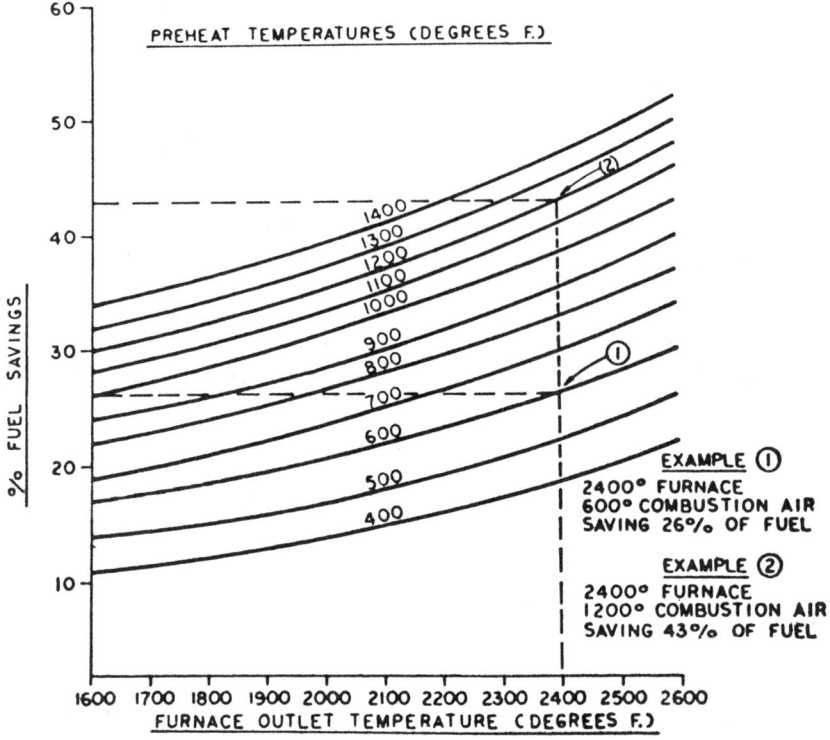

Figure 15-8. Potential savings from heating of combustion air.

In the winter, it is important that plant atmospheres be kept under positive rather than negative pressure so that the building is expelling air rather than bleeding in cold air, which requires makeup heaters to maintain temperature.

Many plants have potential for fuel savings in the heating of combustion air for furnaces, ovens, etc. Furnaces now in operation usually have burners that can utilize preheated air up to 600°F (315.56°C). New furnace manufacturers can make available systems using

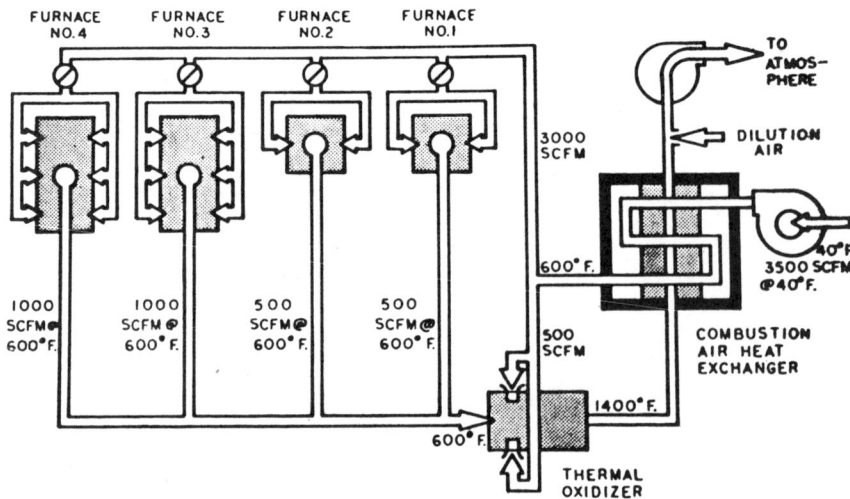

Figure 15-9. Wax and plastic burnout furnaces.

1,000°F (537.78°C) to 1,200°F (648.89°C) combustion air.

Potential savings are shown in Figure 15-8. A typical plant layout showing small furnaces is indicated in Figure 15-9. Special design of controls is necessary. This must be included with the heat exchanger at the time of installation.

VI. RECOVERY BACK TO PROCESS

Incineration of hydrocarbon fumes from paint baking ovens is required to meet air pollution requirements. In a typical system, the oven and coater discharge solvent-laden air to a thermal oxidizer (see Figure 15-10). A primary heat exchanger preheats the incoming air from 270°F (132.22°C) to 860°F (460°C). The fumes then pass over a burner that raises the temperature to 1,400°F (760°C). It is retained at this 1,400°F (760°C) for .5 second, and the combustible hydrocarbons are converted to carbon dioxide and water. The heated 1,400°F (760°C) gases then pass over the primary heat exchanger, reducing gases to 760°F (404.44°C). Next, the

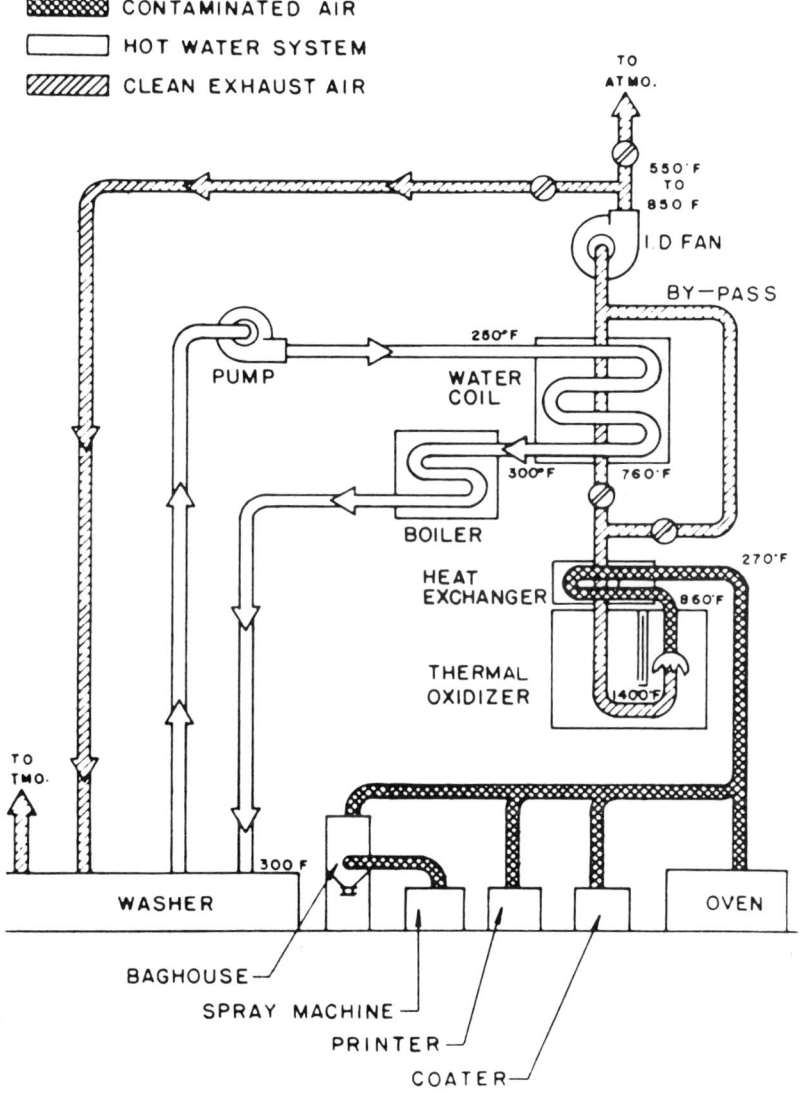

Figure 15-10. System for incineration of fumes from paint-baking oven.

gases pass over the secondary exchanger, which provides heat for high pressure hot water systems and pumps the gases in a closed loop system to heat the washer tanks. The clean gases that are left are then sent to the dryer section of the washer.

Systems of this type can be incorporated into most waste gas streams. The main determining factor is to decide how and where to make the best use of the heat.

VII. COOL GASES WITH HEAT EXCHANGER RATHER THAN BLEED-IN AIR

Figure 15-11 shows a typical furnace and baghouse system with two methods of reducing air temperature before the baghouse. The use of bleed-in air (shown in Figure 15-11A) doubles air flow to 100,000 SCFM (47,200 l/s) and requires 900 hp (671.4 kW) for 400°F (204.44°C) air at 18″ SP (4484 Pa). Installing a heat-dispersing exchanger or exhaust-gas-cooling exchanger (Figure 15-11B) adds only 2″ SP (498 Pa) yet maintains original air flow at 50,000 SCFM (23,600 l/s), reducing horsepower to 300 (223.8 kW). By adding 15 hp (11.19 kW) each for the (4) cooling fans, the total horsepower becomes 360 (268.56 kW), a savings of 540 hp (402.84 kW) over the traditional bleed-in air system. Annual savings, estimated at $200 per hp ($268.1/kW) (per Southern California Edison), become 200 × 530 = $106,000 ($268.1 kW × 395.38 = $106,000).

For a new installation, the baghouse size can be reduced 50% from 100,000 CFM (47,200 l/s) to 50,000 CFM (23,600 l/s). At $3.00 per CFM ($6.36 per l/s), a capital savings of $150,000 ($150,096) will more than pay for the heat exchanger. Industrial estimates based on bag life and 20¢ per CFM (42¢ per l/s) per year maintenance indicate an additional savings of approximately $10,000 ($9,912) per year.

This same type of system can be applied to melting furnaces as well as other heating processes such as lime and cement manufacturing.

VIII. IRON FOUNDRY CUPOLA

An iron foundry cupola is shown in Figure 15-12. Its function is to melt scrap iron, and it is an excellent candidate for heat recovery. In recent years, foundries have had serious prob-

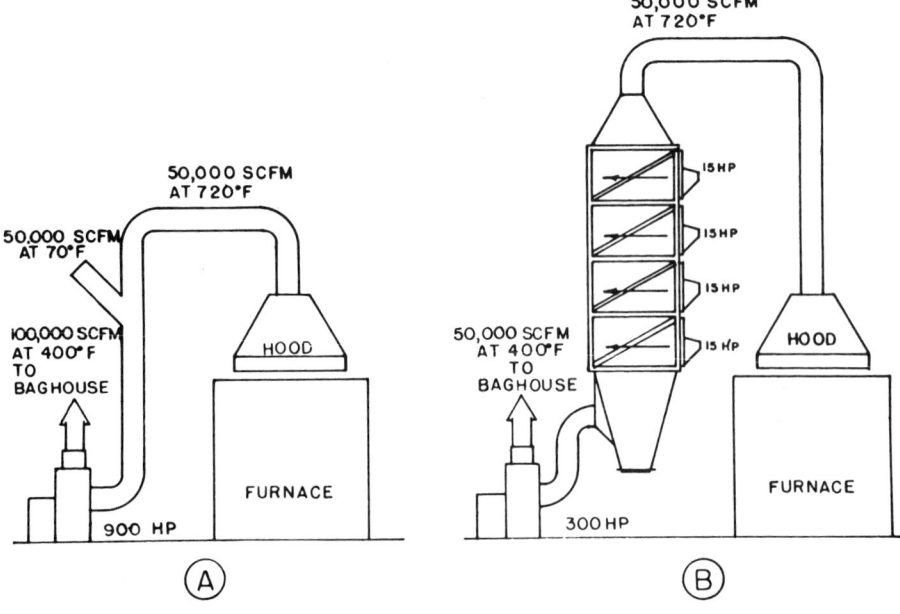

Figure 15-11. Reducing air temperature before the baghouse. A, Bleed-in air. B, Heat-dispersing-exchanger.

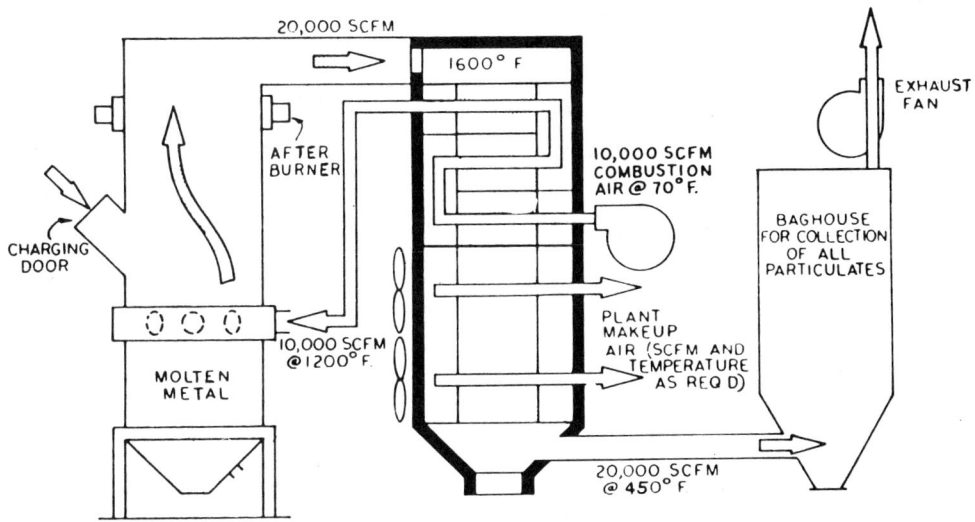

Figure 15-12. Iron foundry cupola.

lems with air pollution, hydrocarbons, carbon monoxide, and particulates. Correcting these problems necessitated the use of additional Btu's of fuel. A method used to recover this heat and also to cool the exhaust air from the cupola before entering the baghouse is the installation of an air-to-air tube-type exchanger in the system. Proper design can raise the combustion air from ambient to 1,200°F (648.89°C) before discharging it into the tuyeres in the lower section of the cupola. Resultant savings occur by reducing the quantities of natural gas and coke required.

The process flow shows a unit producing 20,000 CFM (9,440 l/s) of hot exhaust gases and using 10,000 CFM (4,720 l/s) of "hot blast" air. By passing 10,000 CFM (4,720 l/s) ambient air over the heat exchanger and thereby raising it from 70°F (21.11°C) to 1200°F (648.89°C), the savings will be 10,000 × 1130°F × 1.16 (4,720 l/s × 627.78 × 1.296) = 13,108,000 Btu/hr (3,840,206 W) in coke and natural gas. Assuming $3.00 for the combined price of natural gas and coke per 1,000,000 Btu (1,055,000 kJ) a savings of approximately $40.00 per hour of operation will result. A side benefit can also be realized; as the melting rate increases, productivity can be increased as well.

IX. PREHEAT AIR PRIOR TO INCINERATOR

In heavily populated areas, plants producing offensive odors or exhausting hydrocarbons must eliminate such problems or risk being closed down. Burning the fumes at a high temperature, 1,200°F (648.89°C) to 1,400°F (760°C), is a most effective method of odor destruction; however, because of the difficulty of collecting the odors, large quantities of air are required to carry the contaminant to the air incinerator or thermal oxidizer.

A typical unit installed recently was designed for 28,000 SCFM (13,216 l/s) of air with an inlet temperature of 70°F (21.11°C). Without heat recovery, the gas required to raise the contaminated air from 70°F to the odor-destruction temperature of 1,200°F would have been 28,000 × 1,130° × 1.16 (13,216 × 627.78° × 1.296) = 36,700,000 Btu/hr (10,753,000 W). It was decided to install an 80% preheat heat exchanger in this system with an hourly savings of 29,300,000 Btu (8,584,900 W). Assuming $3.00 per 1,000 cu ft (28,300 l) of gas, the savings were 29.4 × 3.00 = $88.20 per hour. Operating 6,000 hours per year, the annual savings averaged $529,200.00. For details of this system, see Figure 15-13.

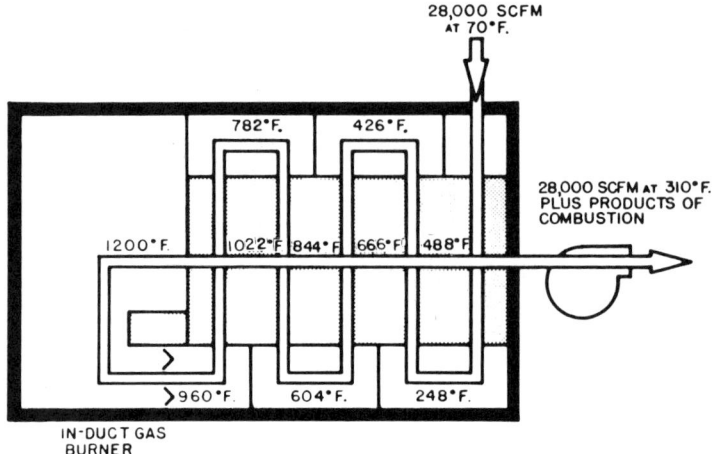

Figure 15-13. Thermal oxidizer with 80% heat exchanger.

X. WASTE HEAT BOILERS

Another method of using available waste heat is in firing a waste heat boiler. Here, we use the exhaust from several furnaces and dilute the temperature to a range more acceptable for standard boiler construction. Each furnace has damper controls so as not to increase the exhaust flow with the induced draft fan at the boiler exhaust. The fan is equipped with a vortex damper that can maintain static pressure control throughout the system.

The process flow shown in Figure 15-14 indicates that five furnaces have an exhaust flow of 9,750 SCFM (4,602 l/s) at 2,300°F (1,260°C). With dilution with ambient air, there is 15,100 SCFM (7,127.2 l/s) of 1,500°F (815.56°C) air available to the boiler.

When we transfer these Btu's to the boiler, assuming approximately an 80% effective transfer into steam, we have a boiler capable of producing 583 hp (434.92 kW).

Relating this cost back to the cost of firing the boiler with natural gas at a cost of $3.00 per 1,000 cu ft, (28,300 l) this is equivalent to $73.46 per hour or, at 2,000 hours per year, $146,920.00 per year.

XI. FUEL COST SAVINGS FOR HEAT RECOVERY ON OPEN FLAME FURNACE

Furnace exhaust:

1,950 SCFM (920.4 l/s) at 2,300°F (1,260°C) × 5 furnaces = 9,750 SCFM (4,602 l/s)

Dilution air required:

$$\frac{[9{,}750 \text{ SCFM } (4{,}602 \text{ l/s}) \times 2{,}300°F (1{,}260°C)] + [5{,}450 \text{ SCFM } (2{,}572.4) \times 70°F (21.11°C)]}{9{,}750 \text{ SCFM } (4{,}602 \text{ l/s}) + 5{,}450 \text{ SCFM } (2{,}572.4 \text{ l/s})}$$
= 1500°F (815.56°C)

Exhaust to boiler:

$$15{,}200 \text{ SCFM } (7{,}174.4 \text{ l/s}) \text{ at } 1{,}500°F (815.56°C)$$

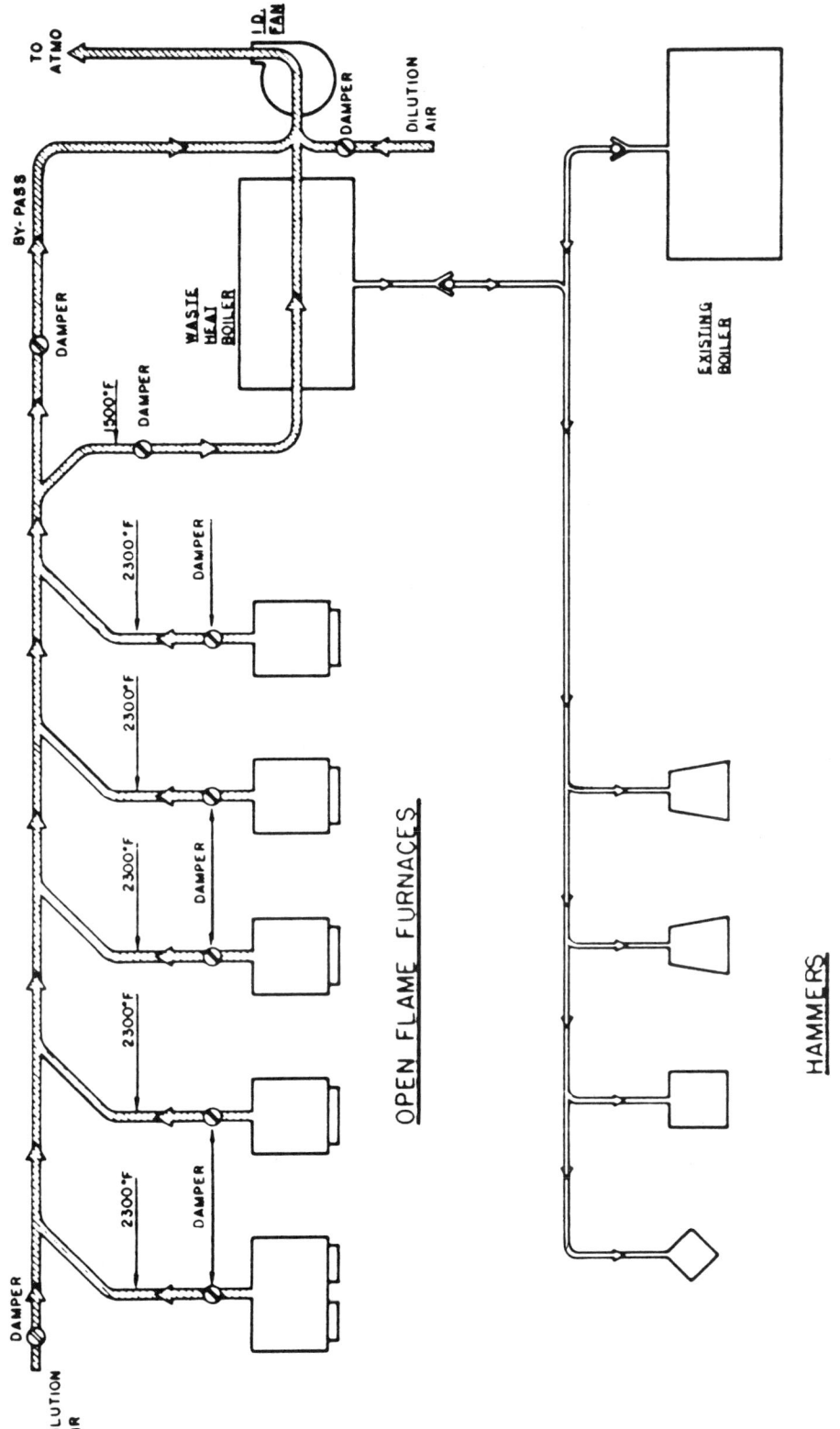

Figure 15-14. Heat recovery from open flame furnace.

Exhaust from boiler:

$$15{,}200 \text{ SCFM } (7{,}174.4 \text{ l/s}) \text{ at } 420°F \ (215.56°C)$$

Heat content in 1,500°F (815.56°C) air:

15,200 SCFM (7,174.4 l/s) × 1.1 (4.52)
    × 1,500°F (815.56°C)
    = 25,080,000 Btu (26,459,400 kJ)

Heat content in 420°F (215.56°C) air:

15,200 SCFM (7,174.4 l/s) × 1.1 (4.52)
    × 420°F (215.56°C)
    = 5,544,000 Btu (5,848,920 kJ)

Btu transferred:

25,080,000 (26,459,400 kJ)
− 5,544,000 (5,848,920)
= 19,536,000 Btu (20,610,480 kJ)

Boiler horsepower:

$$\frac{19{,}536{,}000 \text{ Btu } (20{,}610{,}480 \text{ kJ})}{33{,}475 \text{ Btu/hp } (47{,}389.13 \text{ kJ/kW})} = 583 \text{ bhp } (434.92 \text{ kW})$$

Assuming natural gas cost is $3.00 MCF:

$$\frac{420°F/hr \ (215.56°C/hr) \times 583 \text{ hp } (434.92 \text{ kW}) \times \$3.00}{1{,}000 \ (28{,}300)} = \$73.46 \text{ per hour}$$

Assuming only 40 hr/wk × 50 wk/yr = 2,000 hr/yr, yearly savings:

2,000 hr/yr × $73.46 per hour
= $146,920.00/yr

16. Case Study of a Commercial Retrofit Project: East Texas State University, Commerce, Texas

Morris Backer, P.E., Senior Vice President and J. Phillip Upton, P.E., Department Manager

Bovay Engineers, Inc.
Houston, Texas

	Page
I. Introduction	265
II. Proposal	265
III. Work Planning	265
IV. Building Surveys	266
V. Discussion	266
A. Energy Management Control System	267
B. Reduced Utility Cost Through Limiting Demand	269
C. Electrical Light Relamping	269
D. Utility Distribution	269
E. Building Mechanical System Energy Reductions	270
F. Control Settings and Outside Air Balance	270
G. Architectural Recommendations	270
H. Remodeling of Mechanical Systems	270
VI. Recommendations	271
VII. Conclusions	271
APPENDIX A. Inventory Forms	273
APPENDIX B. Glossary	295

I. INTRODUCTION

East Texas State University is a small, multi-purpose, regional, state university of approximately 12,000 students, located in Commerce, Texas. The map of its physical plant identifies 100 buildings constructed since 1894. The recent rapid escalation of energy costs prompted a study to establish a conservation program. The buildings are served by almost every type of lighting and HVAC system marketed, and in many cases are old enough to be potential maintenance headaches.

II. PROPOSAL

The proposal suggested studying, initially, the 24 buildings using over 85% of the university's total energy expenditure. The scope of the work includes a survey of the mechanical and electrical systems to recommend modifications to energy usage and maintenance requirements.

III. WORK PLANNING

Larry Hardaway, the lead engineer assigned the work, planned his attack by creating a flow chart describing tasks, problems, and expected results. His chart, which follows, may be useful in other applications:

1. OBJECTIVE
   a. Investigate and analyze energy usage.
   b. Identify energy saving opportunities.
   c. Recommend and evaluate energy conservation measures.
   d. Develop appropriate funding.
2. PROBLEMS
   a. Meter usage not available by buildings or systems.
   b. Limited billing histories.
   c. Limited equipment documentation.
   d. Limited building plans.
   e. Limited system flexibility.
3. INITIAL SOLUTION PLANNING
   a. Initial meetings and survey to prepare general study approach.
   b. Develop plans directed to meet objective and overcome problems.
      (1) Gather past billing histories.
      (2) Prepare survey sheets and collect data.
      (3) Conduct on-site investigations.
4. REDEFINE SOLUTIONS
   a. Review resource data and refine study direction.
   b. Collect additional data necessary.
5. COMPUTE SOLUTIONS DATA
   a. Using resource data, prepare energy and cost computations.
   b. Prepare implementation cost estimates.
   c. Evaluate payback periods.
6. DECISIONS
   a. Recommend solutions and associated cost and savings.
7. PRESENT SOLUTIONS
   a. Write report.
   b. Print and prepare copies.
8. ADJUST AS NECESSARY
   a. Incorporate owners' input and adjustments.

IV. BUILDING SURVEYS

The university's buildings are of all ages. Few have plans and specifications or even inventories of their mechanical and electrical systems. Information had to come from physical surveys and operators of the systems. There were limited utility billing histories to provide operating profiles. Metering devices were not conveniently arranged. It was decided a set of inventory forms to be completed by operations personnel could minimize the survey team's time at the site. (Example forms are included in Appendix A of this chapter.)

The forms were given to the university's facilities section with the following instructions:

1. Fill out "Building description form" for each building as listed in the Study Scope.
2. Select the appropriate equipment form for each type of equipment used in the building. Use as many equipment forms as there are pieces of equipment in use.
3. Add additional information in supplementing information sheets as necessary.

V. DISCUSSION

The university uses two forms of primary energy, electricity and natural gas, to operate the building systems. Electrical power, supplied by Texas Power and Light Company, enters the campus at a primary voltage of 12,470 volts through seven metering points, from which it is distributed to all buildings. Three of those points were the sources for the 24 buildings studied. Natural Gas, supplied by The Lone Star Gas Co., is also metered at seven points, two of which serve the buildings studied.

Since the metering points serving the study buildings also serve other buildings, billing history did not provide accurate fuel costs for each particular building. Therefore, the portion of energy used for each building was estimated from an individual building analysis based on equipment inventories and load profiles developed by operators. These data, collected on the survey forms, listed each component in the building mechanical systems along with types of electrical service and lighting systems. Field measurements were taken to determine the average rate of energy used by each system component. By utilizing this rate

Table 16.1. Use of Energy in a Building.

CONTRACTOR	SOURCE	TO DECREASE ENERGY
EXTERNAL	AMBIENT CONDITIONS SOLAR LOADS	MODIFICATION TO BUILDING ENVELOPE
INTERNAL	LIGHTING PEOPLE EQUIPMENT OPERATING PROFILES	CHANGE LOADS OR PROFILES
INHERENT TO SYSTEM	ADDITIONAL ENERGY USED FOR NORMAL OPERATION	MODIFY INSTALLED SYSTEMS

and known operating schedules, energy usage was computed for each building. This computed consumption was then compared with the composite billings to ensure realistic estimates.

The survey forms also documented specific system component details, including age, operating hours, previous operating history, present condition, estimated reliability, type of system, capacity, and other information. Each building was then examined by visual inspection, and interviews were arranged with the operating staff. All information was used to determine which buildings and systems could be retrofitted without producing secondary maintenance and/or environmental problems. Savings opportunities were listed and evaluated for cost-effectiveness. Table 16-1 describes how energy is used in a building. It also defines approaches to savings.

The survey identified many needs in each of the categories examined. There were leaky buildings, buildings that emptied but remained lighted, or buildings being furnished too much air. It became obvious that the two elements needing the most consideration were: (1) the system components, which needed better control in order to match their operation to the building needs; and (2) outdated equipment requiring excessive operating and maintenance costs.

A. Energy Management Control System

The effects of internal contributors as well as the inherent system contributors can be programmed to optimize energy efficiency. The extent of energy savings is dependent on how closely the system components can be controlled to match the needs. Computerized automation systems, often called energy management control systems (EMCS), are manufactured for this specific application. They have varied capabilities and must be selected to match requirements. In this application it was necessary that the control system be capable of start–stop functions, temperature resetting, temperature monitoring, flow monitoring, and data acquisition, as well as recording. Additionally, the system had to be easily programmed and capable of controlling a variety of devices so that all of the campus buildings might be included.

Modular systems, which enable the user to arrange components to meet his present requirements and expand later, were examined. These systems may also be used to monitor equipment and print instructions as part of an equipment maintenance program, which may include logging of operating hours to produce printed notices of needed lubrication, filter maintenance, excess vibration, extreme temperature conditions, high water alarms, etc. Such maintenance can improve scheduling, to maintain more equipment with less manpower, while improving maintenance efficiency.

Technology has been developed that allows microprocessors in the field units, as shown in Figure 16-1, to handle most of the routine processing of data and control of equipment. Programs for the field units may be written in Fortran at the main computer and then downloaded to the field unit. This allows the user considerable freedom after the system is in-

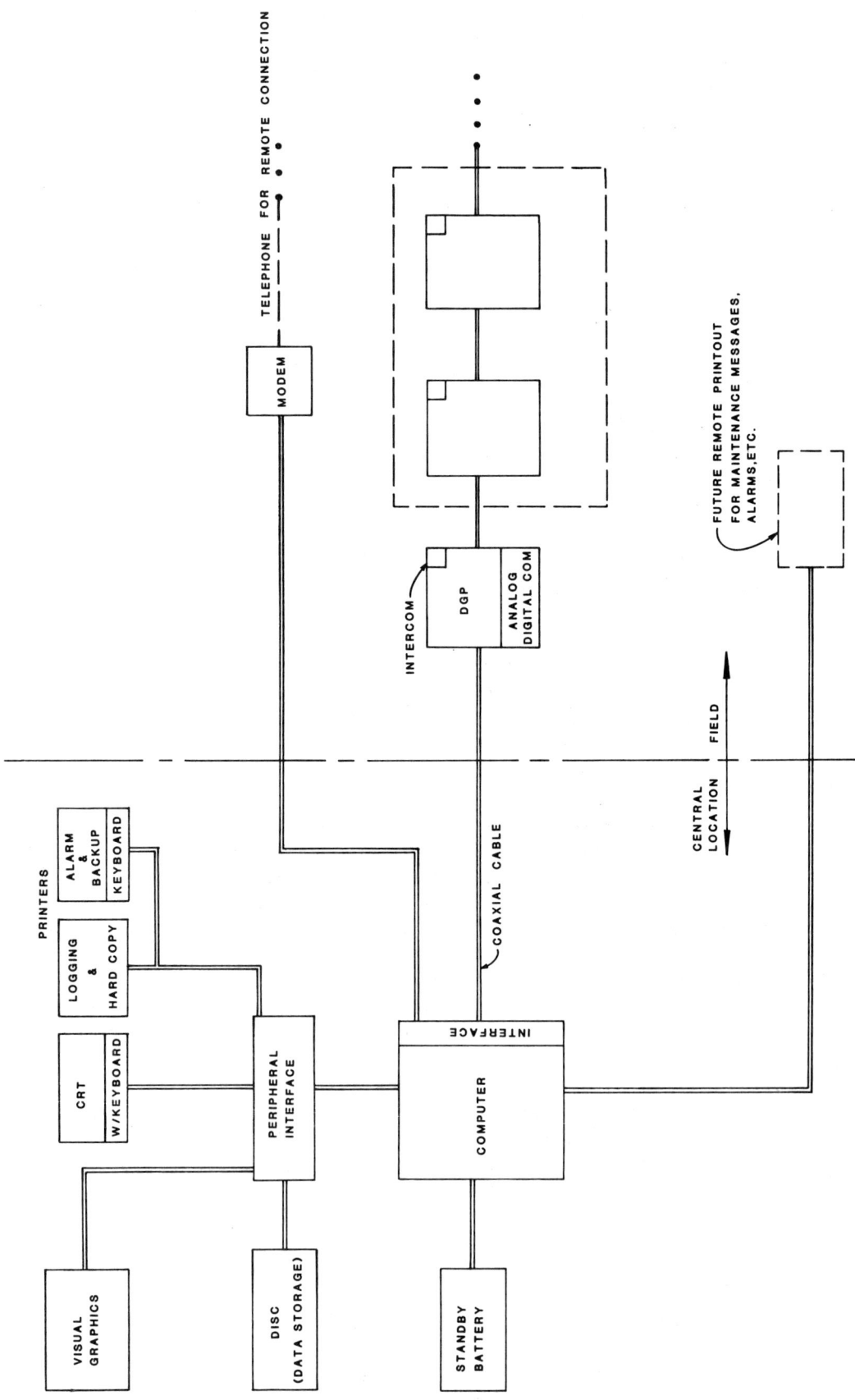

Figure 16-1. System of main computer and field units. (Courtesy of East Texas State University, Commerce, Texas.)

stalled. The field units can be programmed to accomplished start–stop functions, automatic control, and alarm functions even if the central computer is not operating. This type of system is a reliable, flexible, and useful tool in managing energy usage in any installation.

To determine the optimum level of system sophistication, it was necessary to compute the expected energy savings. It was estimated that a savings of $171,700 could be realized by programming various system components to match building requirements. The expected cost for such an automation system would be $1,104,000, resulting in a 6½-year simple payback.

B. Reduced Utility Cost Through Limiting Demand

An analysis of the billing history revealed that the university was paying more per kilowatt-hour for electricity during the low usage months of the year because of the conditions of the Texas Power and Light Company Service Agreement. The billing is based upon actual energy used and maximum rate of demand, but the minimum billing demand quantity will be at least 65% of the maximum demand recorded in a 15-minute interval of the past June, July, August, and September. The demand during several of the past winter months fell below this cost, and the minimum demand provision was exercised.

As a large portion of the utility cost was demand charges, it was found effective to install demand-limiting systems. These systems monitor demand at all times and are capable of turning off equipment on a preselected priority schedule to establish a maximum acceptable level. The equipment is cycled on again when the demand level permits. Using this device, all equipment on the system is assigned a priority so that vital equipment always remains in operation. The estimated cost for this system at the time of our study was $80,000. It was estimated the system would save approximately $59,500 annually after a 1½-year payback period.

C. Electrical Light Relamping

The building lighting and electrical systems were surveyed and light level measurements taken. Computations were then made to determine the savings that could be derived by revising the lighting in the building areas. The latest Illumination Engineers Society criteria were used for each task area. It was determined that a 13% reduction in the present lighting energy and demand was possible without any noticeable effect on illumination. The savings were to be produced by revising the lighting in stairways and corridors, and by reductions in lamp wattage used for aesthetic effect. Significant energy savings appeared possible in the field house and gymnasium by replacement of low-efficiency incandescent lighting with high-efficiency, high pressure sodium units.

The estimated cost for these modifications was $288,400. Expected savings were $149,500 annually after a two-year payback.

D. Utility Distribution

The majority of the university's buildings are served by individual mechanical systems. However, two of the buildings' mechanical plants are capable of being expanded into central plants. The present individual systems vary from small window installations with electric heating coils, to large centrifugal liquid chillers and steam boilers. The total combined cooling capacity of these systems is more than 2,000 tons (7,040 kW) of refrigeration. Central plant installations with distribution mains and building connections offer advantages other than energy conservation, and higher inherent efficiencies of the system components used in central plants can save up to 50% of the energy required for individual systems. A lower initial investment required because less backup machinery is needed, the ability to take advantage of campuswide load diversities, reduced maintenance costs, decreased inventory through standardization of repair parts, improved equipment reliability, and added flexibility for alterations in buildings are all benefits of a central plant.

In analyzing the various buildings and the locations of the existing central plants, a scheme was developed by which most of the independent building systems could be connected. The plan included the addition of new chilled water and steam piping to the existing distribution systems, with connections to those buildings. The capacity of these plants was also to be increased. The cost for this conversion was estimated at $625,000. It was estimated that the improved efficiency in maintenance, utility conversion, operation, control, and life expectancy would save $50,200 per year.

E. Building Mechanical System Energy Reductions

Many different types of air-moving systems were being utilized on the campus. Depending on the type, the amount of fan and heating–cooling energy varied significantly. Systems that vary the supply air quantity to maintain room conditions were comparatively low energy users. Those that supply constant volumes of air but vary the air temperature by controlling heating or cooling coil output were moderate energy users. Multizone and double duct systems, installed in about half of the buildings, were, however, larger users. Such systems maintain comfort conditions by supplying hot and cold air streams that are mixed to produce desirable air temperatures. Cold air must be used to cool hot air in order to produce the desired supply air temperature. This requires that both refrigeration and boiler energy be generated simultaneously and used at all times regardless of actual building loads.

In order to improve the performance of these systems, additional controls may be added to establish cold and hot air temperature settings that result in the least amount of wasted energy. Computations for the ten buildings using these types of systems indicated the cost for modifying the systems was approximately $100,000. The annual savings for these improvements were estimated at approximately $21,800, with a four-year payback period.

The moderate-energy-using systems could be converted to the least-energy-using systems by the addition of terminal control devices. These devices vary the air volume to satisfy room conditions and therefore save fan energy. The cost of retrofitting these systems was estimated at $78,000. The annual energy saved was estimated to reduce utility cost by approximately $13,400, with a six-year payback period.

F. Control Settings and Outside Air Balance

Many building systems were equipped with control cycles that utilized outside air for cooling when possible. Some buildings used fixed outside air quantities to maintain ventilation rates and odor control. Although the correct portions of outside air were set properly at the time of their installation, many dampers were no longer at their original settings. This increased energy consumption; recalibration was required. The improper adjustments increased the amount of cooling and heating needed annually to compensate for undesirable outside quantities. The recalibration was estimated to cost approximately $72,000. Reductions in cooling and heating energy were estimated to save $9,300 annually. An eight-year recovery period was envisioned.

G. Architectural Recommendations

The survey identified improvements that could be made to decrease loading for several air conditioning and heating systems. The improvements included the addition of solar screening devices, the addition of wall insulation, and modification to roofing systems in order to reduce mechanical system energy consumption.

For a cost of $300,000, a reduction in utility consumption of $28,700 per year, with a 10½-year payback period, was estimated.

H. Remodeling of Mechanical Systems

Several building systems called for complete remodeling in order to achieve a reliable op-

CASE STUDY OF A COMMERCIAL RETROFIT PROJECT 271

Table 16.2. Recommendations.

NUMBER	ITEM	CONSTRUCTION COSTS	ANNUAL SAVINGS	YEARS PAYBACK
(1)	ENERGY MANAGEMENT COMPUTER SYSTEMS	$1,104,000	$171,700	6½
	DEMAND LIMITING	80,000	59,500	1½
(2)	RELIGHTING	288,400	149,500	2
(3)	DISTRIBUTION	625,000	50,200	12½
(4)	BUILDING MECHANICAL SYSTEMS	100,000	21,800	4½
	VARIABLE VOLUME	78,000	13,400	6
(5)	OUTDOOR AIR ADJUSTMENTS	72,000	9,300	8
(6)	ARCHITECTURAL	300,000	28,700	10½
	TOTALS	$2,647,400	$501,100	5½
	OTHER			
(A)	REMODELING MECHANICAL	500,000	—	—

Source: Courtesy of East Texas State University, Commerce, Texas.

erating condition. Most of these systems had served their expected useful lives, and their maintenance costs were very high.

VI. RECOMMENDATIONS

Table 16-2 is a summary of the recommended improvements that were described in Section V. The savings indicated are annual amounts based on the projected utility costs for January 1980. Cost figures include estimated construction, priced on January 1980 dollars.

VII. CONCLUSIONS

East Texas State University received the report of this study in July of 1978 and proceeded to obtain funding. In mid-1979 part of the funding became available; design is being completed on the first phase, which encompasses recommendations numbers (1) and (3). The first portion will include the use of the university's CDC computer and its in-house capability to provide the software for the energy management system. The design work on that portion includes the sensory and control elements needed to input the data required. The second portion of this first phase includes the addition of the distribution piping and central plant capabilities. Following phases will convert building systems and connect them to these distribution systems, as well as the energy management computer system.

APPENDIX A. Inventory Forms

SYSTEM NO.		SYSTEM DESCRIPTION	IP/KW	PE		AMPS	
				EFF		VOLTS	
DAY TYPE __ OF __							

		AM												PM										
ORIG.	1	2	3	4	5	6	7	8	9	10	11	12	1	2	3	4	5	6	7	8	9	10	11	12
OPER.																								
% LOAD																								
KWH																								
REV.																								
OPER.																								
% LOAD																								
KWH																								
ENERGY TOTALS																								
OPER.																								
% LOAD																								
KWH																								

DAILY ENERGY USAGE	DAILY ENERGY SAVINGS	ANNUAL ENERGY USAGE	ANNUAL ENERGY SAVINGS

METER READINGS

		DATE:						
	HR.	MONDAY	TUESDAY	WEDNESDAY	THURSDAY	FRIDAY	SATURDAY	SUNDAY
AM	1							
	2							
	3							
	4							
	5							
	6							
	7							
	8							
	9							
	10							
	11							
	12							
PM	1							
	2							
	3							
	4							
	5							
	6							
	7							
	8							
	9							
	10							
	11							
	12							

INDUSTRIAL EQUIPMENT DATA

SHOP_____ ROOM_____ EQUIP. NO._____

I. Electrical Data:

                              H.P._____ ETF_____
Voltage_____ Phase_____ Hz_____

KVA Input_____ P.F._____ KVA Output_____

II. HVAC Data:

1) Total KW_____ x 3.415 = _____ MBTUH
    EFT x _____

2) Space KW_____ x 3.415 = _____ MBTUH

3) Spec. Air Cooling_____ cfm(1.08)_____ *TD-_____ MBTUH
                          1,000

4) Ind. Water_____ gpm(.5)_____ TD-_____ MBTUH

5) Net Heat to Space = (2-3-4)   _____ MBTUH

6) Heat transfered to space No._____ (1-2)_____ MBTUH

7) Estimated Hours Operated Daily (Diversity)_____ HRS

III. Industrial Equipment Data:

1) Ind. Cooling_____ gpm, _____ EWT, _____ LWT

2) Ind. Gases - Type_____, _____ SCFM _____ PSIG

3) Compressed Air - _____ Peak SCFM, _____ Steady SCFM
    _____ PSIG

4) Steam - _____ Peak lbs/hr, _____ Steady lbs/hr
    _____ PSIG

5) Misc. Req._____

6) Special Treatment:
    a) Ventilation - Scrubbing _____
    b) Industrial Waste Treatment _____

IV. Plumbing - Industrial Waste:

1) Potable Water _____ gpm

2) Sanitary Waste
    a) Hub Drains _____ size
    b) Floor Drains _____ size

3) Industrial Waste
    a) Hub Drains _____ size
    b) Floor Drains _____ size
    c) Flow _____ gph

*Indicate Entering and Leaving Air Temperature

CAMPUS TOTAL INFORMATION

(All information from previous fiscal years) Annual electrical utilities cost previous 5 years_____

Monthly electrical cost immediate past year electrical service contract

Annual Natural gas utilities cost previous 5 years_____

Monthly gas cost immediate past year gas service contract.

Total maintenance budget previous 5 years

Total cost of sub-contracted work

Present energy conservation program

Answers

_____ Yes

_____ No

_____ Does not apply

_____ Information presently not available'

_____ Approximate value

GENERAL INSTRUCTIONS

1. Fill out "Building description Form" for each building as listed in study scope.

2. Select the appropriate equipment form for each type of equipment used in the building. Use as many equipment forms as there are pieces of equipment in use.

3. Add additional information on supplementary information sheets as necessary.

CASE STUDY OF A COMMERCIAL RETROFIT PROJECT 277

	DATE:
BOVAY ENGINEERS, INC. HOUSTON • SPOKANE • BATON ROUGE AUSTIN • ALBUQUERQUE • WASHINGTON, D.C.	BY:

LIGHTING SYSTEMS

Building_____

Fluorescent_____, Wattage_____, No. of Units_____
Lighting task_____
Operation - _____AM to _____PM, Continuous_____
Incandescent_____Wattage_____, No. of Units_____
Lighting task_____
Operation_____AM to _____PM, Continuous
Fluorescent_____, Wattage_____, No. of Units_____
Lighting task_____
Operation _____AM to _____PM, Continuous_____
Incandescent_____Wattage_____, No. of Units_____
Lighting task_____
Operation _____AM to _____PM, Continuous_____
Fluorescent_____, Wattage_____, No. of Units_____
Lighting Task_____
Operation _____AM to _____PM, Continuous_____
Fluorescent_____, Wattage_____, No. of Units_____
Lighting Task_____
Operation _____AM to _____PM, Continuous

Period for Lamp replacement_____
Period for Cleaning lenses and bulbs_____

BOVAY ENGINEERS, INC.		DATE:
HOUSTON • SPOKANE • BATON ROUGE AUSTIN • ALBUQUERQUE • WASHINGTON, D.C.		BY:

MULTIZONE AHU SYSTEM

Building Served_____, AHU NO. _____

MFR. Name_____ Model No._____, Age_____

Design Total Supply CFM_____, SP_____

Design Total Outside Air CFM_____

Motor Nameplate HP_____ Volt_____ Phase_____

Motor Nameplates Amps_____

Actual Amps L_1_____, L_2_____ L_3_____

Actual Voltage L_1_____, L_2_____ L_3_____

No. of Zones_____

Type Damper controls - Pneumatic_____, Electric_____

Hot Deck Reset_____, Cold Deck Reset_____

Type Cooling Coil - Dx_____, CHW_____

Type Heating - Steam_____, Hot water_____, Electric_____

Preheat coil_____

Economizer cycle_____

Humidity control_____

Previous operation problems:

Control failures_____, Zones_____, Coils_____

Zones too cold in winter_____

Zones too hot in summer_____

Zones hot and cold alternately_____

Present operation:

Manual start_____, Time Clock start_____

Operating hours_____AM to _____PM, Continuous_____

Week-end operation required_____

		DATE:
BOVAY ENGINEERS, INC. HOUSTON • SPOKANE • BATON ROUGE AUSTIN • ALBUQUERQUE • WASHINGTON, D.C.		BY:

SINGLE ZONE AHU SYSTEM

Building Served _____, AHU No. _____

MFR. Name _____ Model No. _____ Age _____

Design Total Supply CFM _____ S.P. _____

Design Total Outside Air CFM _____

Motor Nameplate Amps _____

Actual Amps, L_1 _____ L_2 _____ L_3 _____

Actual Voltage L_1 _____ L_2 _____ L_3 _____

Single Zone _____, Terminal ReHeat _____

Type Cooling Coil - Dx _____, CHW _____

Type Heating - Steam _____, Hot Water _____, Electric _____

Preheat Coil _____

Economizer Cycle _____

Humidity Control _____

Previous operating problems: _____

Control Failures _____, Coils _____

Space too Cold in winter _____

Space Too Hot in Summer _____

Space Hot and Cold Alternately _____

Present Operation:

Manual Start _____, Time Clock Start _____

Operating Hours _____ AM to _____ PM, Continuous _____

Week-end Operation Required _____

BOVAY ENGINEERS, INC. HOUSTON • SPOKANE • BATON ROUGE AUSTIN • ALBUQUERQUE • WASHINGTON, D.C.		DATE:
		BY:

DOUBLE DUCT AHU SYSTEM

Building Served_____ AHU No._____

MFR. Name_____ Model No._____ Age_____

Design Total Supply CFM_____, SP_____

Design Total Outside Air CFM_____

Motor nameplate HP_____ Volt_____ Phase_____

Motor nameplate amps_____

Actual amps L_1_____,L_2_____ L_3_____

Actual voltage L_1_____ L2_____ L_3_____

No. of Mixing Boxes_____

Type cooling coil - Dx_____, CHW_____

Type heating - Steam_____, Hot water_____, electirc_____

Hot Deck Temp. Setting_____°F

Cold Deck Temp. Setting_____°F

Hot Deck Reset_____

Economizer Cycle_____

Preheat Coil_____

Humidity Control_____

Previous Operating Problems:

Control Failures_____, Coils_____, Air_____

Spaces Too Hot in Summer_____

Spaces Too Cold in Winter_____

Spaces Hot and Cold Alternately_____

Present Operation:

Manual Start_____, Time Clock Start_____

Operating Hours_____AM to _____PM, Continuous_____

Week-end Operation Required_____

BOVAY ENGINEERS, INC.	DATE:
HOUSTON • SPOKANE • BATON ROUGE	
AUSTIN • ALBUQUERQUE • WASHINGTON, D.C.	BY:

FAN COIL UNITS

Building Served_____ No. of Units_____

MFR..lame_____ Model No._____ Age_____

Design total supply CFM_____, S.P._____

Design Total outside air CFM_____

Motor HP or watts_____, Volts_____

Type Control:

Fan speed_____, Coil Modulating Valves_____

Electric_____, Pneumatic_____

Previous operating problems:

Control Failures_____, Coil Valves_____

Control Signal_____, Fan Speed Switch_____

Space Too Hot in summer_____

Space Too Cold in Winter_____

Space Hot and Cold Alternately_____

Present Operation:

Operating hours_____AM to _____PM, Continuous_____

Week-end Operation Required_____

BOVAY ENGINEERS, INC. HOUSTON • SPOKANE • BATON ROUGE AUSTIN • ALBUQUERQUE • WASHINGTON, D.C.		DATE:
		BY:

HEATING-VENTILATION UNIT

Building Served_____, AHU No._____

MFR. Name_____, Model No._____, Age_____

Design Total Supply CFM_____, S.P._____

Design Outside Air CFM_____

Motor Nameplate HP_____, Volt_____, Phase_____

Motor Nameplate Amps_____

Actual Amps L_1_____, L_2_____, L_3_____

Actual Voltage L_1_____, L_2_____, L_3_____

Type Heating - Steam_____, Hot Water_____, Electric_____

Min-max outside air dampers_____

Previous Operating Problems:

Control failures_____, Coils_____, Air_____

Space to cold in winter_____

Space to hot in summer_____

Present Operation:

Manual Start_____, Time Clock Start_____

Operating Hours_____AM to _____PM, Continuous_____

Week-end Operation required_____

	DATE:
BOVAY ENGINEERS, INC. HOUSTON • SPOKANE • BATON ROUGE AUSTIN • ALBUQUERQUE • WASHINGTON, D.C.	BY:

VENTILATION FAN UNIT

Building Served_____   Fan NO._____

Space Served_____

MFR Name_____ Model No._____ Age_____

Motor nameplate HP_____, Volt_____, Phase_____

Actual Amps L_1_____, L_2_____ L_3_____

Type Drive - Belt_____, Direct_____

Type Blades - Propellar_____, Centrifugal_____

Present Operation:

Manual Start_____, Time Clock Start_____

Operating Hours_____AM to _____PM, Continuous_____

Week-end operation required_____

BOVAY ENGINEERS, INC. HOUSTON • SPOKANE • BATON ROUGE AUSTIN • ALBUQUERQUE • WASHINGTON, D.C.		DATE:
		BY:

EXHAUST FAN SYSTEMS

Building Served_____, Exhaust Fan No._____

System Served_____

MFR. Name_____, Model NO._____ Age_____

Design Total CFM_____, Static Pressure_____

Motor Nameplate HP_____, Volts_____ , Phase_____

Actual motor Amps L_1_____, L_2_____, L_3_____

Type Drive - Belt_____, Direct_____

Present Operation:

Manual Start_____, Time Clock_____

Operating Hours_____AM to _____PM, Continuous_____

Week-end Operation Required_____

BOVAY ENGINEERS, INC. HOUSTON • SPOKANE • BATON ROUGE AUSTIN • ALBUQUERQUE • WASHINGTON, D.C.		DATE:
		BY:

COOLING TOWERS

Building served_____ Unit NO._____

System served_____ Age_____

MFR._____, Model No. _____

Fan HP_____, Volts_____, Phase_____

Motor nameplate amps_____

Actual amps L$_1$_____, L$_2$_____, L$_3$_____

Control temperature setting _____ °F

Control method:

Fan sequenced_____, Water by-pass_____

Present condition:

Poor_____, Fair_____, Good_____

BOVAY ENGINEERS, INC. HOUSTON • SPOKANE • BATON ROUGE AUSTIN • ALBUQUERQUE • WASHINGTON, D.C.		DATE:
		BY:

CENTRIFUGAL WATER CHILLERS

Building name_____ Unit No._____

Tons_____ Age_____

MFR. name_____ Model No._____

Compressor KW_____, Volt_____, Phase_____

Hot gas by-pass option_____

Load limiter option_____

Chilled water control setting_____

Present Operation:

Operating hours_____AM to _____PM Continuous_____

Seasonal Operation - months per year_____

		DATE:
BOVAY ENGINEERS, INC. HOUSTON • SPOKANE • BATON ROUGE AUSTIN • ALBUQUERQUE • WASHINGTON, D.C.		BY:

WATER COOLED RECIPROCATING CHILLERS

Building Served_____ Unit No._____

Mechanical system served_____

MFR. Name_____ Model No._____

Tons_____ Age_____

No. of compressors_____

Compressor KW_____. Voltage_____, Phase_____

Condenser water pump HP_____, Volt_____, Phase_____

Condenser water pump amps L$_1$_____, L$_2$_____, L$_3$_____

Operating hours_____AM to _____PM, Continuous

Manual start_____, Time clock start_____

Chilled water pump HP_____, Volt_____, Phase_____

Chilled water pumps amps L$_1$_____, L$_2$_____, L$_3$_____

Operating hours _____AM to _____PM, Continuous

Manual start_____, Time clock start_____

Chilled water temp. setting_____

		DATE:
BOVAY ENGINEERS, INC. HOUSTON • SPOKANE • BATON ROUGE AUSTIN • ALBUQUERQUE • WASHINGTON, D.C.		BY:

<u>AIR COOLED COMPRESSOR UNITS</u>

Building Served_____ Unit No._____

Mechanical system served_____

MFR. Name_____ Model No._____

Tons_____, Age_____

No. of Compressors_____

Compressor KW_____, Voltage_____, Phase_____

BOVAY ENGINEERS, INC.		DATE:
HOUSTON • SPOKANE • BATON ROUGE AUSTIN • ALBUQUERQUE • WASHINGTON, D.C.		BY:

PACKED ROOF TOP SYSTEMS

Building Served_____, Unit No._____

MFR. Name_____, Model No. _____ Age_____

Tons_____

Design Supply CFM_____, SP_____

Design outside air CFM_____

Supply air fan motor HP_____, Volts_____, Phase_____

Fan motor amps_____

Actual fan motor amps L_1_____, L_2_____, L_3_____

Type heating - Gas_____, Electric_____

Economizer Cycle_____

Previous operating problems:

Control failures_____, Air_____, Compressor_____

Space too cold in winter_____

Space too hot in summer_____

Space hot and cold alternately_____

Present operation:

Manual start_____, Time clock start_____

Operating hours_____AM to _____PM, Continuous_____

Week-end operation required_____

BOVAY ENGINEERS, INC.		DATE:
HOUSTON • SPOKANE • BATON ROUGE AUSTIN • ALBUQUERQUE • WASHINGTON, D.C.		BY:

BOILER SYSTEMS

Building Served_____ Unit No._____

MFR. Name_____ Model_____

Rated Capacity_____

Type fuel - Gas_____, Oil_____, Electric_____

Type boiler - Cast iron_____, Fire tube_____, Water Tube_____

Type generation - steam_____, Hot water_____, Pressure_____

Type burner - ATM_____, Forced draft_____

Forced draft data:

Motor HP_____, Volts_____, Phase_____

Operating Conditions:

Age_____, Poor_____, Fair_____, Good_____

Seasonal Operation:

Continuous_____, No. Months_____

| BOVAY ENGINEERS, INC. | | DATE: |
| HOUSTON • SPOKANE • BATON ROUGE AUSTIN • ALBUQUERQUE • WASHINGTON, D.C. | | BY: |

DOMESTIC HOT WATER SYSTEMS

Building served_____ Unit No._____

Type: Gas_____, Steam_____, Hot water_____

MFR._____, Model_____, Age_____

Circulator pump HP_____, Voltage_____ Phase_____

Pump operation Continuous_____, Week-ends_____

Condition: Poor_____, Fair_____, Good_____

Temperature setting_____°F

Primary Service:

Rest Rooms_____

Showers_____

Kitchens_____

Laboratories_____

Dormatories_____

BOVAY ENGINEERS, INC. HOUSTON • SPOKANE • BATON ROUGE AUSTIN • ALBUQUERQUE • WASHINGTON, D.C.		DATE:
		BY:

BUILDING DESCRIPTION FORM

Building Name_____ No._____

Building Age _____

Primary Occupancy Use_____
_____

Floor Area_____ Number of Floors_____

%Glass in Walls, N_____, E_____, S_____, W_____

Operation Schedule:

    Week day_____AM to _____P M, No. Per Year_____

    Week-end_____AM to _____PM No. Per Year_____

    Holiday_____AM to _____PM No Per Year_____

Approximate No. of peiple at max. occupancy_____

Requirements For Continual Temperature Control.

    a. Laboratories _____Temp. R.H._____%

    b. Computers _____Temp. R.H._____%

    c. Occupancy _____Temp. R.H._____%

    d. Material Storage _____Temp. R.H._____%

    e. Security Office _____Temp. R.H._____%

Type Air Handling Systems:

Multizone_____, Single Zone_____, Fan Coil Unit_____

Double Duct_____, Variable Volume_____, Terminal Reheat_____

Packaged Unitary or Roof Top_____

Type Heating-Cooling Service:

Air Cooled Dx_____ Water Cooled Dx_____

CHW - From Central Plant_____, Local Chiller_____

Steam from Central Plant_____

Hot Water Boiler_____, Electric_____, Gas_____

Electric Heating_____, Gas Furnace_____

Lighting - Fluorescent_____, Incandescent_____, Other_____

Available Building Information:

Plans and specifications_____

Separate metering - Gas_____, Electricity_____, Steam_____

No. of Elevators - Escalators_____

Building Distribution Voltage_____ Phase_____, Wires_____

APPENDIX B. GLOSSARY

Absorption: A process whereby heat extracts one or more substances present in an atmosphere or mixture of gases or liquids, accompanied by physical change or chemical changes.

Absorption chiller: A refrigeration unit based upon absorption refrigeration.

Absorption refrigeration: Cooling caused by the expansion of liquid ammonia into a gas. Heat is the primary source of energy!

ACS: Automatic control systems.

AEB: Annual energy budget.

Air change: The movement of a volume of air in a given period of time in or out of a building or room. Air changes are expressed in cubic feet per minute.

Ambient temperature: The temperature of the outside or surrounding air.

ASHRAE: American Society of Heating, Refrigerating, and Air Conditioning Engineers.

Boiler: A device used to heat water and/or produce steam. Major components include burner, heat exchanger, flue and expansion chamber, and controls. Traditional fuels are oil, gas, coal, and electricity.

Boiler efficiency: The ability of a unit to convert a form of energy (gas, oil, coal, electricity) to heat energy at the highest possible rate; that is, 80 to 85% efficiency should be an attainable goal.

BOMA: Building Owners and Managers Association.

British thermal unit: The amount of heat necessary to raise the temperature of one pound of water one degree Fahrenheit. A unit of thermal (heat) energy approximately equal to the amount of heat given off by burning a kitchen match.

Btu: (See British thermal unit.)

Building envelope: The physical elements that make up the exterior of a building.

Building orientation: Considered in relation to the site orientation, the direction of prevailing winds and the sun. These relationships directly affect the heat loss/gain properties of a building.

Capacity: The usable output of a system or system component.

CFM: Cubic feet per minute.

Chimney effect: The tendency of air or gas in a duct or other vertical passage to rise when heated owing to its lower density compared with that of the surrounding air. In buildings, the tendency toward displacement (caused by the difference in temperature) of internal heated air by unheated outside air owing to the difference in density of inside and outside air.

Comfort zone: The ranges of indoor temperature, humidity, and air movement under which most persons enjoy mental and physical well-being.

Compression refrigeration system: A process by which the pressure and temperature of the refrigerant are increased to allow for greater heat transfer.

Conduction: A process of heat transfer whereby heat is transmitted through a material.

Convection: Transfer of thermal energy (heat) by the movement of a fluid or gas.

Damper: A device used to vary the volume of air passing through an outlet, inlet, or duct.

Degree-day: The difference between 65°F and the average daily temperature (°F); i.e., average daily temperature = 25°F, so the degree-day would be 65 − 25 = 40. The greater the number of heating or cooling degree-days, the higher the energy consumption.

Demand factor: The ratio of the maximum demand of a system, or part of a system, to the total connected load of a system, or part of a system, under consideration.

Dewpoint: The temperature at which water vapor begins to condense.

Dry bulb temperature: The temperature of a gas or mixture of gases indicated by an accurate thermometer after correction for radiation.

ECM: Energy conservation measure.

Economizer cycle: A method of operating a ventilation system to reduce refrigeration load. Whenever the outdoor-air conditions are more favorable (lower heat content) than return-air conditions, outdoor-air quantity is increased.

EMCS: Energy management control system.

Energy: The capacity for doing work, taking a number of forms that may be transformed from *one into another, such as thermal (heat), mechanical (work), electrical, and chemical; in customary units, measured in kilowatt-hours (kWh) or British thermal units (Btu); in SI units, measured in joules (J) where 1 joule = 1 watt-second.*

Energy control system: A system designed and operated to control the energy-consuming equipment of an institution or installation of buildings, usually automatically, to optimize energy conservation.

Energy cost savings: Energy savings times energy unit (therm, 10^3 ft^3, kWh). Note: The results indicate a cost avoidance based upon anticipated reduction in energy consumption.

Enthalpy: For the purpose of air conditioning, enthalpy is the total heat content of air above a datum, usually in units of Btu/lb. It is the sum of sensible heat and latent heat and ignores internal energy changes due to pressure change.

EPRI: Electric Power Research Institute.

Flow rate: Velocity at which a fluid travels, usually through an opening or duct.

Flue gas analysis: A test procedure whereby the relationships between air, fuel, and stack temperatures are monitored, thus indicating the apparent transfer of energy during combustion of the fuel.

Foot-candle: Energy of light at a distance of one foot from a standard (sperm oil candle).

Heat, latent: The quantity of heat required to cause a change in state.

Heat, sensible: Heat that results in a temperature change but no change in state.

Heat capacity: Sometimes called the thermal capacity, a measure of how much heat is required to raise the temperature of a specific quantity of given material by a given amount.

Heat exchanger: A device used to transfer heat from one medium to another.

Heat gain: As applied to HVAC calculations, it is that amount of heat gained by a space from all sources, including people, lights, machines, sunshine, etc. The total heat gain is the quantity of heat that must be removed from a space to maintain indoor comfort conditions.

Heat loss: A decrease in the amount of heat contained in a space, resulting from heat flow through walls, windows, roof, and other building envelope components.

Heat pump: A reversible refrigeration system that delivers more heat energy to the end use than is put in the compressor. The additional energy input results from the absorption of heat from a low-temperature source—the combination heating and cooling unit. It operates like a normal air conditioner in summer, and in winter operates in reverse, ejecting warm air indoors and cool air (or water) outdoors.

Heat wheel: A device used in ventilating systems that tends to bring incoming air into thermal equilibrium with exiting air. As a result, hot summer air is cooled, and winter air is warmed.

HVAC: Heating, ventilating, and air conditioning systems.

IES: Illuminating Engineering Society.

IGT: Institute of Gas Technology.

Infiltration: The process by which outdoor air leaks into a building by natural forces, especially through cracks around doors and windows, etc. (usually undesirable).

Insulation, thermal: Any material high in resistance to heat transmission that when placed in the walls, ceilings, or floors of a structure will reduce the rate of heat flow.

Life cycle cost: The cost of equipment over its entire life including operating and maintenance costs.

Luminaire: Light fixture designed to produce a specific effect.

Mass: The property of density of a material. The use of mass by itself or in combination with insulation gives building structures the capacity to store thermal energy.

Moisture content: The relative quantity of water in a volume of air expressed in grains of moisture; a grain of moisture equals approximately $\frac{1}{7000}$ lb.

OSES: On-site energy system.

Payback period: The time required for anticipated energy savings to recover the cost of the investment; also termed recovery rate.

Peak load: Maximum predicted energy demand of a system.

Psychrometric: Pertaining to the device comprising two thermometers, one a dry bulb, the other a wet bulb or wick-covered bulb, used in determining the moisture content or relative humidity of air or other gases.

Reflectance: A property of a material indicating the percentage of light that is reflected when a certain amount of light strikes the surface of the material or is transmitted through it.

Relative humidity: The ratio of the amount of water vapor at a given temperature to the maximum amount of water vapor that could be held as vapor.

Setback: Lowering of the thermostat setting. The technique is used to reduce the amount of energy required to heat a space.

TES: Thermal energy storage.

Thermal lag: The ability of materials to delay the transmission of heat; can be used interchangeably with time lag.

U value: A coefficient that indicates the energy (Btu) that passes through a component for every degree Fahrenheit of temperature difference from one side to the other under steady-state conditions.

Vapor barrier: A component of construction impervious to the flow of moisture or air.

Ventilation: Air introduced to an occupied space or building to accomplish comfort and odor control. Commonly refers to outside-air quantities mixed with room air.

Weatherstripping: Foam, metal, or rubber strips used to form a seal around windows, doors, or openings to reduce infiltration.

Wet bulb temperature: The lowest temperature attainable by evaporating water into the air without the addition or subtraction of energy.

17. Case Study of a School Building Retrofit*

Frank T. Carroll

Product Line Manager, Environmental Control Division
American Air Filter Company, an Allis-Chalmers Company
Louisville, Kentucky

	Page
I. Introduction	298
II. Overview	298
III. History and Its Impact	299
IV. Beginning the Renovation	299
V. Solar System	300
VI. Water Source Heat Pump	301
VII. Conclusion	303

I. INTRODUCTION

Then there is the story of the school district that became enamored with energy conservation. Thermostats were set back for a few percentage points of energy savings. Bus routes were made more efficient. Insulation was beefed up. A more efficient heating–cooling system was installed.

The list of energy-saving methods and materials went on and on, with 15% savings here, 7% there, and 12% elsewhere. Adding all the percentages together, the district soon saved well over 100% of its energy. It then went into business selling this excess energy and never had to worry about Proposition 13 again.

There may be a slight exaggeration there somewhere, but the point is that energy conservation opportunities can seem almost unlimited and go well beyond a few percentage points. Hand Middle School, in Columbia, South Carolina, provides an example of what may be the extreme in making an old school building maximum energy-efficient.

II. OVERVIEW

Two-thirds of the school's window area was replaced with insulated panels. Remaining single-glazed, wooden double-hung windows were replaced with a fenestration arrangement having a thermal efficiency equal to one-inch insulated glass. Roof insulation was added to reach a .05 (0.284 W/m^2°C) U factor. Walls were re-insulated and ceilings were lowered.

Old, cast iron, steam radiators were replaced with EnerCon® water source heat pump system made by American Air Filter, an Allis-Chalmers company, for year-round comfort control without open windows. This system is assisted by solar collector panels mounted on the roof and a 15,000-gallon (56,780-l) water tank that stores excess heat from daytime cooling and recycles it for nighttime heating.

---

*This chapter is reprinted from the *CEFP Journal* (Council of Educational Facility Planners), July/August, 1979.

Room thermostats are placed at seven-foot heights to prevent tampering. There are automatic night setback controls with two-hour override switches on all heating–cooling units. The setback override controls are operated from a central office.

Ventilation air fans have controls that allow them to operate only during the hours the school requires the air to meet state ventilation codes. Heating–cooling unit fans are thermostatically actuated to shut off when adequate temperature is reached. This prevents the units from recirculating room air and provides double energy savings. Moving air requires more heat to maintain comfort, and, of course, fan energy is consumed.

Architects William Fulmer & Associates, along with consulting engineers Durlach, O'Neal, Jenkins & Associates, apparently turned every stone in their quest for energy economics.

An instantaneous domestic water heater was installed so that it is no longer necessary to hold large amounts of heated water in readiness for showers and lavatories.

During summer months, when it is anticipated 90°F (32°C) water will be available from the EnerCon system pipe loop, shower water will bypass the domestic hot water system and go through a heat exchanger with the cooling system water. This will provide warm-enough water for showers and, at the same time, take some of the load from the system's heat rejection component, thereby increasing energy savings.

Finally, it was determined the school's offices would be in use 40% more hours per week than its classrooms. Because it would be energy-wasteful to activate the entire HVAC system to meet the requirement of this relatively small area of the building, a separate condenser water pump was installed in this section of the building. It circulates water from the main pipe loop and back to provide optimum comfort in offices, while the rest of the school remains on night setback and the main, 10-hp (7.46 kW) circulating pump stays off. All of these efforts aimed at energy efficiency were incorporated in a $2,322,500 (private communication) renovation of the school.

III. HISTORY AND ITS IMPACT

Built in 1929, the school received additions in 1938 and 1950. Another 2,600 sq ft (241.5 m^2) of space were added to the school during its renovation in 1977–78. The total rehabilitation cost of the school, which now encompasses 98,000 sq ft (9,104 m^2), was $23.70/sq ft ($255.11/m^2). Hand School had been closed for two years following a fire which resulted in an insurance claim payment of $750,000.

Studies showed the school was situated at the heart of a growing residential area of Columbia, and, while it might have cost little more to build a new school, this was not a feasible alternative. Either the cost of tearing down the existing school or of acquiring land elsewhere would have had to be added to the building cost. This would have made a new school of equal size cost much more than the renovation of Hand.

The Hand School, despite the fire and despite its age, was still a sound structure. There was no aspect of its arrangement of spaces that indicated the renovated school would be significantly less functional over the next decades than a new building. (See Figure 17-1 for a view of the renovated school.)

IV. BEGINNING THE RENOVATION

The first phase of the renovation was to bring the school in line with current fire and safety codes. This meant providing smoke doors at appropriate locations and constructing stairways and exits at the ends of dead-end corridors. Lesser modifications to the structure were also initiated.

Next, a tight "envelope" was provided. Both rigid and fiber insulation was applied to the inside of roofs and perimeter walls to double the thermal efficiency of the walls and bring all roof areas to a .05 U (0.284 W/m^2·°C) factor.

Approximately two-thirds of all the window area was replaced with an arrangement of brick veneer on the outside to be aesthetically compatible with the remainder of the brick structure. This was followed by an insulative dead air space and U.S. Gypsum thermal sheathing mounted on 3⅝-inch (92-mm) metal

Figure 17-1. Hand Middle School, Columbia, South Carolina.

studs. The final, inside surface was lath and plaster to match existing walls.

Windows that remained were replaced with dual glass windows. These differ from conventional double-glazed windows by providing a thicker air space, or thermal break, between the panes of glass. They are rated at the same thermal efficiency as one-inch insulated glass.

Getting rid of the old radiator heating system was next on the renovation schedule. Ceilings were dropped, and horizontal water source heat pump units were installed above the suspended ceilings to provide filtered, tempered air to each classroom.

Each classroom now has its own thermostat that senses the input of heat from students and lights and uses all of the heat in the room. The system can recirculate excess heat to rooms that require it or store the excess heat from daytime cooling and recycle it for nighttime heating.

V. SOLAR SYSTEM

All HVAC units in the system are interconnected by a single closed pipe water loop that, in turn, is tied in, by means of heat exchangers, to a roof-mounted solar collector, a 15,000-gallon (56,780-l) water storage tank, and to the domestic hot water system.

The role of the solar collector in reducing energy costs is more academic than actual at this point, with only 462 sq ft (42.9 m^2) of collector in place. (See Figure 17-2 for a view of the installation.) Even with the water source heat pump system, which requires water in a temperature range of 60°F (15.4°C) to 90°F (32°C), compared with a hot water system needing 150°F (65°C) to 180°F (82°C), the solar collector's contribution to energy savings is minor.

However, there is provision to add another 2,500 sq ft (232 m^2) of collector to the system as its benefits become apparent and funds become available. It is estimated that 3,000 sq ft (279 m^2) of collectors would supply 30 to 40% of the school's heating needs.

Meanwhile, readout instruments showing the hourly input of the collectors are located in the school's science classrooms; students will be able to study solar energy first hand as school officials study the same data with an eye toward expanding the solar system at Hand School and the prospect of using solar collectors at other district schools.

Figure 17-2. Solar collector panels and American Air Filter's water source heat pumps.

VI. WATER SOURCE HEAT PUMP

Project engineer J. L. Jenkins, of the firm that designed the mechanical systems at Hand School, points out that the water source heat pump, backed with solar collectors and water storage, is the most energy-efficient kind of system possible for a school the size and shape of Hand.

The EnerCon system (Figure 17-3) utilizes a series of unitary, water-to-air reverse cycle air conditioners. These heating and cooling units are connected to a nonrefrigerant central water system that is maintained within an approximate temperature range of 60°F (15.4°C) to 90°F (32°C) by a supplementary water heater and a heat rejector.

"During occupied hours," Jenkins says, "highly insulated schools require more cooling than heating—even on very cold days. Students and lighting give off heat that will soon become intolerable in a structure where this heat gain exceeds the building's heat loss. Conventional HVAC systems simply waste this excess heat into the atmosphere.

"Schools change from cooling to heating very quickly in late afternoon when students go home, lights go out and the sun goes down," Jenkins continues. "At this point the school begins to need heat to maintain night setback temperatures."

Hand School doesn't throw away heat, except in summer when the cooling load is high during the day and little or no heat is required at night. It transfers heat from air conditioning units on cooling cycle, through the water pipe loop, to units on heating cycle as long as any heating is needed anywhere in the building.

If the system is still capturing heat, water in the pipe loop and in the storage tank absorbs the excess heat. Meanwhile, storage tank water is being heated by the solar collector as well as by rejected heat from areas of the building being cooled.

Throughout the night, the system draws on this stored heat to maintain setback temperatures. It continues to take heat from the stored water in the piping and tank until water temperature throughout the system drops to 60°F (15.4°C). The school's original boiler takes over at this point to maintain the 60°F (15.4°C) water temperature.

Conversely, when the system can no longer reject heat into the pipe loop and tank without forcing water temperature over 90°F (32°C), a heat rejector in the form of an evaporative

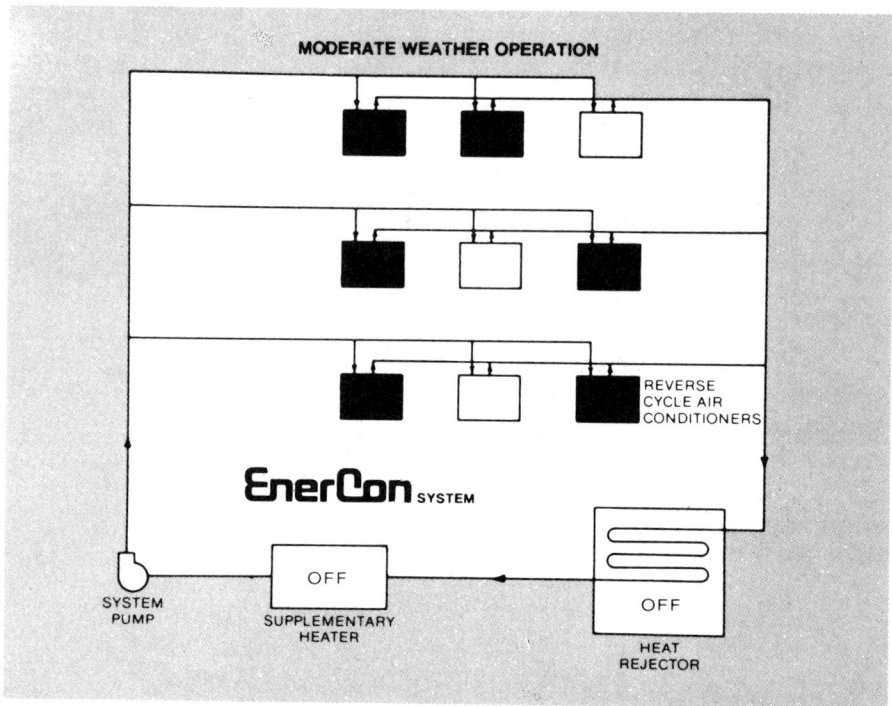

Figure 17-3. EnerCon system in moderate weather operation. Some units are cooling, some are heating, and some aren't needed at all. The heat put into the water loop by units cooling is being utilized by those heating. Neither the evaporative cooler nor the water heater is operating. (From "Enercon: How it Works," American Air Filter Company, Louisville, Kentucky.)

cooler located outside of the building takes over to maintain the temperature range for most efficient operation of the heat pump system.

The air conditioners thus frequently start the day with cool water for most efficient cooling, and start the night with warm water for most efficient heating. The net result of this inherent and effective heat recovery is to reduce power input to the air conditioners, and reduce

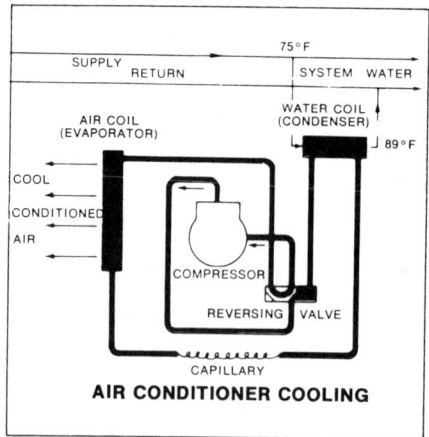
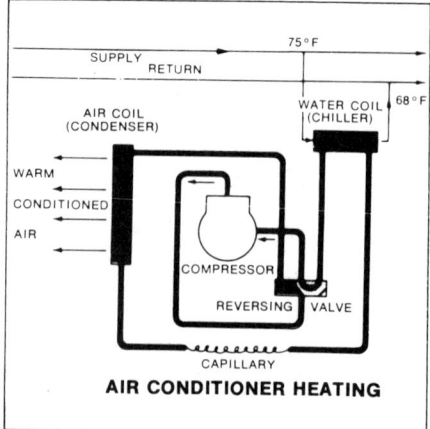

Figure 17-4. Schematic of the EnerCon system. (From "Enercon: How it Works," American Air Filter Company, Louisville, Kentucky.)

water heater and heat rejector loads and hours of operation. Figure 17-4 is a schematic illustration of the heating–cooling system.

The supplemental water heater and the heat rejector are, of course, the least energy-efficient components of the system. The less they are used, the lower the total energy costs remain.

VII. CONCLUSION

American Air Filter Company water source heat pump sales literature promotes a more than 25% energy savings over other kinds of heating–cooling systems. Insulated-window and insulation manufacturers have claimed savings as high as 40% when their products are used on old buildings. Add to this the energy-conserving benefits of night setback controls, solar collectors, heat recovery from exhaust air, tinker-resistant thermostats, and preheating of domestic hot water with excess heat from the comfort cooling system, and savings become extremely significant. Hand Middle School may still be far short of the mythical 100% energy savings. But it surely has shown that there are few limitations when energy economy receives a high priority in school renovation projects.

18. Retrofitting the Astrodome for Energy Conservation

Dale S. Cooper, P.E.[*]

Energy Consultant, Houston Sports Association
Houston, Texas

		Page
I.	Introduction	304
II.	Preliminary Findings	305
III.	Control Solutions	307
IV.	Other Projects in Progress or Completed at the Astrodome	313

I. INTRODUCTION

In the retrofitting of large air conditioning systems for energy conservation there has developed a logical procedure in which the stages of development are arranged in the order of their economic feasibility. It is to the advantage of all concerned that changes showing the greatest possibility for improvement with the least expenditure of funds be accomplished first.

In many cases, the savings may appear to be so dramatically large that the owner does not believe them to be possible. At this point the engineer's credibility is being challenged; he should therefore start out to accomplish at least *one project* that will show a meaningful improvement before proceeding to the next. Once he has gained the confidence of management in this manner, the remaining items will fall into place with very little opposition.

Unfortunately, retrofitting involves the expenditure of rather substantial amounts of money, which has to be capitalized and written off over a period of time. Management tries to avoid such expenditures, but if it can be shown that the return on the investment is three years or less, there will usually be very little opposition.

Before proceeding to explain the details of the first major step in the retrofitting operation at the Astrodome, it might be useful to explain what led to its consideration.

Houston's Astrodome was designed in 1963, at which time all utilities were considered to be inexhaustible and of little consequence in the economics of operation.

The stadium was originally planned to take care of about 100 baseball games and 10 football games per year. Expected expenditure for basic fuel was about $100,000 per year, based upon electricity at 5 mills/kWh and natural gas at about 35¢/1,000 cu ft (1.24¢/1,000 l).

In January 1976, the cost of these fuels was running about $115,000 per month, based upon electricity at 2¢/kWh and gas at $1.95/1,000 cu ft (6.9¢/1000 L) a 550% increase. From that date to August 1978, there was a steady month-to-month increase. Electricity rose to 3¢/kWh, and gas approached $3.00/1,000 cu ft (11¢/1,000 l).

These rates included fuel adjustments, kVA demand charges, cost of service charges, state sales taxes, etc. In comparison to other areas of the country, the rates were still considered to be quite low.

---
[*]Current affiliation: Dale Cooper Consulting Engineers, Inc., Houston, Texas.

The overall increment of cost, from $100,000 per year in 1963 to $115,000 per month in 1976, was not all due to utility rate increases, however. In that period, the Astrohall and the Astro Arena were added to the complex and connected to the stadium chilled water system to take care of many more revenue-producing events not suitable to the dome. Addition of these large-area expositions—horse shows, stock shows, boat and recreational vehicle shows, homebuilders expositions, offshore technology conferences, etc.—tended to double the number of events to be accommodated.

Within the two-year period between January 1976 and January 1978, the usage of all facilities had stabilized, and most of the events had become repetitive on a year-to-year basis. Thus the use of energy could reasonably be compared on a month-to-month basis.

II. PRELIMINARY FINDINGS

Although the physical plant had functioned to everyone's satisfaction in maintaining the desired inside conditions, the preliminary survey showed it to be operating very inefficiently, because of the following:

1. The designing engineers were told that the systems were to maintain 72°F (22.22°C) and 50% relative humidity, with an occupancy of 66,000 people, on a design day in August, with the sun streaming through the translucent dome with sufficient intensity to permit the growth of grass on the field.
2. The original operating engineers were told to maintain these conditions on a 24-hour-per-day basis, summer and winter, because of the then "questionable effects" of temperature change upon the dome structure.
3. None of the main stadium air-handling units was designed to control from return air temperature. Control of the discharge air temperature from the central control panel (manual reset) was to be made in response to temperature sensors in the area.
4. Terminal reheat was employed on all units to accomplish the 50% R.H. condition. However, the stadium was quite loose in construction, and multiple doors opening to the four points of the compass at five different levels made this requirement a practical impossibility.
5. The dewpoint controls were set to maintain 50% R.H. at 72°F (22.22°C) and were hardly ever satisfied, with the result that reheat was called for all the time to offset the effects of the wide-open chilled water valves. This means that the large high pressure steam boilers had to be fired continuously, along with steam turbine operation, to produce the chilled water.

When it was found that the natural grass had to be replaced with what is now called "Astroturf," and the translucent dome panels had to be partially painted to stop the glare, there were appreciable changes in the components of heat gain. Watering of grass was discontinued, and field lights had to be burned during all sporting events.

These and other changes resulted in an appreciable increase in the sensible heat load and a reduction in latent heat gain, which in effect made it unnecessary to attempt to maintain positive control of humidity.

Twelve years later, the systems were being operated as outlined above, which is shown graphically by Figure 18-1.

These were the operating conditions on a cold day in October 1976 when the temperature was 45°F (7.22°C) outside and there was nothing taking place in the stadium.

Both steam boilers were operating at less than half capacity, using about 40,000 cu ft (1,132,800 l) of gas per hour to produce about 27,000 lb (12,247. kg) of steam per hour at an efficiency of 70%.

The steam was divided almost equally, between the steam turbine, producing about 1,140 tons [4008 kW] $= \left(\dfrac{12000 \times 1140}{3413}\right)$ of refrigeration, and the high temperature hot water converters, circulating some 500 gpm

306 6/RETROFITTING FOR INDUSTRY

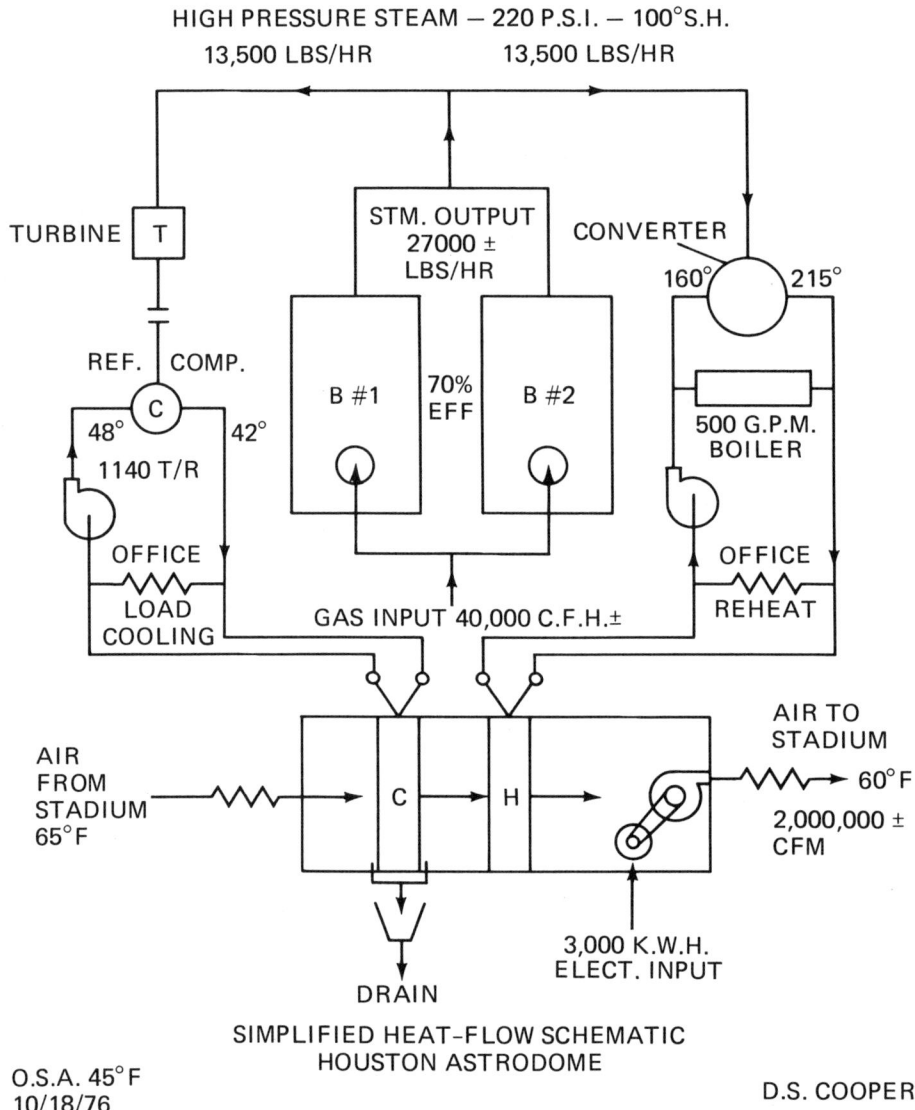

Figure 18-1

(31.55 l/s) of hot water through the heating coils at the temperature shown.

All stadium air-handling units were running and maintaining about 65°F (18.33°C) (average) in the stadium with 60°F (15.56°C) air leaving the units. Office heating and cooling loads were also canceling each other out, as the internal heat from lights, appliances, and occupants just about equaled the heat loss of the spaces.

Here was a system with large heating and cooling capability, running on a low-load condition, burning gas and electricity without limitation and producing no appreciable effect, except moisture removal, which was not needed or fully attainable.

This situation is not an isolated case peculiar only to a domed stadium. Precisely the same problem exists today in hundreds of large office buildings, department stores, hospitals, etc.

Whenever a boiler or other heating device is found to be in operation in the middle of the hot summer season, there will probably be a terminal reheat system that cannot maintain its desired temperatures without the use of *two* mediums using prime energy simultaneously—one to cool and the other to reheat. These types of systems are *gross wasters of energy* and should be outlawed as a means for providing human comfort, except in special applications.

Energy conservation standards, such as ASHRAE 90-75, are rapidly being adopted by states and municipalities as enforceable law. They place serious limitations upon such systems.

From careful observation of the manner in which the system was being operated and controlled, it became obvious that large quantities of gas and power could be saved without in any way affecting the function of the stadium during an event.

III. CONTROL SOLUTIONS

The most important problem to solve was that of control, and after careful study of the original air-handling unit control systems and capacities, the following procedure was decided upon:

1. Increase the "event" temperature to 74°F (23.33°C), and eliminate the "reheat" feature on all air-handling units within the stadium by making the cooling and heating valves operate in sequence to maintain temperature. (Sequencing of valves is commonly understood to mean the closing of one valve before the opening of the other, thus preventing the simultaneous usage of the two mediums being employed.) Figure 18-2 shows the arrangement of components in a typical, seventh-level, air-handling unit. Chilled and hot water coils are now operated in sequence, but with different "dead bands," depending upon the mode of operation.
2. Establish a "non-event mode" between the temperatures of 60°F (15.56°C) and 80°F (26.67°C), in which neither heating nor cooling could take place. In Houston there are about 3,000 hours in this temperature zone; however, sun effect through the dome would reduce this time to about 2,500 hours.
3. In the non-event mode, shut down all air-handling units except those on the seventh level and the peripheral offices. The seventh-level units were chosen to maintain the stadium in non-event status because of their size and location. Figure 18-3 shows the location of the seventh-level units around the stadium, providing excellent air coverage at the higher elevation above the field. These six units were found to have sufficient capacity to hold the stadium within the desired limits—heating or cooling—in non-event status.
4. Provide a means for manually switching all the seventh-level units from event mode to non-event mode and vice versa when desired by the operators.
5. Let a contract to the original control contractor to make the necessary changes and additions to the existing control system. This would accomplish the above items and, at the same time, provide for the application of an electronic system to encompass the entire complex at a future date.

This was designated Phase I of the new control system, shown graphically in Figure 18-4 above the dashed line. The boxes marked P are outside air pretreatment units delivering some 17,000 CFM (8,024 l/s) of outside air at 60°F (15.56°C) into the recirculated air stream of the large seventh-level units.

Phase II, below the dashed line, was the next major increment to the control system. It has now been completed, but Phase I was the improvement that made most of the savings possible.

Phase I was accomplished for the very nominal cost of $25,000, because it did not require any new control wiring between the control room and the seventh level.

By starting and stopping the outside air pretreatment units (P), the main return temperature control could be switched from the event

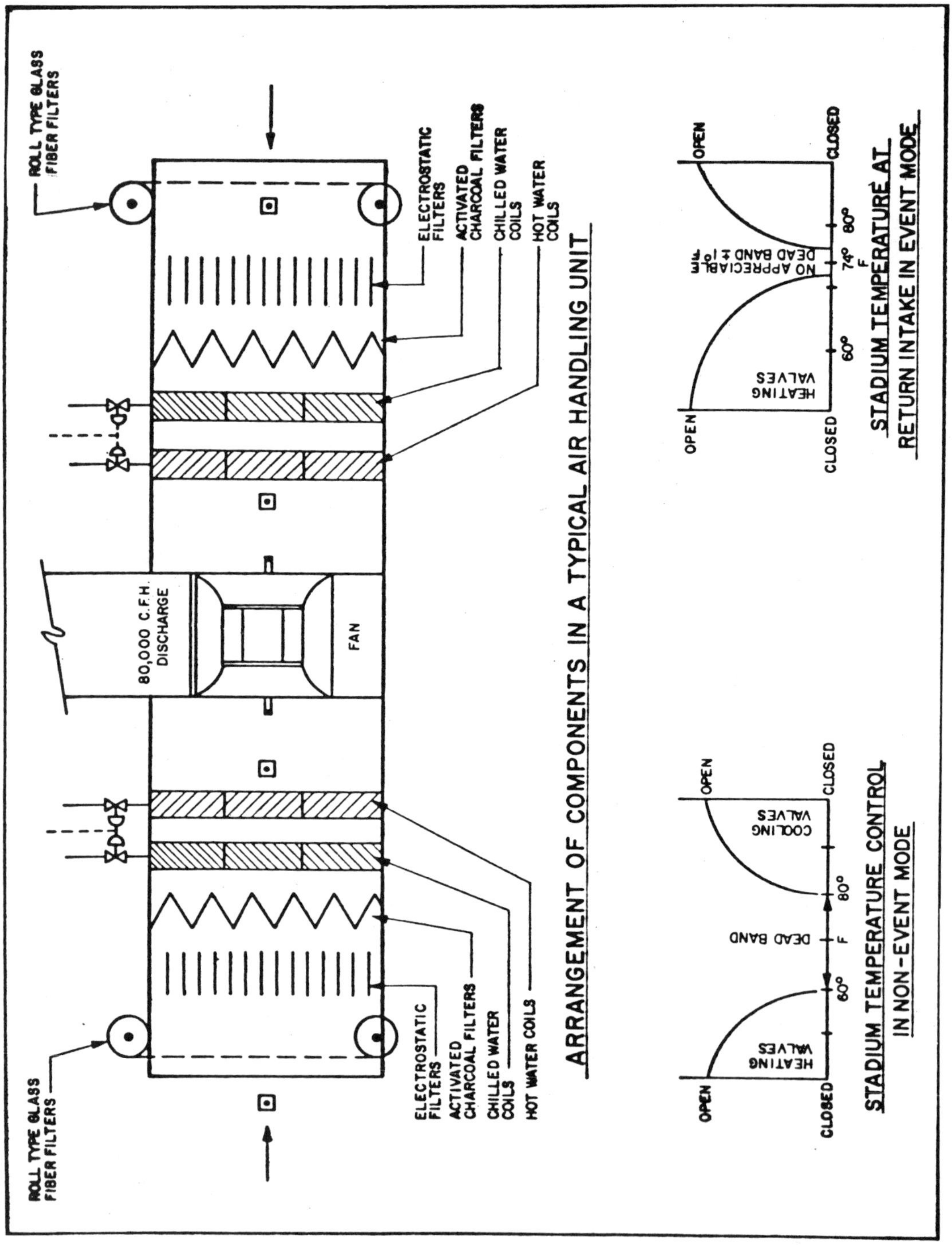

308 6/RETROFITTING FOR INDUSTRY

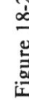

Figure 18-2

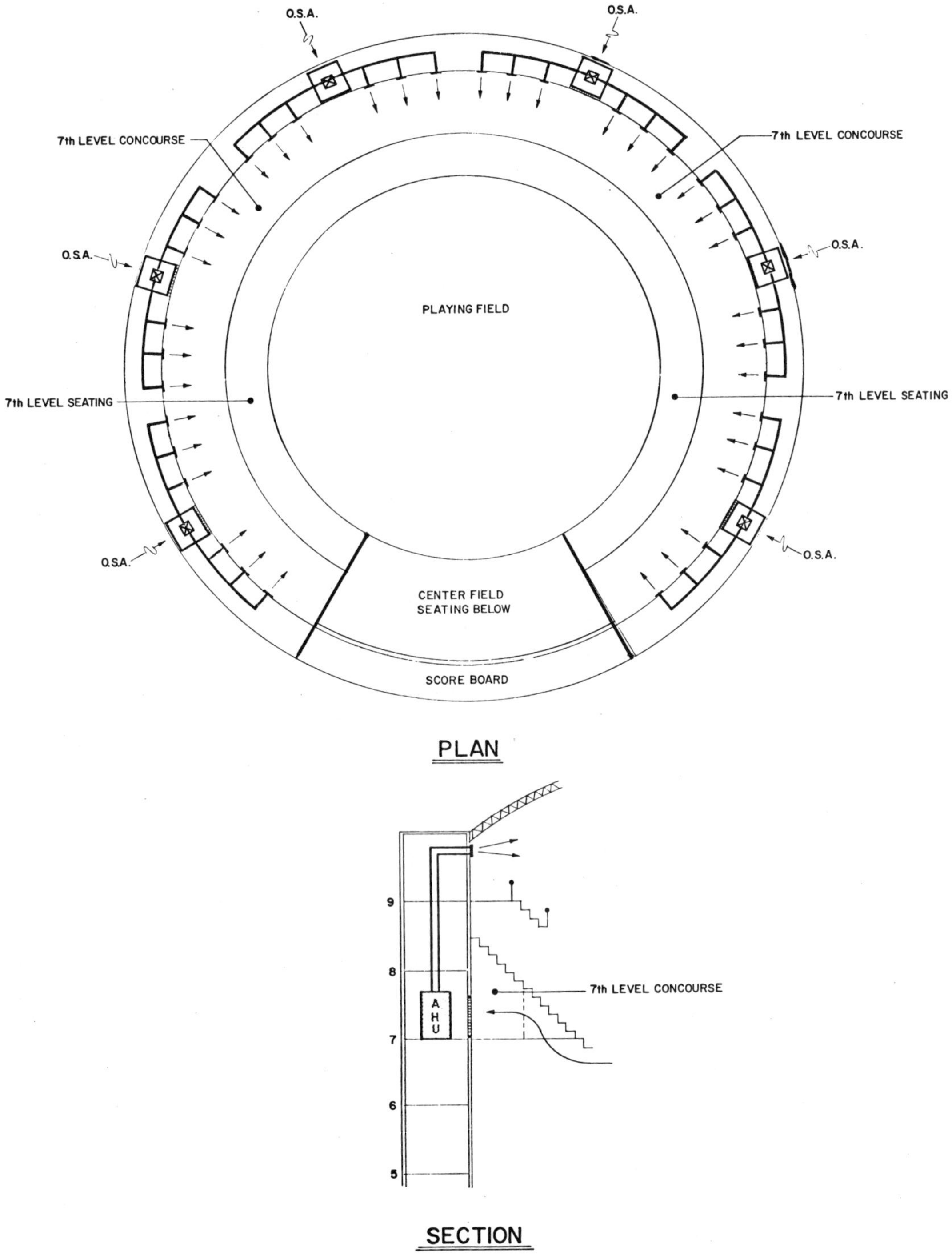

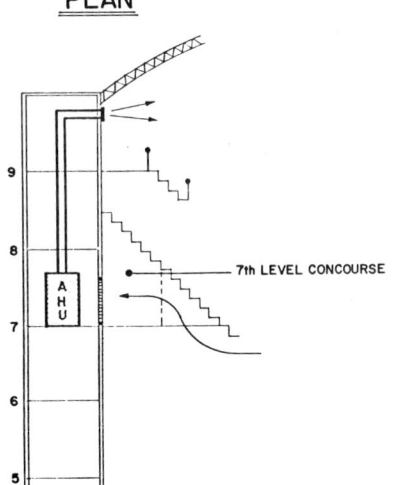

PLAN

SECTION

Figure 18-3

310 6/RETROFITTING FOR INDUSTRY

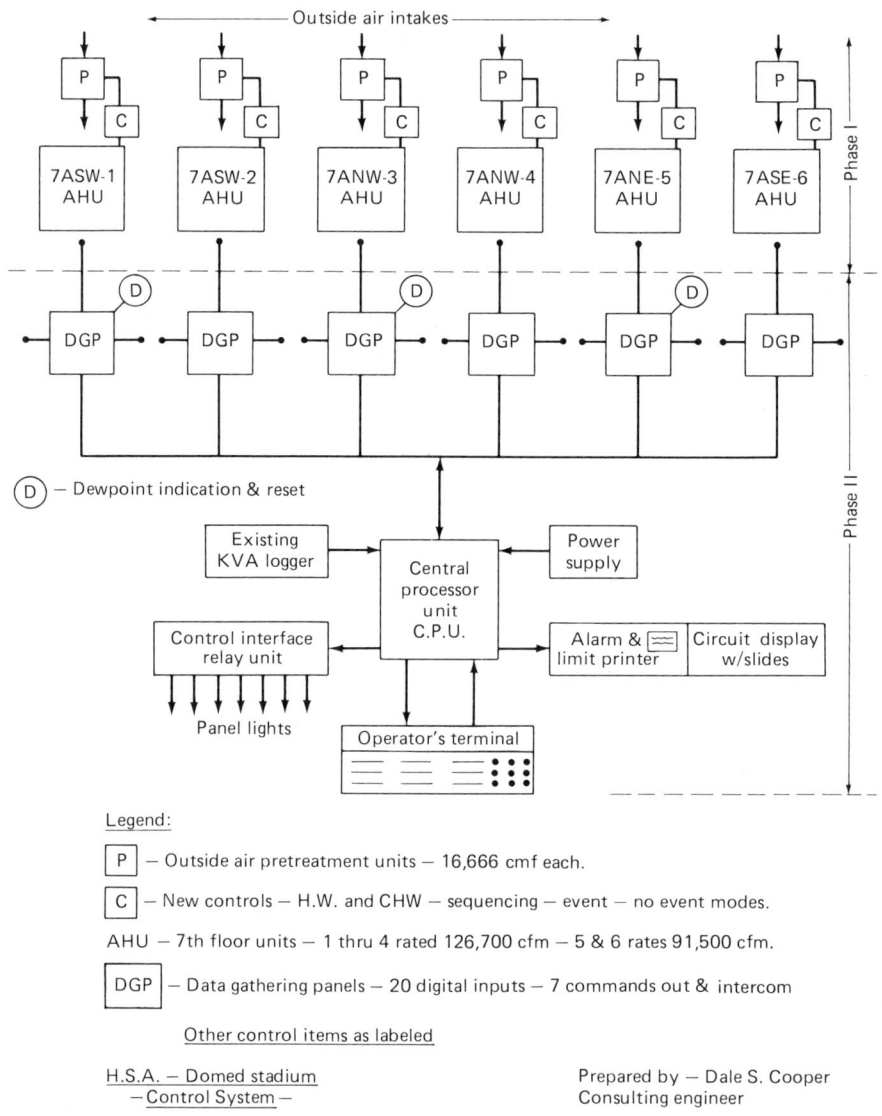

Figure 18-4

to non-event mode with hot and cold valves sequencing as shown in Figure 18-2.

It was now possible to shut down the large boilers and turbines, which previously ran continuously. The necessary cooling required for the peripheral zones could be supplied by the electric chillers, and the service hot water for cooking and showers could now be handled by the small hot water boiler originally installed for that purpose.

The results of the above changes in control and methods of operation are shown by the bar charts designated in Figures 18-5 and 18-6.

Figure 19-5 shows the total power used in the Astrodome complex, in thousands of kilowatt-hours, comparing similar months on a year-to-basis.

Unfortunately, there are no submeters on the air conditioning system, which would allow a more accurate analysis of the retrofitting previously described. The field lights, concourse lighting, general lighting, parking lot

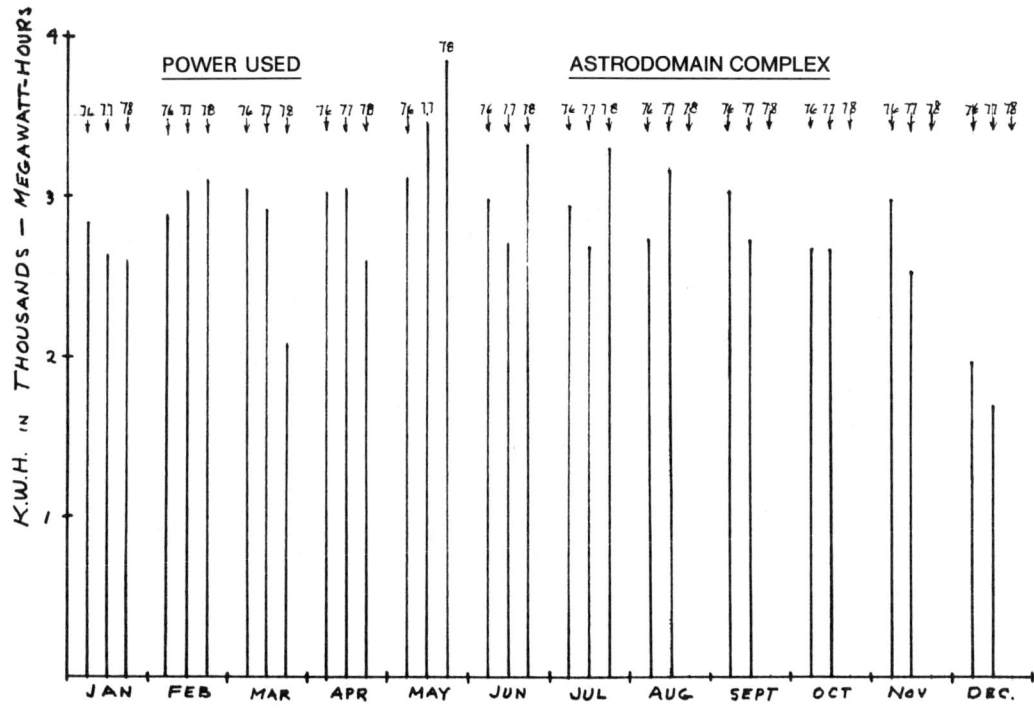

Figure 18-5

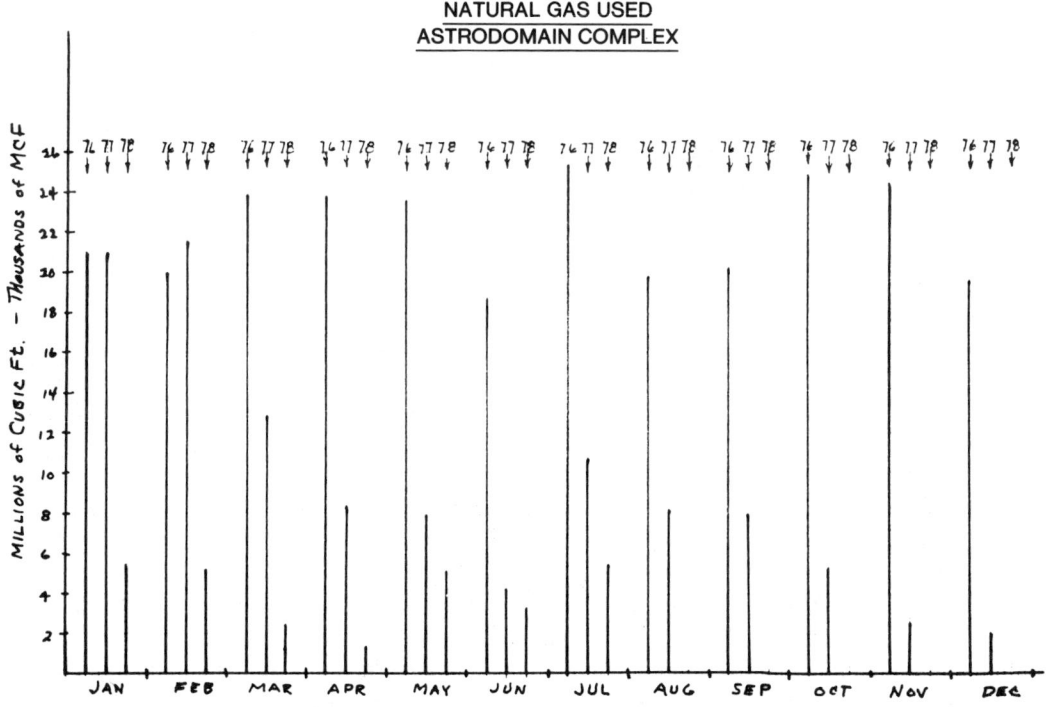

Fugure 18-6

lighting, and all other electrical usage in the complex, with the exception of the Astro Arena, are included in the monthly totals shown.

For these reasons there seems to be no set pattern for electrical savings, particularly when it is realized that, in shutting down the turbines, all non-event cooling was produced with the electric centrifugal machines.

Even so, the average monthly usage of power was reduced from 1976 to 1977, by 73,000 kWh per month.

Attention is directed to the power usage in May 1978, which exceeded all previous peaks because of one large event (an offshore technology conference), which lasted about one week and required the full capacity of all systems on a 24-hour-per-day basis.

Figure 18-6 shows the total gas used in the Astrodome complex in thousands of MCF, comparing similar months on a year-to-year basis.

In January and February of 1976 and 1977, gas usage was stabilized at an average of about 21,000 MCF (594.3×10^3 m^3). In March of 1977 the gas usage fell off to about 13,000 MCF (367.9×10^3 m^3). This was the beginning of the energy reduction program; the boilers and turbines were shut down.

From that time forward, to date, the monthly usage of gas has diminished. By 1978, monthly gas usage averaged less than 4,000 MCF (113.2×10^3 m^3), mainly attributable to service water heating, showers, dishwashers, lavatories, etc.

The full impact of these reductions can only be realized by a look at the figures with the month-to-month rate increases applied to the actual usage.

Instead of a detailed tabulation of these month-to-month figures and their cumulative effect upon the economics of operation, a summary of the results may convey the same information in simpler form.

Table 18-1 shows a summary of the two basic fuels used in the Astrodome complex, prepared from monthly billings before and after the retrofit. The year 1976 was a typical year, conforming to the system operation that had been established in prior years. Figures to the right of the double line represent the 20-month period following the retrofitting program. This period started early in 1977 and continued 8 months into 1978.

Table 18-1 reveals several interesting things:

1. In reversing the operation from base load gas with peak load electric, to base load electric with peak-loading gas, an appreciable increase in electrical energy was anticipated. However, this did not take place. There was an average reduction, due to fewer large A/C fans in operation and elimination of boiler auxiliaries such as forced draft fans, boiler feed pumps, cooling tower fans, sump pumps, etc. The savings from this reduction were nearly canceled out by rate increases averaging nearly $15,000 per month.

2. In the 20-month period following the retrofit there was an appreciable increase in the size and number of revenue-producing events. These have increased not only the kilowatt-hours used, but also established new demand charges that affect the rates over a 13-month period. As a result, it may be said that there has been very little monetary savings in electricity, although the system is operating with better efficiency with the increased load factors and power factor. The appreciable savings that might have been shown, had the event usage been stable, have no doubt shown up in increased profits.

3. The real savings, which are rather spectacular, show up in the reduction of gas usage by reason of eliminating the dewpoint control along with the reheat feature. These, in turn, permitted the shift from gas to electricity for maintaining the stadium between wide limits in the non-event mode.

The reduction in gas usage along with the percentage rate increases and taxes applied thereto have shown savings that are best called "payments avoided," which are determined by taking the actual differences in MCF between the corresponding months in 1976 (base year) and the month in 1978 and applying the overall 1978 rate to that difference. This incorporates the effect of rate increases, taxes, etc., into the economic picture and becomes the dol-

Table 18.1. Power and Gas Summary, Astrodome Complex, Houston, Texas.

	BASE YEAR 1976, JAN. RATES	AFTER RETROFIT 1977	8-MONTHS JAN.-AUG. 1978	20 MONTHS, TOTALS
Electric Power Usage				
Total power used (kWh × 1000)	33,240	33,364	24,324	57,688
Electric energy cost $	717,009	699,710	507,312	1,207,022
Rate increases and taxes	31,786	138,532	152,580	291,112
Total power cost $	748,795	838,242	659,892	1,498,134
Average monthly cost $	62,399	69,853	82,486	74,906
Energy saved (net kWh)	—	876,000	(759,000)	117,000
*Payments avoided $	—	22,057	(15,744)	6,313
Natural Gas Usage				
Total Gas Usage (MCF)	262,334	112,259	31,368	143,627
Gas energy cost $	510,827	217,848	61,700	279,548
Rate increases and taxes	86,345	78,639	30,565	109,204
Total gas cost $	597,172	296,487	92,265	388,752
Average monthly cost $	49,764	24,707	11,533	19,437***
Energy saved (MCF)	—	150,077	142,245	292,322
*Payments avoided	—	411,161	422,154	833,315
Combined Gas and Electric Energy Costs				
Gas cost	510,827	217,848	61,700	279,548
Electric cost	717,009	699,710	507,312	1,207,022
Total energy costs	1,227,836	917,558	569,012	1,486,570
Rate increases and taxes	118,131	217,171	183,145	400,316
Total cost	1,345,967	1,134,729	752,157	1,886,886
Average monthly cost	112,164	94,560	94,019	94,344
*Payments avoided $: Gas	—	411,161	422,154	833,315
Elec.	—	22,057	(15,744)	6,313
		$433,218	$406,410	$839,628

*Payments avoided are the dollars that would have been spent had the 1976 energy usage been continued through the subsequent months without retrofit.
***Note savings of $30,327 per month for 20 months = $606,555.
Source: Monthly billings, data courtesy of Houston Sports Association, Inc., Houston, Texas.

lars that would have been spent had the stadium continued to operate on the 1976 usage pattern without retrofit.

Actually, the increased usage of the facilities in the 20-month period would have caused even more expenditures than are shown.

The energy conservative program at the Astrodome seems to be a never-ending opportunity for saving money. Unfortunately, retrofitting costs money, which must be capitalized and amortized over a relatively few years. Cash flow considerations and mounting expenses in other areas put a damper on how much can be accomplished.

IV. OTHER PROJECTS IN PROGRESS OR COMPLETED AT THE ASTRODOME

A new field lighting system using metal halide fixtures has been installed at a cost of about $175,000. It doubled the intensities on the field with half the number of fixtures, with expected payout in about two years.

A new concourse lighting system using high pressure sodium fixtures indicates a payout in less than three years, but at a cost of about $1,000,000.

Extensions to the new electronic control system will show an energy savings of about $150,000 per year, but at a cost of about half a million dollars.

These and other equally interesting projects are in progress or being designed for future application. But unless there is some tax help or other incentive that removes the entire burden from the owners and operators, there will be slow improvement in the energy situation.

19. Heat Reclaim System Project for a Chemical Laboratory

Fritz A. Traugott

Robson & Woese, Inc.
Syracuse, New York

		Page
I.	Basic Data	315
II.	Initiation of Energy Conservation Program	315
III.	Construction Costs	319
IV.	Energy Savings by Use of Actual Meters	319
V.	Economic Review of Construction Cost versus Energy Savings	326

I. BASIC DATA

The biochemistry building, located at the University of Rochester River Campus, was occupied in 1972 and provided with an air conditioning system for various occupancy and environmental usages (Figure 19-1). The center core of the building is occupied by laboratories and is five stories high. There are 17 building bays, approximately 40 ft × 19 ft (12.19 m × 5.79 m), on each side of a central pipe space. The air distribution for this building is provided by 17 air-handling units located in a penthouse, providing vertical air distribution through the pipe space into two reheat boxes per floor and bay. The original design made air available through laboratory hoods in each bay on the first four floors and through laboratory hoods in each bay on the top floor (Figure 19-2). Individual exhaust fans were installed in the penthouse for all the aforementioned hoods, totaling 425 (Figure 19-3).

Environmental criteria dictated that the laboratory space be kept under negative pressure in relation to the corridor. The exhaust capacity for each fan was designed to be 1,390 CFM (656.08 l/s). Exhaust air, then, equaled 425 × 1,390 CFM (425 × 656.08 l/s) or 590,750 CFM (278,834 l/s). One hundred percent of the outdoor air for the entire center core of this building was required, encompassing some 500,000 CFM (236,000 l/s), or approximately 3 CFM per sq ft (15.24 l/s per m^2).

From the above information, it was obvious that a considerable amount of heating was expended to maintain proper indoor environment conditions when ambient conditions of 93°F (33.89°C) during the summer and −5°F (−20.56°C) during the winter needed to be considered. A supply temperature of 55°F (12.78°C) was selected for year-round use. This air was then reheated by the individual reheat boxes before delivery to the laboratories.

II. INITIATION OF ENERGY CONSERVATION PROGRAM

The utility invoices in 1975 amounted to one million dollars for electric power and $350,000 for central steam supplied by the university's own plant. With these inflated bills, it was obvious that a study was needed to investigate

316 6/RETROFITTING FOR INDUSTRY

Figure 19-1. Typical laboratory (supply and exhaust). (All figures in this chapter courtesy of the University of Rochester, Rochester, New York.)

the ways in which energy could be saved in this building. Several ideas were considered and put into practice:

1. Rebalance supply air quantities in spaces, consistent with calculated loads, to reduce preheat requirements.
2. Provide a means to manually shut off exhaust fans from hoods that are not being used and, at the same time, reduce the supply air quantity to match the requirements of that space.
3. Reclaim energy from the exhaust air and transfer it to the supply system (Figure 19-4).

To prove the effectiveness of the energy conservation study, we instituted the changes on three of the existing systems and metered both the modified systems and the nonmodified system to compare operating costs.

To permit the described procedure, the spaces conditioned by these three selected systems were surveyed to discover which could be effectively shut off. Provision was made to do this and to reduce the air flow to these spaces. The cooling load calculations for the spaces were reselected for the lower air quantities, and the reheat boxes were adjusted downward for the new capacities.

A coil run-around system (Figure 19-5) was designed and implemented with a coil installed in the outdoor air intake for each of three selected air-handling units. This coil was installed in front of the existing preheat coil and behind the filters. The existing exhaust fans for each hood of space were stacked three high in the penthouse, with individual outlet stacks projecting through the roof in rows of three on

Figure 19-2. Typical fume exhaust fan system.

Figure 19-3. Typical exhaust ports on roof without energy recovery.

Figure 19-4. Typical exhaust on roof with energy recovery units.

each side of the central pipe space. Because of space limitations in the penthouse, the exhaust fan discharges needed to be discharged through the roof via an exhaust reclaim coil. A unique design combining the three exhaust fan outlets discharging through a single coil with appropriate baffles to separate the air flows was prepared. These coils were installed above the roof at a 45° angle, with the air flow making a 90° offset before being discharged vertically (Figures 19-6 and 19-7). A glycol solution, 40% by volume, was used in the piping distribution system to connect with each coil. Appropriate meters were installed to measure the heat removed by the chilled water coils, the heat added to the preheat coils, and heat transferred in the reclaim system (Figure 19-8). Electric meters were installed to measure power usage of the hot water circulating pumps, the supply air fans, and the reclaim heat circulation pump. The power usage of the exhaust fans was calculated from one-time

Figure 19-5. Typical energy recovery unit piping on roof.

HEAT RECLAIM SYSTEM PROJECT FOR A CHEMICAL LABORATORY

meter readings. (For details on the coil system, see Figures 19-9, 19-10 and 19-11.)

III. CONSTRUCTION COSTS

The contractor making the system construction changes determined that the cost for energy conservation amounted to $52,000 per air-handling system. Of this, $1,477 was allotted for system upgrading maintenance items and air side balancing.

IV. ENERGY SAVINGS BY USE OF ACTUAL METERS

The energy usage was metered for a 2½-month period from November 11, 1976 to January 31, 1977. The net savings per system was calculated to be $7,787. This was based on electrical costs of $.025/kWh and steam costs of $3.92/1,000 lb ($8.63/1,000 kg) of steam. These costs were the actual costs paid by the university for the 1975–76 period. The number of degree-days for this period was 3,371, which

Figure 19-6. Singular energy recovery unit detail.

Figure 19-7. Exhaust fume connection to energy recovery unit.

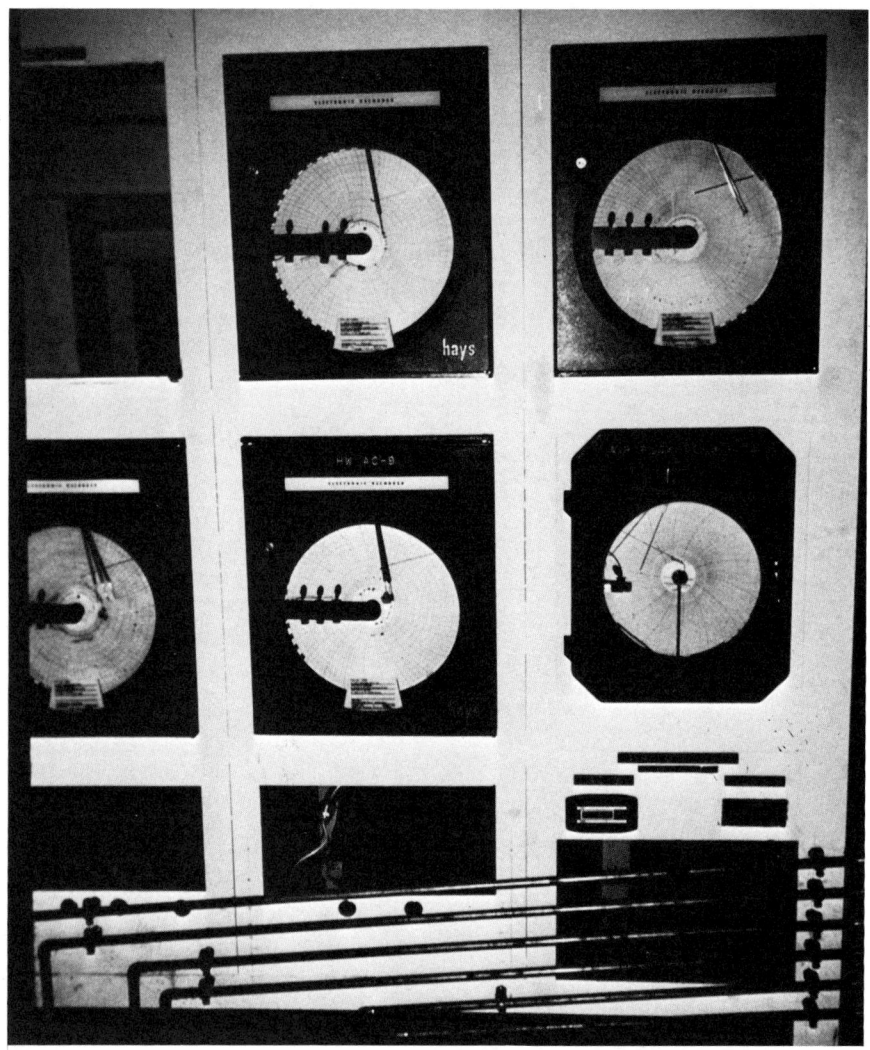

Figure 19-8. Test instrumentation control board.

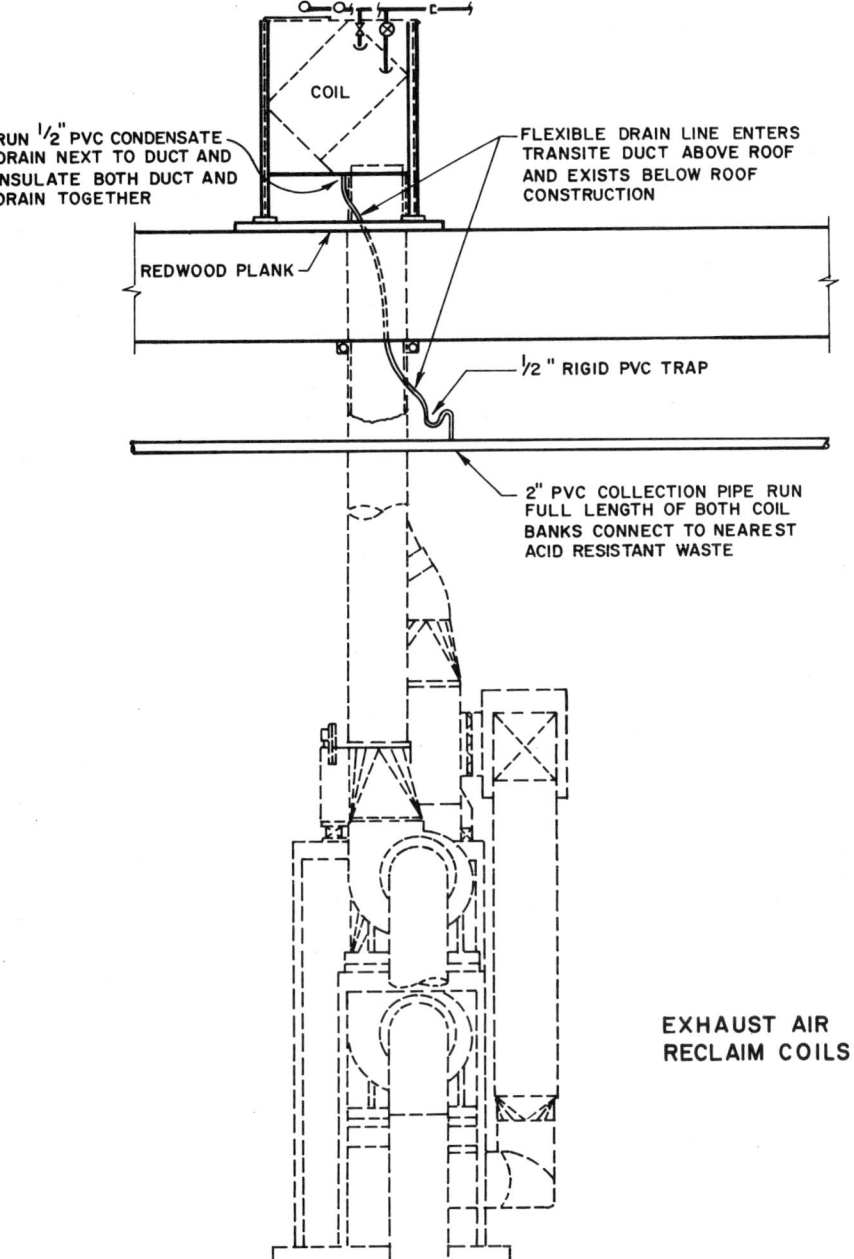

Figure 19-9. Exhaust air reclaim coils, side view.

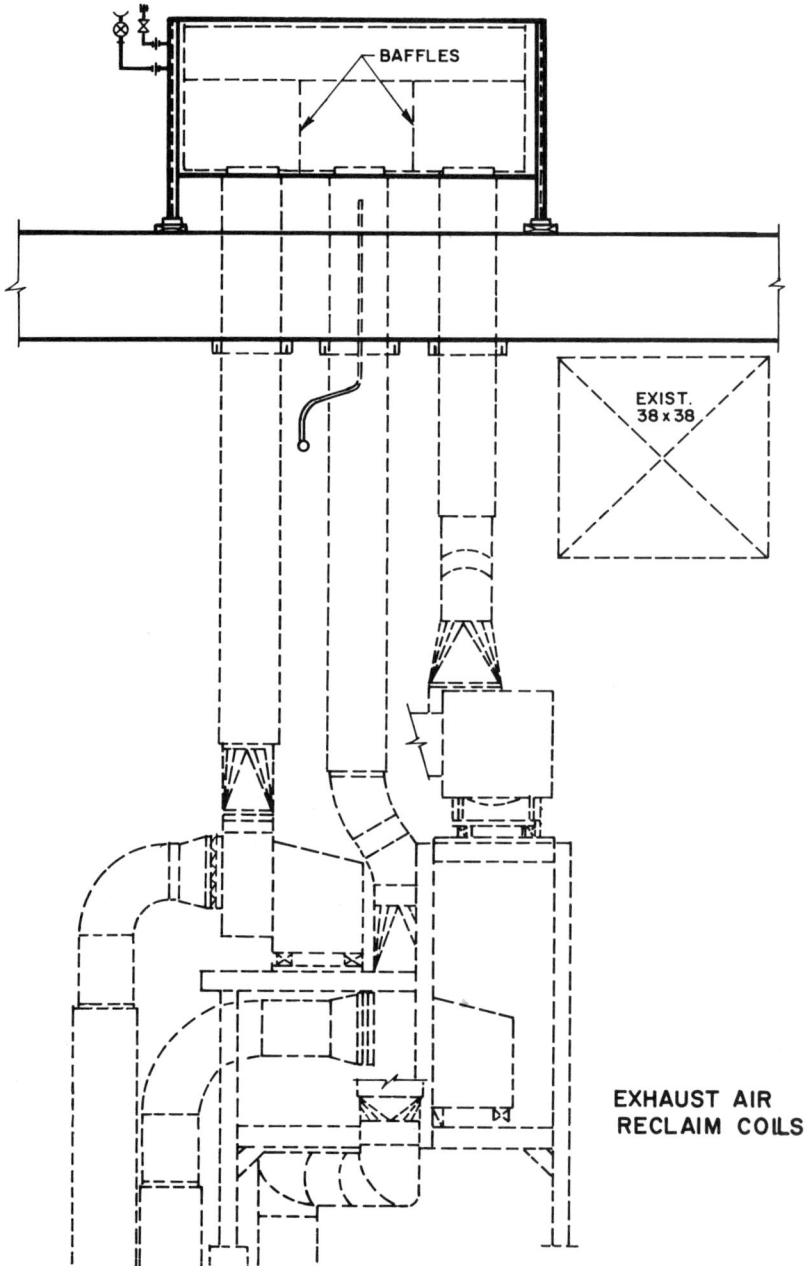

Figure 19-10. Exhaust air reclaim coils, front view.

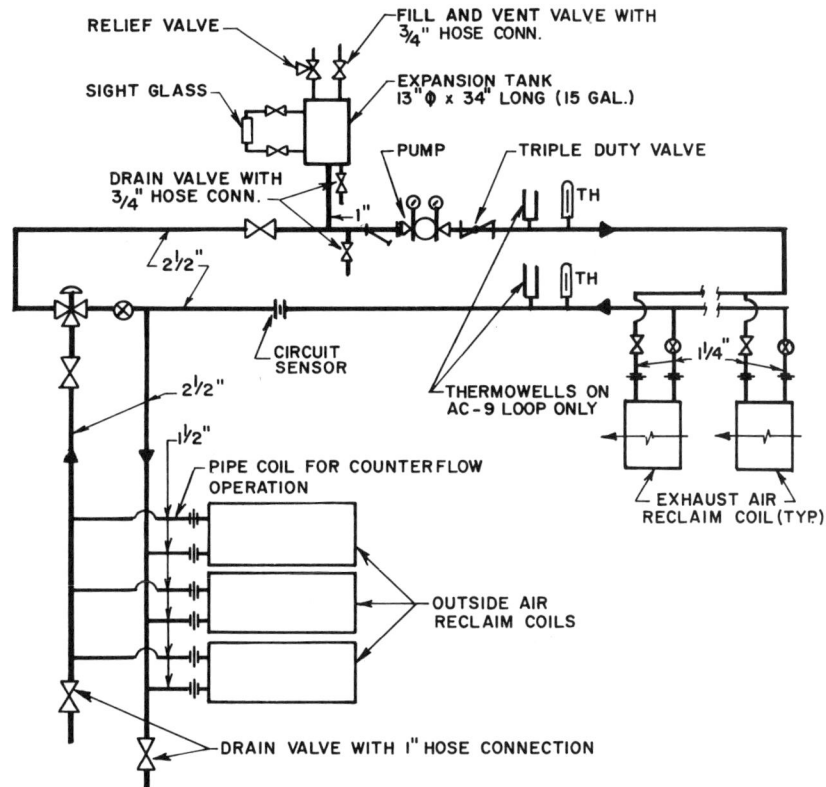

Figure 19-11. Heat transfer system piping schematic, no scale.

Table 19.1. Energy Calculation Sheet.

MONITORED ITEM	RUN #1 METER READINGS	RUN #2 METER READINGS	DIFFERENCE	METER MULTIPLIER	OTHER MULTIPLIER	QUANTITY SAVED	ENERGY RATE	COST SAVINGS LOSS	REMARKS
Date of reading	11-11-76	1-31-77							
Time of reading	8:30	10:30	1,946 hr						
Outdoor air temperature	32	13							
C.W. Btu's AC11 & AC12	—	—	—	15,000	—	—	—	—	
C.W. Btu's AC8 & AC9	—	—	—	15,000	—	—	—	—	
4 − 5 difference						—	—	—	Not applicable
H.W. Btu's AC12 (kJ)	039544	278797	239253	10,000	—	$2,392 \times 10^6$	2.52×10^9		
H.W. Btu's AC9 (kJ)	170471	190427	19956	6,000	—	120×10^6			$8,910 @ $3.72/10^6$ kJ
7 − 8 difference						$2,273 \times 10^6$	3.46/10.6 Btu	$ +7865	$3.92/10^6$ Btu
Reclaim Btu's AC9	NA	NA	—	3,000	—	—			
Supply fan kWh AC12	6163	9409	3246	10	—	32,460			
Supply fan kWh AC9	4886	9155	4269	10	—	42,690			
11 − 12 difference						— 10,230	.025/kWh	$ − 256	

H.W. pump kWh AC12	0488	1429	941	10	—		9,410	
H.W. pump kWh AC9	0268	0991	723	10	—		7,230	
14 − 15 difference							2,180	
Reclaim Pump kWh AC9	0299	0805	506	10	—		5,060	
16 − 17 difference						.025/kWh	2,880	$ − 72
Supply air avg. CFM (l/s) AC12	398,297 843,850	915,667 1,939,974	517,370 1,096,125	2,400	1/2c/× 60		22,530 (10,634)	
Supply air avg. CFM (l/s) AC9	199,456 422,577	599,700 1,270,552	400,244 847,975	2,400	1/2c/× 60		17,430 (8,227.0)	
19 − 20 diff. (CFM)							5,100	
Avg. sup. air temp. AC9 & 12	50°F 10°C/ 58°F 14.4°C	50°F 10°C/ 58°F 14.4°C						
Reheat savings (Btu) (kJ)						$3.46/10⁶/ $3.28/10⁶ kJ	— 164.4 × 10⁶	−569	−$644 @ $3.72/10⁶ kJ 3.92/10/6 Btu
Exhaust fan kWh AC9	Calculated		15.6 kW		1946		30,358	
Exhaust fan kWh AC12	Calculated		12.5 kW		1946		24,325	
24 − 25 difference						$.025/20⁶	— 6,033	$ −15i	131
Net savings for 1,946 hr							$ +681T	One system only

Source: Courtesy of the University of Rochester, Rochester, New York.

resulted in a savings of $2.31 per degree-day. The average heating degree-days for Rochester is 6,255, which equated to an annual savings of $14,450 per unit. The estimated cooling savings was $2,200 per year, resulting in a total annual heating/cooling operating cost savings of $16,650 per unit. (See Table 19-1 for specific details on savings.)

V. ECONOMIC REVIEW OF CONSTRUCTION COST VERSUS ENERGY SAVINGS

Using the projected annual savings of $16,650 per unit and the previously indicated construction cost of $52,000 per unit, the capital cost recovery can be obtained in less than four years, using an 8% interest rate. It is obvious that if we use today's higher energy costs, the capital recovery costs period will go down to probably between one and two years. Therefore, projects of this type become very lucrative in cost management savings. This type of retrofit permits capital recovery of first costs at a much greater rate than could be obtained by improvement of the building envelope and related techniques. This does not imply that the building envelope should not be improved; it simply points out that system retrofit modifications in general have a shorter payback than other related construction energy savings.

Part 7

Solar Retrofitting Potentials

20. Engineering Solar Retrofit Projects

Frederick H. Kohloss

*Frederick H. Kohloss and Associates, Inc.
Honolulu, Hawaii*

		Page
I.	Use of Solar Energy in Projects	329
II.	Preliminary Assessment	329
III.	Selection of Solar Systems	330
IV.	Solar Collectors	330
V.	Collector Location and Orientation	331
VI.	Other Considerations	333
	A. Pipe and Fittings	333
	B. Valves	334
	C. Water Circulation and Pumps	334
	D. Storage	338
	E. Insulation	338
	F. Controls and Instruments	338
	References	340

I. USE OF SOLAR ENERGY IN PROJECTS

Retrofit projects involving solar energy fall into two general building classifications: (1) applying solar energy to an existing building's use, or (2) modifying an existing building, then applying solar energy to its new configuration and use.

The classifications of *active* and *passive* solar systems are useful. Active systems use solar energy in mechanical systems for heating and cooling, while passive systems admit, store, or repel solar heat by natural means without mechanical systems.

Nearly all solar retrofit installations will benefit from some passive techniques. Most frequently, economics will limit the amount of investment in passive techniques, and augmentation by active techniques is then the preferred approach.

In engineering a retrofit project, the design engineer must keep the desired result foremost in his thinking. He must know what capital cost limitations exist, for even the most attractive design in life cycle cost must be initially affordable. It is easy to be carried away by enthusiasm for energy saving, to the extent that incremental means of energy saving included in a project may not be truly economical.

II. PRELIMINARY ASSESSMENT

Solar energy use in active systems may include the direct conversion to electricity by photovoltaic cells; the collection of heat to be used for space heating, industrial process heating, domestic water heating, or swimming pool heating; the collection of heat for thermal power cycle or absorption refrigeration systems; or the collection of heat to distill water.

In active solar systems now, photovoltaic systems or solar stills are economic only for a relatively small load remote from a conventional energy source. For thermal power or absorption refrigeration, there must be unusually favorable circumstances. For space heating, solar energy is sometimes economic, and for domestic water and pool heating, it is quite often economic. Often solar systems become economic when supplementing a source of waste heat.

Early in the project's concept stage, the design engineer must define the areas of solar energy application that appear to offer attractive economic prospects, and then proceed to study them in concert with other design disciplines.

Solar energy application is still somewhat in its "chicken or egg" period. There is a lack of a proven market for many products that are useful in solar projects, but that cannot be economic unless produced in quantity, with good quality control. The lack of market hampers development. Fortunately, this problem is being addressed by manufacturers, design professionals, and promulgators of codes and standards.

Retrofit solar projects can, of course, use many standardized components, but offer a challenge to the engineer's ingenuity in coping with problems of installation space, access for maintenance, structural support, and appearance.

"Staging" solar work can be considered. For example, if reroofing is currently needed, and other construction is contemplated for the near future, roof penetrations and anchor bolts for solar collectors should be provided during reroofing to save cost later.

III. SELECTION OF SOLAR SYSTEMS

Basic to the selection of any solar system is the location of the sun relative to the building, diurnally and annually. Legal uncertainties exist as to solar rights. Some state laws and municipal ordinances address the problem. If insolation can be affected by future neighboring construction, an easement for solar rights might best be negotiated before the solar system is built, in the absence of clear legal solar rights.

Having confirmed when and where insolation is available, conceptual design can proceed with establishing the scope of the active solar system. In the course of this determination, prospective passive measures evolve, and decisions can be made relative to their inclusion.

IV. SOLAR COLLECTORS

The type of collector to be used depends upon the required temperature range of the solar fluid to be heated. Use of air as the collector fluid avoids water freeze-up problems, but may require higher power consumption and difficult flow balance, particularly in more extensive systems. Liquids have been more frequently used.

The typical collector efficiency plotted versus the collector parameter $\Delta T/I$ (fluid inlet temperature in collector minus ambient temperature, divided by insolation) is shown in Figure 20-1. Efficiency of collection of solar heat is highest at low values of $\Delta T/I$.

For warming of swimming pools or other low temperature service (70 to 100°F, 21 to 38°C), unglazed collectors of plastic or metal are usually most cost-effective. For service hot water or space heating hot water (100 to 200°F, 38 to 93°C), the most usual collector type is single-glazed or double-glazed metallic flat-plate. Where space cooling is to be achieved (180 to 250°F, 82 to 121°C), evacuated tubular collectors, concentrating collectors, sun-tracking collectors, and high-efficiency flat-plate collectors have all been used. For thermal power cycles, fluid temperatures up to 600°F (316°C) are desirable, for which concentrating, tracking-type collectors are needed. Figure 20-2 illustrates the typical temperature range for various applications, and relative efficiencies of various types of collectors at these temperatures. Evacuated tube-type collectors are relatively unaffected by ambient temperature, and can collect diffuse radiation as well as direct radiation. Concentrating collectors must "see" the solar disc.

The high cost of solar collectors requires that they be carefully selected and located to effect economic application.

While it is possible to use collectors mounted as an integral part of a structure, in

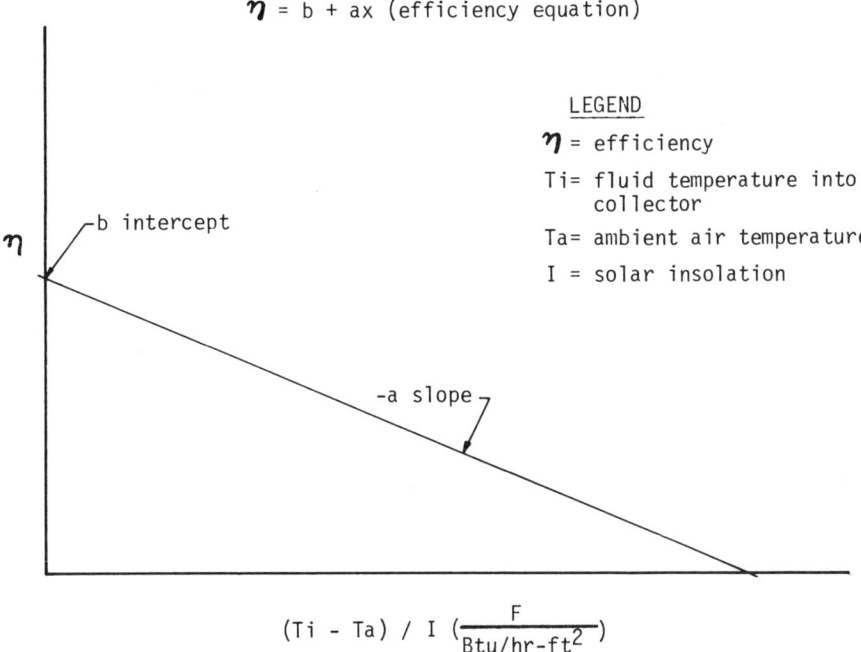

Figure 20-1. Solar collector performance.

most solar retrofit projects the collectors are mounted on racks or other supports. This permits more ready access for collector or pipe connection repair or replacement, plus access to the structure for its maintenance.

Ideally, in either the Northern or the Southern Hemisphere, flat-plate collectors should face the equator (due south for Northern Hemisphere), tilted at an angle near their latitude. For space heating, the tilt should consider the sun's lower winter path, and thus be 10° or 15° more than the latitude angle. For domestic water heating, and for cooling and heating, the tilt should equal the latitude. There is little loss in efficiency for a 10° variation up or down from optimum tilt, and a 20° variation from the due south orientation. Local climate, adjacent masking or reflecting surfaces, or existing structural configurations can all cause some deviation from the generalized "ideals."

Nearly all solar energy is collected between about 8:30 A.M. and 3:30 P.M. sun time. If adjacent rows of collectors must be fitted into available space, self-shading of back rows by front rows should be avoided between those hours. The back slope of flat-plate collector racks can be clad with a reflective surface to increase radiation falling on the collectors. This is effective, of course, only when the back slope is at an angle less than about 45° from horizontal.

V. COLLECTOR LOCATION AND ORIENTATION

Collector orientation deserves careful consideration. In retrofit installations, the available space, prospective angle of elevation, and prospective orientation all may be limited; so it is desirable to be able to visualize the effect of less-than-optimum situations on performance.

Year-round performance of solar collectors varies markedly with the seasons in higher latitudes, but exhibits far less seasonal variation nearer the equator. For example, clear-sky incident solar radiation on a horizontal plane for a locality with high values in summer, Grand Junction, Colorado, latitude 39° 07′ north, is 2,645 Btu (348 W/m^2) per day per sq ft for June, and 793 (104) for December [1]; while for Honolulu, Hawaii, latitude 21° 24′ north,

332 7 / SOLAR RETROFITTING POTENTIALS

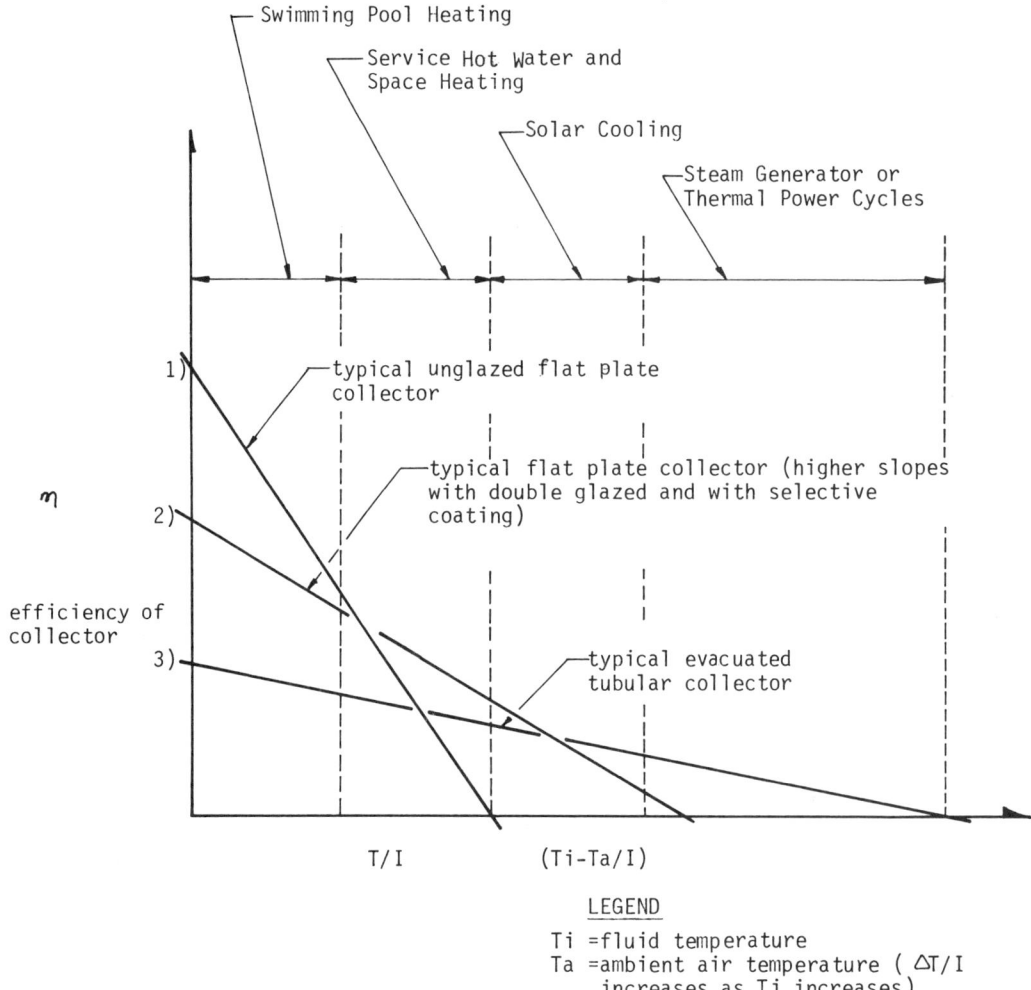

Figure 20-2. Effect of $\Delta T/I$ range of application (or, required temperature of heated fluid) on type of solar collector used.

it is 2,310 Btu (304 W/m^2) per day per sq ft for June, and 1,354 (178) for December [2]. Annual total insolation is 20,022 Btu per day per sq ft (2,633 W/m^2) for Grand Junction; 22,996 (2,932) for Honolulu. Thus, there is a great difference in availability of solar energy as well as in heating and cooling needs: Grand Junction has 5,641 heating degree-days a year and an average winter temperature of 39.3°F (4.1°C), while Honolulu has no heating degree-days and an average winter temperature of 74.2° F (23.4°C). Further, Grand Junction has 1,140 cooling degree-days compared to Honolulu's 4,221.

Since economic building use of solar energy usually involves domestic water heating and/or space heating, the drop in available insolation in cold weather is a factor in retarding wider use of active solar systems.

Should significantly less-costly solar cooling systems be developed, the annual variation in insolation fairly closely tracks cooling load requirements for all parts of the United States. Currently, solar cooling's economics are questionable, although a sharp rise in energy costs could change this picture.

With the sharp drop in available heat from the sun in northern winters, collector efficiency and collector orientation are very critical for space heating applications where heating load is highest when available solar energy is low. It is thus necessary to consider handling only a portion of the total heating task by solar means. A popular technique is the "F-chart" method of Klein, Beckman, and Duffie [3], which is well presented in the U.S. Department of Housing and Urban Development's 1977 Intermediate Minimum Property Standards Supplement, "Solar Heating and Domestic Hot Water Systems."

VI. OTHER CONSIDERATIONS

Other than the solar collectors, careful selection and application engineering is required for the remainder of liquid type solar heating systems components such as:

Pipe and fittings
Valves
Pumps
Tanks
Controls and instruments
Insulation

A. Pipe and Fittings

Pipe material should be compatible with the fluid. Some waters are soft, with a high tendency to corrode, often containing dissolved air whose oxygen accelerates possible corrosion. Some local water sources will produce hard scale on pipe interiors, and usually this is accentuated if water is heated above about 140°F (60°C). In most climates, the design engineer must be concerned with freezing of fluids in solar collector piping. Where this is a problem, practical choices are draining or, more often, use of a secondary heat transfer fluid in the collector piping, often a water–glycol solution, with a heat exchanger to a separate domestic water circuit. Many codes require double-walled heat exchangers, often with further special requirements for testing and certification. In the tropics, solar water heating systems will generally use water in the collector piping, without need of draining, since freezing temperatures will not be encountered.

Ferrous metal piping, even zinc-coated (galvanized), is not appropriate for most domestic water piping. Many plastic piping materials are not suitable for hot water. Piping for solar domestic water heating is subject to plumbing code requirements that restrict use of plastics. Suitability should be checked at building departments before planning the use of any particular plastic piping or tubing.

The most common material is copper water tube, which meets all codes, is fairly easy to join, and gives good service with hot domestic water. If copper piping is used, it is desirable that water passages in collectors, pump impellers, and valves be of copper, or at least of bronze. Aluminum or ferrous metal will corrode if connected to copper in the presence of moisture. Dielectric fittings are imperative where copper piping connects to dissimilar metal, but their use will not preclude all corrosion.

Hard copper tubing with brazed joints, or at the least 95/5 tin/antimony solder should be used. The technique of making good solder joints is simple, but shortcuts will not give good results. Tube should be cut square, burrs removed, tube end and fitting socket cleaned, and flux applied to both, the joint assembled and excess flux removed, and heat applied. Solder is then applied to allow it to fill the socket by capillary action; then the joint is allowed to cool.

Soft copper tubing may be used, with flare joints, not compression fittings; cut tube, remove burrs, slip coupling over end, flare end precisely in a flaring block, hand-assemble to fitting, and tighten without tube turning on flared fitting face. No soft-tube, flared-joint system is as good as a hard-tube, properly braced or hard-soldered system.

High temperatures can be encountered in solar water heating systems, particularly under stagnation (no-flow) conditions. Piping materials must be able to withstand high temperatures at operating pressures without leaks or excessive deformation. Provision for expansion and contraction is essential, as is proper support of piping.

Flexible connections and quick-disconnecting means are often desirable for collector piping systems. This is particularly true at connections to collector panels and other locations where items must be removed on occasion. Collectors on roofs should be installed so as to permit reroofing or repair of roofing. By swaging the pipe ends, synthetic polymer sleeves can be clamped to connect two pipe ends to permit easy disconnecting. The material of the sleeve should be tested and rated to withstand the 350 to 400°F (177 to 204°C) temperatures possible in no-flow conditions in the sunlight. The clamps should be of corrosion-resisting metal. Wherever flexible sleeve-type joints are used, they should be located where they can be visually inspected for leaks, and where leaks will not cause damage. A strainer should be provided at an accessible location on the suction side of each pump.

B. Valves

No special requirements for valves are presented by solar water heating systems. Valves and piping specialties commonly used include gate valves, ball valves, butterfly valves, globe valves, plug valves, check valves, balancing cocks, gage cocks, balancing devices, air vents, and relief valves.

For shut-off, gate valves have been traditionally used. Ball valves are suitable, and, usually in larger sizes, butterfly valves may be used. For modulating flow to any rate from shut-off to full open, globe valves have been traditionally used. Ball valves and plug valves are also suitable for modulating flow. To adjust a flow rate initially, but not for frequent operation over a wide range of flow, balancing cocks may also be used. All of these valves are available in bronze in small sizes, and in iron bodies with bronze trim for larger sizes. All valves used should be pressure-rated for the service and should have ends for the proper type of joints.

Open piping systems, in particular, have a fairly large amount of air dissolved in the water, which will separate at no-flow and low-flow conditions or locations, particularly at high points and particularly if the water is heated. Piping should be planned to minimize pockets in which air can be trapped. Air venting should be provided. Automatic air vents may be used where discharge is not objectionable, but some engineers prefer to use manual air vents.

As water is heated, it expands. Provisions for expansion and contraction of all piping runs must be made. A temperature and pressure relief valve should be installed on a solar collector piping loop at the highest point, with discharge arranged to a safe location.

Low points in piping should be provided with drain valves or hose bibs to permit flushing out foreign material or sediment.

Outdoor air frequently is corrosive to ferrous metal. Valves and operating wheels or handles usually should be constructed of nonferrous materials if they are installed outdoors.

C. Water Circulation and Pumps

Circulation of water may be by gravity ("thermosyphon") or by pumping. The thermosyphon system is particularly useful for smaller systems, in nonfreezing climates. It requires no power input and avoids moving parts in the entire solar water heating system.

Since water becomes less dense when heated, hot water will rise and colder water will drop in a piping system. A half-century and more ago, many U.S. residences were heated by gravity hot-water space heating systems. Data developed by ASHRAE for that purpose are equally applicable to solar domestic water heating. See Figure 20-3 for the head developed by a temperature difference, and Figure 20-4 for a pipe sizing graph.

Since the solar-heated water from the collectors is the system's hottest water, the storage tank for hot water in a thermosyphon system must be at a higher elevation than the collectors. This will create the thermal circulation of the colder water down, hotter water up. The entire system is normally pressurized at supply water main pressure.

Manufacturers of collectors have published data on circulation rates for optimum performance. A typical figure is about 1.75 gallons per hour per sq ft of flat-plate collector aperture surface. Within a fairly wide range around this flow rate, there is little difference in typical collector performance. If it is assumed that about 250 Btu/hr per sq ft (788 W/m^2) are received by the water in the collec-

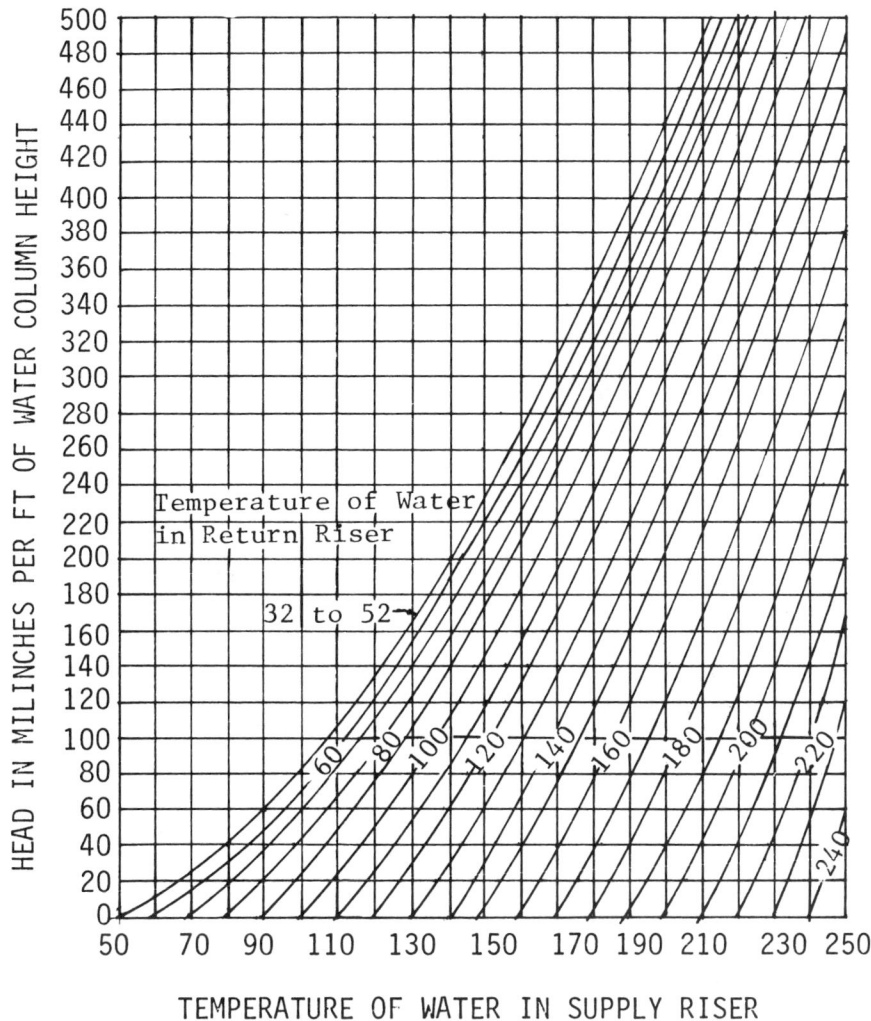

Figure 20-3. Heads from temperature difference. (Data from American Society of Heating, Refrigerating and Air-Conditioning Engineers, Inc., "Applications," in *ASHRAE Guide and Data Book,* New York, 1962, Chapter 8.)

tor of a solar domestic water heating system under favorable conditions at sunlight near midday, it can be seen that the expression:

$$\frac{250 \frac{\text{Btu}}{\text{h-ft}^2}}{1.75 \frac{\text{gal}}{\text{hr-ft}^2} \times 8.33 \frac{\text{lb}}{\text{gal}} \times 1 \frac{\text{Btu}}{\text{lb-F}}}$$

$$\left(\frac{0.788 \frac{\text{kJ}}{\text{s-m}}}{0.0198 \frac{1}{\text{s-m}^2} \times 0.9996 \frac{\text{kg}}{\text{l}} \times 4.19 \frac{\text{kJ}}{\text{kg-°C}}} \right)$$

gives a temperature rise of 17°F (9.5°C). Typically, 15 to 30°F (8.3 to 16.7°C) is the temperature rise in actual systems.

In Figure 20-3, if water in the return riser is 100°F (38°C) and in the supply riser is 74°F (23°C), there is developed a gravity head of 52 milinches (0.052 inch, 1.32 mm) for each foot of water column height. As the water is heated in the collector, available head increases. With supply at 174°F (79°C) and return at 200°F (93°C) the head has increased to 115 milinches (2.92 mm).

The total available head for circulation must be ascertained under design conditions, then,

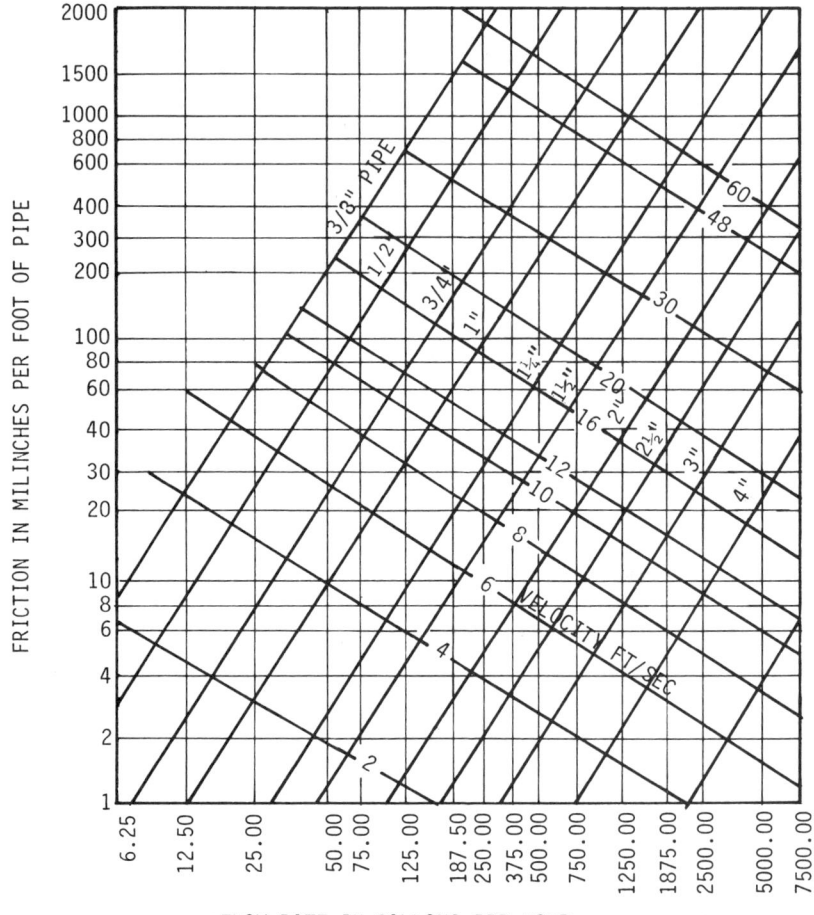

Figure 20-4. Graph for hot water pipe sizing. (Data from American Society of Heating, Refrigerating and Air-Conditioning Engineers, Inc., "Applications," in *ASHRAE Guide and Data Book,* New York, 1962, Chapter 8.)

and the friction in the collectors subtracted from it, and then the piping sized so the desired flow rate can be achieved with the remaining available head. Usually the measured length of piping times two, to allow for friction losses in valves and fittings, is used. This need only be run through once to get a feel for the sizing, after which rules of thumb can be used for sizing small systems, such as the data in Figure 20-5, from an Australian publication [4], *Solar Water Heaters,* CSIRO Bulletin No. 2, 1964. In Figure 20-5, the typical Australian solar water heater's tank is at atmospheric pressure, fed by float valve makeup from the cold water piping. The only hot water pressure available is that due to elevation of the storage tank above the plumbing fixtures using hot water. Note that in Figure 20-5 the effect of greater height between bottom of the tank and top of the collector is to increase the available head for circulation; so with greater height smaller pipes can be used.

Friction in copper tubing is slightly less than that in ferrous piping. For pressures encountered in solar domestic water and space heating, Type M copper is adequate, although thicker-wall Type L and even thicker Type K are still used. For a given nominal size, the outside diameter of copper tube is the same for all types, so that Type M has a larger internal cross section (with its thinner wall), and thus less friction than Types L or K. In extensive commercial systems with large-diameter pipe, the pressure rating of Type M tubing may not be adequate.

Thermosyphon systems will flow in reverse

COLLECTOR AREA, SQ FT	100			50		
↓ Y.FT X. FT ⟶	5	10	25	5	10	25
1	1.25	1.25	1.25	0.75	0.75	1.00
5	1.00	1.00	1.00	0.75	0.75	0.75
10	0.75	0.75	1.00	0.75	0.75	0.75

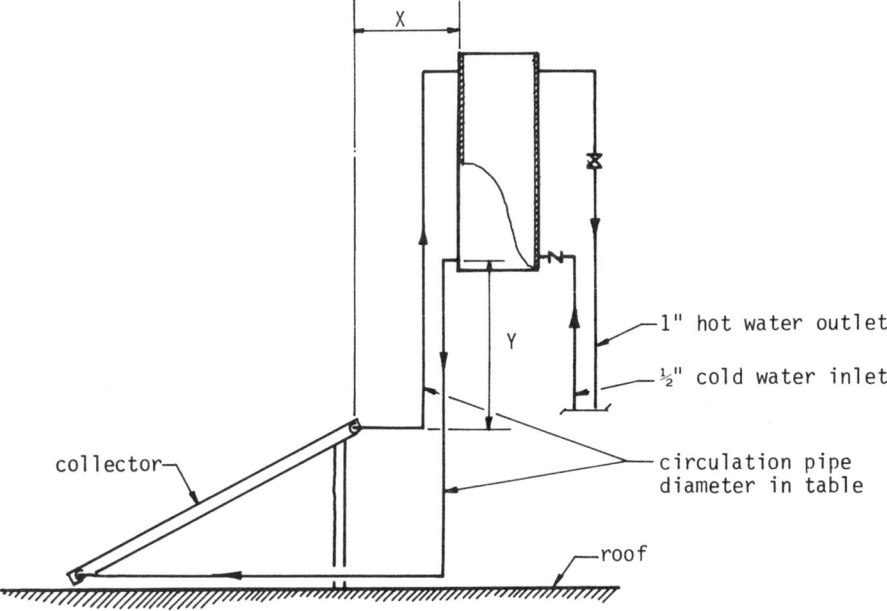

Figure 20-5. Thermosyphon. (Adapted from Ref. 4.)

if the water in the collectors is cooled by radiation and convection to the night sky. A check valve should be placed in the return line from the tank to the collectors to prevent reverse circulation.

Whenever the storage tank must be installed below the elevation of the collectors, a thermosyphon system cannot be used, and there must be a circulating pump. Because of the inconvenience of elevating a storage tank, most solar water systems are pumped. Where existing water heaters are used as part of a solar system for an existing building, they are usually on the ground floor, below the collectors. On new projects, the structural cost of elevating a tank is significant. Remember that a 100-gallon (379-l) tank has over 800 lb (363 kg) of water plus the weight of the tank.

In most parts of the United States, a solar domestic water heating system is considered similar to other means of water heating in that it is subject to plumbing code requirements. Thus a temperature and pressure relief valve is mandatory, the materials of the system are usually subject to review by plumbing code checkers or inspectors, and the system as a whole is subject to building code requirements.

In most cases the head-capacity characteristics of available pumps are not precisely suitable for desired collector flow rate. It is desirable to minimize the power consumed by circulating pumps, but, of even more importance, it is desirable to select a pump with probable long service life. Pump impellers and waterways should be all-bronze. Some small pumps have magnetic drive, eliminating the need for a shaft seal. For larger systems, pump selection should be made with more care to ensure optimum circulation for heat transfer while matching the system resistance of the piping, collectors, and other components. Most pump manufacturers do not manufacture their motors; so in selecting a pump, consider the manufacturer's reputation to be reasonably sure of proper matching of motor, drive, and pump. Sizing of the piping and selection of the pump follow conventional practice for hot water systems.

Consideration should be given to the temperature at which the hot water will be used. At times a properly designed and installed solar water heating system can deliver water too hot for safe delivery to fixtures, particularly for occupancies with children or the elderly. A mixing valve can blend cold water in, to supply water at reasonable temperature, which certainly need be no higher than 120°F (49°C) for residential use.

D. Storage

Several publications have suggested 1½ to 2 gallons (5.7 to 7.6 l) of water storage per square foot of collector as the desirable quantity for liquid solar systems, assuming that the solar flat-plate collector area is sized for collecting approximately 1½ to 2 average days' usage. The size of storage tanks and the area of collectors are interrelated by factors of insolation, ambient temperature, collector efficiency, temperature, and variability of usage, among others. Tank materials can be copper (often too expensive), copper-lined ferrous metal, glass-lined ferrous metal, or even concrete with an impervious liner not toxic to humans. Tanks may be baffled internally to assist stratification. Tank supports are very important, to avoid any strain on piping connections.

Water is a preferred thermal storage means, having high heat capacity, being readily available and inexpensive, and being easy to handle and move. Sometimes an antifreeze such as a glycol is added to the water.

Two other storage media are fairly common. Particularly in air systems, beds or boxes of pebbles or rocks are used; and latent-heat (phase-changing) storage media are now being manufactured. One phase-changing product manufactured in 6-ft (1.83-m) rods, 3½ inches (89 mm) in diameter, weighing 32 lb (14 kg), has 2,720 Btu (2,869 kJ) storage, comparable to 400 lb (182 kg) rock and 16.3 gallons (61.8 l) water, in a 70 to 90°F (21 to 32°C) temperature range. The phase-changing rod occupies only about 15% of the volume of the equivalent water, and 5% of the volume of the equivalent rock bed. It changes from solid to liquid at 81°F (27°C).

E. Insulation

Efficiency of a solar water heating system is affected by heat loss from storage, as well as heat loss from collectors and piping. No matter how much insulation is applied, there will be some heat loss through it, if the stored water is hotter than the surroundings.

Where storage tanks are located so they cannot create any smoke hazard to building occupants, foamed urethane type insulation can be used, which is very effective thermally. Fiber glass, polystyrene, foamed rubber-like plastic and other materials have all been used. If water under pressure is to be stored at temperatures above 250°F (121°C), it is desirable to use a layer of fiber glass insulation next to the tank surface; then more efficient foam insulation can be added as a second layer, where it cannot be damaged by contact with the higher-temperature pipe wall.

Economic thickness of insulation is a matter for design. It will depend on the temperature difference of the fluid and surroundings, and to a lesser extent on the diameter, in the case of piping. Smaller pipes require less insulation thickness than larger pipes for the same heat loss. Metal jackets are ideal but expensive. Frequently an all-service insulation jacket with thick mastic coating is used.

Whenever insulation can become wet, for example, in all exterior piping, it must have a weatherproof jacket.

F. Controls and Instruments

The small domestic type solar water heating system needs relatively simple controls, but larger systems may need additional instrumentation and possibly more controls. A thermosyphon system needs no controls, but it is de-

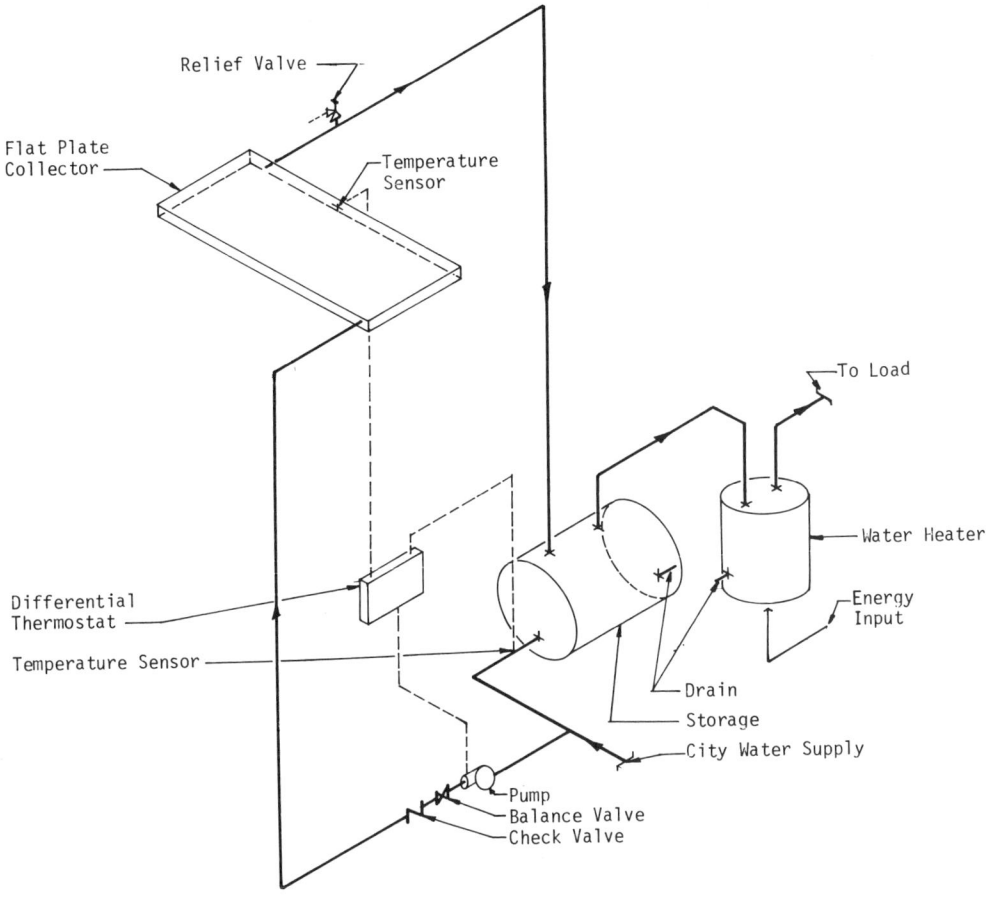

Figure 20-6. Typical application of controls and sensors.

sirable to provide thermometer wells to check performance.

Most frequently the primary system control for a pumped system is a differential temperature sensor (Figure 20-6). Whenever water temperature in the collector is sufficiently higher than stored water temperature to make pumping worthwhile, the controller will start the pump, usually at above a 20°F (11°C) difference. When the difference drops to about 5°F (3°C), the pump is stopped (see Figure 20-7). Some sensors have adjustable settings and differential. Similar controllers are available with proportional output, to vary pump speed with temperature difference.

Accessories are available in control packages including the following: high-limit, to turn off the pump if the storage tank temperature must be limited; low-temperature sensing, for activating drain systems in freezing climates; dual outputs, to control a second pump or other design feature; dual-thermistor sensor—one to control, the other for providing a signal for instrumentation, or for an auxiliary control.

Relatively inexpensive photovoltaic pyranometers to indicate rate of insolation are available. Microprocessors can be provided to receive pyranometer data, fluid temperatures, and flow rates, and continually process the system's output. They can be combined with printed output. These costs cannot be justified for smaller systems, but for large commercial systems they are sometimes desirable.

Automatic draining of water is feasible in mild winter locations where freezing is infrequent or of short duration. One control means for this is to provide solenoid valves, energized

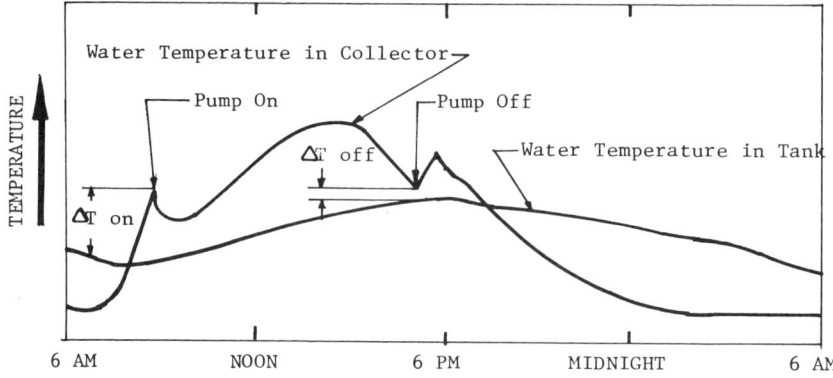

Figure 20-7. Differential thermostat control. (Adapted from "Solar Controls and Instrumentation," Rho-Sigma Corporation, Van Nuys, California.)

to keep the drain line closed and the water makeup line open, which will be de-energized by a drop in outdoor air temperature below the set point (usually about 37°F, 3°C) of a thermostat. Where freezing is a frequent problem or one of long duration, a non-freeze solution should be used in the collector piping, with an indoor heat exchanger to potable service water.

When the solar system's basic capacity and configuration have been determined and the system components have been selected, the solar system must be integrated with the conventional supplemental heating or heating-and-cooling systems, for proper operation and control.

Plans and specifications must be prepared that clearly set forth the design, for negotiations or bids, with appropriate installers. As in other building systems engineering, the ultimate success of a solar retrofit depends on clear plans and specifications, careful installation, and proper testing, balancing, and placing in operation. The designer should observe these phases, and assist the owner in achieving actual results commensurate with the design.

In the year ahead, solar retrofit projects will, no doubt, become increasingly feasible, particularly in the Sun Belt states. As energy costs rise, attitudes in commercial and industrial institutions will become more flexible, and new projects will be evaluated and initiated. Already government support has increased solar retrofit activity in many institutional settings. It is hoped that the expertise and cost efficiencies gained from such projects will spread throughout the economy, enabling solar retrofitting to assume a position of true significance in the overall energy picture.

REFERENCES

1. Robert C. Jordan and Benjamin Y. H. Liu, eds., *Applications of Solar Energy for Heating and Cooling of Buildings,* ASHRAE Publication GRP 170, New York, 1977.
2. Takeshi Yoshihara and Paul C. Ekern, *Assessment of the Potential of Solar Energy in Hawaii,* The Hawaii Natural Energy Institute, Honolulu, 1978.
3. William A. Beckman, Sanford A. Klein, and John A. Duffie, *Solar Heating Design by the F-Chart Method,* John Wiley & Sons, New York, 1977.
4. Commonwealth Scientific and Industrial Research Organization, *Solar Water Heaters,* CSIRO Bulletin No. 2, Melbourne, 1964.

21. Case Study of an Energy and Solar Heat Study

Billu Merraro

Vice President, The Meckler Group
Encino, California

	Page
I. Introduction	341
II. Space Heating System	342
III. Solar Energy	343
A. Weather Data and Location of Solar Panels	343
B. Types of Solar Panels	344
IV. Domestic Hot Water Solar Heating System	345
A. Standby Auxiliary Heater	345
B. Solar Storage Tank	345
V. Space Solar Heating System	345
VI. Combining Space Heating and Domestic Hot Water Solar Heating Systems	346
A. Solar Panel Installation	346
B. Solar Storage Tank	346
C. Auxiliary Standby Heater	346
VII. Life Cycle Cost Analysis	346
A. Electricity	347
B. Natural Gas	347
VIII. Life Cycle Costing	347
IX. Recommendations	349

I. INTRODUCTION

There are many types of hot water heating systems being marketed today, and more than a few of these systems feature some degree of integration with a solar system, usually located on the roof of the building using the hot water. More and more, such integrated systems are being put into place in a wide variety of residential, commercial and industrial applications. Yet despite the continued popularity of the idea of "free" solar heating, the realities are more complex; in order for such an integrated system to be truly viable, payback periods must be determined, current solar technology must be evaluated in terms of practicality, and the individual characteristics of the building, its location and its hot water requirements must be determined. Only when all of these factors have been taken into account can an integrated hot water system be weighed in terms of its actual viability.

The process of quantifying and evaluating these factors, then combining this data to arrive at an informed and correct decision is not simple, and it warrants a detailed, in-depth look. It is the purpose of this article to examine such a solar hot water heating project from its beginning to its conclusion, so that methods for evaluating integrated hot water heating systems will become more clear to the reader.

Our examination will center on the hot water heating system of a USDA laboratory located in Pasadena, California; it represents a typical institutional candidate for such an integrated system, and it is situated in a typical Southwestern setting, where solar heating is commonly feasible. As such it serves as an example for many similar structures in sunbelt locations.

As with any building being evaluated for an integrated solar hot water heating system, this laboratory had already in place an existing water heating system, and it is against this that any new systems or modifications have to be compared. The system in place at the beginning of this study featured a steam-to-water heat exchanger, with a 375-gallon (1,420 l.) capacity. Located in the basement, near the entrance to a tunnel that connects the building to the boiler pit, this exchanger used steam from the boiler to heat the domestic hot water. During the summer months, when space heating was not required, the boiler had to be kept in operation to satisfy continuing domestic hot water needs.

Information concerning this exchanger and other pertinent data was obtained, of course, during a site survey, as well as through interviews with USDA maintenance personnel. This survey determined that 14 hot water faucets were located on the premises, in the laboratories, shops, and in the pilot plant. One additional hot water faucet was being used extensively by the janitorial staff. Measurements were taken, which determined the total maximum hot water demand to be 20 gallons (76 l.c.l.) per hour. Interviews with staff personnel revealed, however, that this demand was unlikely to be reached with any frequency at the facility. The hot water was, in fact, being used mainly for washing hands or for the occasional washing of laboratory utensils.

Before proposing an integrated solar hot water heating system, it seemed appropriate to examine more conventional alternatives to the exchanger system then in place. To this end, the installation of a gas heater with a capacity of 40 gallons (150 l.) was investigated. Such a plan presumed the removal of the existing heat exchanger, which would mean that a net gain of usable space would be obtained for other purposes. And, of course, the installation of a conventional gas heater would allow the existing steam boiler to be shut down during the summer months, when there was no demand for space heating. This would reduce annual operating and maintenance costs.

II. SPACE HEATING SYSTEM

Further site investigations were undertaken to determine space heating requirements for the facility and to evaluate the system already in place. At the time of the survey, the only area heated during the winter was the northwest portion of the main laboratory building. Heating was accomplished via steam convectors located along the perimeter of the building; however, the unit heaters were disconnected and not functioning.

The total steam capacity of the existing convectors was measured at 155.5 MBH (164 MJ). The steam, at 30 psig (207 l.c.k.) was being supplied through a 2-inch (5 centimeter) main line, with a ¾ inch condensate return steam line. Plant personnel informed the survey staff that plans were in effect to relocate the existing main steam lines to the outside of the building, where both the steam and condensate lines were to run alongside the building walls.

At the time of the site survey, the existing convectors were being controlled automatically, through steam valves located at each convector.

Steam was supplied from a 125 psig (862 l.c.k.) steam boiler connected to the steam and condensate lines through an underground tunnel. The steam supplied by the boiler passed through a 30 psig (207 l.c.k.) pressure reducing station located in the basement of the main building.

Changing the existing steam space heating system to a system that heats space through circulating hot water in the convectors was briefly considered at this point, but it was soon determined that there would be no advantage to such a system except that it would be new and would therefore presumably require less maintenance. Certainly, energy consumption would not have been reduced significantly by such a system, so other alternatives were sought that would reduce gas consumption by a significant amount.

The existing steam converters had a capacity of 155.5 MBH (164 MJ), with a supply of 30 psig (207 l.c.k.) steam. However, further calculations revealed that the area being heated required only 83 MBH (87.6 MJ), assuming an indoor space temperature of 72° (22°C).

Using the existing converters with 150° (66°C) water would have only supplied 63 MBH (66.4 MJ) to the space. Therefore, it was determined that additional converters or other heating equipment were needed to maintain the desired space temperature.

After further inspection, it was decided that portions of the existing steam and condensate piping could be utilized in a conversion from steam heating to hot water heating. The existing main condensate return line would have to be increased, however, to a 1½-inch (4 cm) line in order to provide for adequate water flow in the hot water circuit. The existing 2-inch (5 cm) main steam line was determined to be capable to handling the required hot water flow to heat the space.

Provisions were made so that the water heating temperature could be readjusted, based on the outside temperature, to provide hot water with a temperature ranging from 140 to 180°F (60 to 82°C), in which case the existing convectors would be adequate to supply the 83 MBH (87.6 MJ) needed at the outside design temperature of 36°F (2°C).

III. SOLAR ENERGY

After the existing systems had been examined and evaluated, along with conventional solutions to the space and hot water heating problems of the facility, we began to study the potential advantages and drawbacks inherent in a solar system at this location.

It is important at this point to recall that the amount of solar energy that can be collected at any given building site is dependent upon a large number of external variables, such as ambient temperatures, wind temperatures and velocities, and cloud cover. Then there are the building-related variables such as the size, frequency, and variations of the building loads, and the amount of storage capacity required to retain the collected energy for periods when solar energy collection temporarily can not be attained. Finally, any solar heating system must be evaluated on its own features: collector efficiency, the tilt and orientation of the collector panels, whether reflectors are used, minimum system operating temperature and the like.

The most desirable loads for solar applications are those that remain relatively constant throughout the year and that peak during the daytime hours when peak solar energy collection is available. Any building used primarily during the daytime hours, in which hot water is needed for kitchen, restaurant, or especially for laundry applications, is ideal.

While such buildings are common, space heating applications are much more difficult to justify economically, since in most areas of the country the requirement for heating is seasonal, and generally the need for space heating is greatest during periods when solar collection efficiency is lowest.

Moreover, space heating is usually less desirable than domestic hot water heating because much of the space heating load occurs during the nighttime hours when peak solar input is not available; therefore, more storage is required for space heating than would be required for hot water heating. Also, the amount of heat needed for space heating is much larger than the amount required for domestic hot water heating.

A. Weather Data and Location of Solar Panels

In performing an evaluation of solar systems for a given locality, weather data from a nearby location must be incorporated. For the USDA laboratory at Pasadena, data from the Los Angeles Civic Center was used.

As a proposed location for solar panels, the roof of the laboratory building was chosen. This location proved the most satisfactory in terms of the following requirements:

1. Availability of adequate space for the solar panel installation.
2. No obstruction in the path of direct solar flux falling on the panels.
3. Minimization of the shadow-casting hours for the solar panels.
4. Short distance between solar panels and solar storage tank, thus reducing heat loss from the piping system.

B. Types of Solar Panels

A great and sometimes confusing variety of solar hardware exists on the market today. There are different types of solar hot plate panels. Solar panels can be flat-plate, single glazed, double glazed, or unglazed. The panels can be constructed with copper, plastic, or aluminum tubing for the waterways. The casings may be insulated or uninsulated, the solar collecting plates coated or uncoated. There are concentrating solar panels available, with or without provisions for sun tracking. And among those that do track the sun, there are collectors that track in two directions (east-west or north-south), and those that track in four directions, compensating for the changing seasons.

For each of these items or systems, there are sales claims, exaggerated and otherwise. Yet the simple truth remains that each item has its strengths and its shortcomings. The concentrating collectors are used mainly for collecting energy at higher temperatures of 200° to 600°F (93° to 316°C). The single glazed, flat-plate panels are capable of supplying hot water

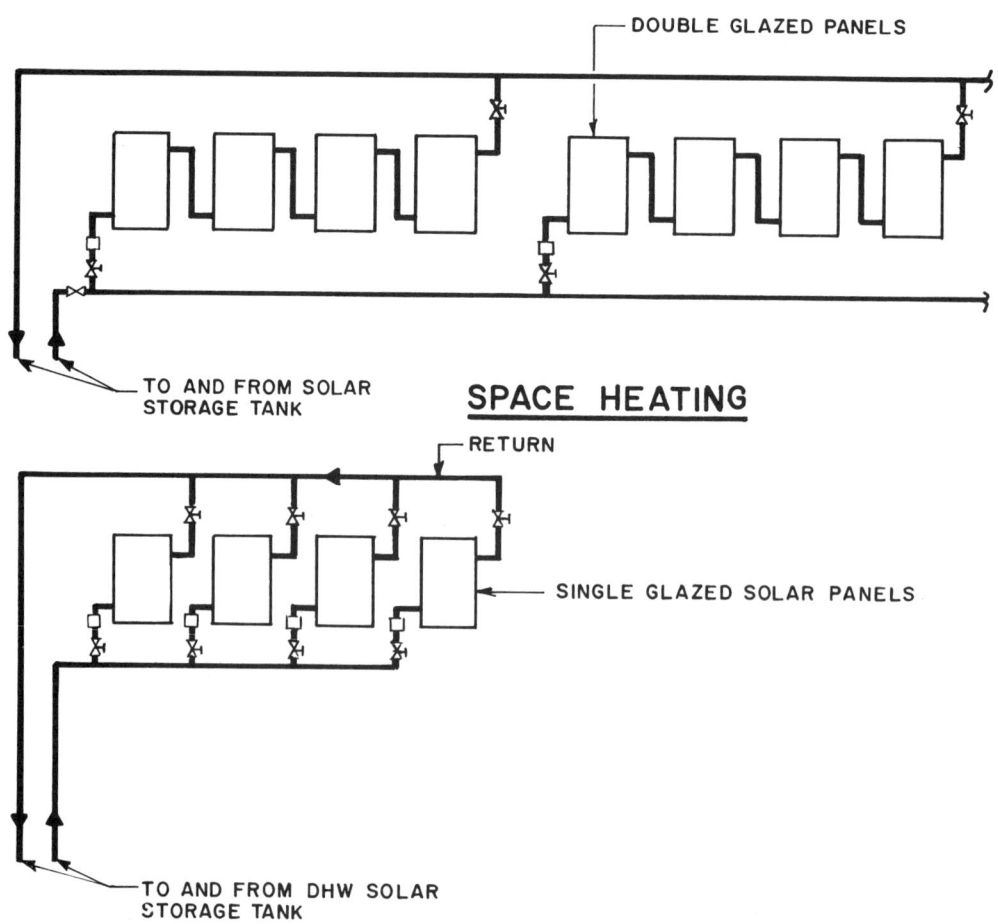

Fig. 21-1

at temperatures between 110° and 125°F (43° and 51°C); the unglazed panels are more economical in applications where the desired water temperature is 110°F (43°C) or lower. The double glazed panels are used for applications requiring water temperatures of 140° to 160°F (60° to 71°C). Furthermore, all metal panels have insulated casings. And although plastic panels are lower in price than the aluminum ones, they tend to become brittle after long exposure to the sun's rays and thus have a shorter life expectancy.

After a comprehensive evaluation of all major solar systems available, the single glazed aluminum fin copper tubing with insulated casing was deemed most appropriate, since the domestic hot water temperature required by the building was only 120° (49°C). The optimum tilt for the proposed solar panels was determined, based on location, to be 54°, measured from the horizon. A decision was made to install the panels facing due south. The tilt angle chosen maximizes the solar energy collection in the winter months, when the solar flux is the lowest during the year.

The next decision which had to be made concerned the number of solar panels that would be required to satisfy the domestic heating requirements. Taken into account were the average monthly insolation available in Pasadena, the average monthly cloud cover, and the efficiency of the single glazed solar collectors. With these factors in mind, it was determined that four collectors, would provide a total solar collection area of approximately 70 ft^2 (6.5 m^2). (For a more detailed look at the arrangement of the panels, please see Figure 21-1.) Provisions for draining the system at, or close to, freezing temperatures were to be incorporated into the system.

IV. DOMESTIC HOT WATER SOLAR HEATING SYSTEM

A. Standby Auxiliary Heater

It was determined at this point that, in addition to the solar system, an auxiliary system would be needed to ensure a domestic hot water supply throughout the year with the appropriate water temperatures. The auxiliary water system would be capable of maintaining the specified water temperature even if the solar system were shut off for any reason—such as maintenance—or if there was not enough sun to maintain the necessary temperature of the domestic hot water.

Also at this juncture, engineering personnel made provisions for a supporting structure for the solar panels. An investigation of the existing roof on the laboratory facility revealed that the roof was adequate for the load and would require no reinforcement to support the solar collectors and panels.

B. Solar Storage Tank

Next came the decision concerning the location of the solar storage tank associated with the domestic hot water system. Several possible locations were discussed, and it was concluded that the best place for the tank would be above ground near the building's west entrance, adjacent to the tunnel connecting the boiler pit with the building. Using measurements from the size of the solar panels as well as data concerning the building's hot water needs, the storage tank was sized; it would have a 120-gallon (454 l.) capacity. And, of course, it would be insulated to minimize heat loss to the environment. The proximity of the storage tank to the existing hot water distribution system would further reduce the overall heat loss from the entire system.

V. SPACE SOLAR HEATING SYSTEM

Once the initial options concerning the solar hot water heating system had been explored, the engineering staff turned its collective attention to the area of space heating; the feasibility of converting the existing steam space heating system to a solar hot water heating system was investigated. Six different alternatives were considered:

1. Convert the heating of the area heated by steam to heating via a 150° (66°C) hot water system using the existing boiler

as an auxiliary heater for the solar heating system or for standby.
2. Convert the heating of the area heated by steam to heating via a 150° (66°C) hot water system using an instantaneous electric auxiliary heater for the solar heating system or for standby.
3. Provide enough heating capacity to heat the entire building with the system described above in Number 1.
4. Provide enough heating capacity to heat the entire building with the system described above in Number 2.
5. Convert the heating of the area heated by steam to heating via a 150° (66°C) hot water system using a hot water boiler as an auxiliary heater for the solar heating system or for standby.
6. Provide enough heating capacity to heat the entire building with the system described above in Number 5.

An in-depth investigation was performed for alternative no. 2 at this point, because the engineering staff believed that this option had the best potential to yield the most favorable payback.

VI. COMBINING SPACE HEATING AND DOMESTIC HOT WATER SOLAR HEATING SYSTEMS

The next possibility that was examined concerned the actual combining of the existing steam space heating system with a solar hot water system that would supply all the heating needed for the heated area. Once again, a 72° (22°C) inside maintained temperature was assumed, along with a heating season lasting from October to April. It was determined that such a combined system would require:

1. Twenty-eight solar collector panels with copper waterways, double glazed and coated with metal insulated casing.
2. A 1,000-gallon (3,785 l.) solar storage tank and a separate electric auxiliary standby heater.
3. All the necessary piping, pump, and controls needed to operate the solar system.

A. Solar Panel Installation

The 28 solar panels, providing approximately 500 ft^2 (46.5 m^2) of collector area, were examined in detail at this juncture. It was decided that the panels would have the same arrangement as the panels for the domestic hot water heating solar panels and would be located on the roof of the main building as well.

B. Solar Storage Tank

The solar storage tank mentioned above would be located above or just below grade near the existing boiler room. It would be insulated to minimize heat loss into the surrounding environment.

C. Auxiliary Standby Heater

A standby heater would be needed for such a system to ensure that the required temperature for heating the space would be supplied at all times. Using the existing steam boiler for this purpose was quickly ruled out because of the following factors:

1. Steam boiler pressure would have to be maintained whenever the boiler was in operation, regardless of the demand or lack thereof.
2. The steam and condensate lines would have to be checked annually and all connections would have to be maintained in proper condition.
3. All steam traps and strainers would have to be serviced regularly.
4. The steam boiler would have to be approved annually by the insurance underwriter.

Because such factors would automatically and significantly raise maintenance and servicing costs, it was decided that a small gas-fired, instantaneous hot water heater might be used in lieu of the existing boiler.

VII. LIFE CYCLE COST ANALYSIS

The option of combining the hot water heating and space heating systems seemed on the surface to be a logical possibility. However, upon closer examination it became apparent that a new combined system would be both more complicated and more costly, for the following reasons:

1. The liquid circulating through the solar panels used for space heating would have to be enclosed in a separate system and could not be mixed with the water circulated through the convectors and/or the domestic hot water heating tank. This was due to state and county regulations regarding water treatment.
2. The heat exchanger used for the domestic hot water would have to be of double wall construction to ensure that the potable water would not become contaminated in the event of a rupture in one of the walls of the heat exchanger.
3. More sophisticated controls would be needed to divert the flow through the solar panel circuit and through the respective solar storage tanks, based on the temperature of the liquid in the solar panel circuit.

For the above reasons, the combining of the two systems was deemed economically unfeasible.

The next step in the cost analysis phase of this project entailed an analysis of the energy costs. The USDA laboratory uses two types of energy: electricity and natural gas.

A. Electricity

Like electrical bills in many parts of the United States, the monthly bill for the laboratory facility consists of a combination of two major items, the kilowatt demand and the kilowatt per hour consumption. The demand and the consumption are each calculated at a different rate; thus when establishing the total electrical cost, every month's bill has to be calculated separately. Once the yearly cost is known, an average cost per kilowatt hour can be obtained.

The laboratory facility pays for electrical energy according to two different rate schedules as well:

1. Pasadena Department of Water and Power's G-2 schedule for three-phase power.
2. Pasadena Department of Water and Power's G-1 schedule for single-phase electrical power.

A detailed analysis determined that the three-phase power cost an average of $0.026 per kilowatt hour, and the single-phase power cost an average of $0.045 per kilowatt hour. These costs were based on actual electrical bills for the facility during 1977 and 1978.

B. Natural Gas

The natural gas costs at the facility are charged as a fixed amount plus a direct straightforward rate; that is, the cost increases in direct proportion to the consumption (with the addition of $3.10 to the total bill as a monthly customer service charge). The cost used in our analysis was $0.00197 per ft^3 ($0.06956 shillings per m^2) or $0.197 per therm ($0.187 per MJ), assuming a high heating value of 1,000 Btu/ft^3 (37 MJ/m^2).

VIII. LIFE CYCLE COSTING

The various energy conservation alternatives examined during the course of this project were studied from the vantage point of four basic elements of cash flow:

1. Initial investment
2. Annual maintenance costs
3. Annual energy savings
4. Salvage value at the end of the useful lifetime (This was indeterminate and assumed to be zero in our study.)

In order to devise a mathematical formula for the various elements involved, the following values were assigned:

I = Estimated initial investment in equipment and installation

Table 21.1

NO.	ALTERNATIVE	INITIAL INVESTMENT	ENERGY SAVINGS ($/YR)	MAINT. SAVINGS ($/YR)	DEPRECIATION* ($/YR)	PAYBACK (YRS)
1.	D.H.W. solar heating system	$ 9,300	165	-0-	465	19.6
2.	Space solar heating system	$90,000	2,636	-0-	3,220	16
3.	D.H.W. Gas heater	$ 2,200	31	1,670	110	1.2

*Assume 20 years life expectance.

F = Estimated annual energy savings at present costs
f = Estimated annual energy cost escalation rate
d = Expected serviceable lifetime of the equipment
M = Estimated annual savings due to reduced maintenance costs, directly attributable to the implementation of the energy conservation alternatives
m = Estimated annual inflation rate of maintenance cost
i = Estimated investment discount rate (assumed to be zero as advised by USDA staff)
n = payback in years

Using these values, the formula which follows determined the Net Present Value:

Payback is calculated from the life cycle cost analysis expression, using a tabulation of values to find the correct value on "n" when the net present value is zero.

In our analysis, a 15% annual energy cost escalation rate (for both electricity and natural gas) was assumed, along with an 8% annual escalation for maintenance costs.

The proposed solar system would tend to increase the overall maintenance costs, but the elimination of the existing boiler would decrease these same costs. For the purposes of this study, therefore, we assumed no changes in the annual maintenance costs; thus there were no maintenance savings.

In Alternative 3 (discussed in Section 5.), we assumed an annual savings of $1,670 by shutting down the boiler for the summer months (June through September, inclusive). See Table 21-1. for further details.

As can also be seen in Table 21-1., the solar-augmented space heating system discussed in this article would pay back its cost throughout the life-time of the equipment. The solar heating domestic hot water system would pay back at the end of its life. However, solar space heating seemed not to be a particularly attractive investment at the time of this study because of prevailing energy costs and the small amount of energy saved.

The domestic gas water heater heating system has a 1.2-year payback period and thus is a very attractive option. Based on the first cost and payback in years, our engineering staff felt that Alternative 3 of Table 21-1. should be implemented. However, when we considered the greatest amount of energy saved as the pri-

Table 21.2

NO.	ALTERNATIVE	RATIO INITIAL INVESTMENT/ENERGY SAVINGS
1.	D.H.W. solar heating system	56.4
2.	Space solar heating system	34.14
3.	D.H.W. gas heater	71.0

mary goal of any proposed changes, then Alternative 2 seemed the most attractive. See Table 21-2. for alternative rankings based on the ratio of initial investment to energy savings.

IX. RECOMMENDATIONS

It was the recommendation of our engineering staff that the conversion of the existing domestic hot water heating system from heat exchanger to gas water heater be executed as soon as possible, in view of the very attractive payback and the relatively low estimated initial investment of $2,200.

The domestic hot water solar heating system, with an estimated investment of $9,300, would pay back its cost toward the end of its estimated economic life, but this would not eliminate the need for conventional domestic hot water heating. We estimated that about 80% of the energy needed to heat the domestic hot water would come from the solar system. In view of the results of the life cycle analysis and the small quantities of energy saved, we were unable to recommend this solar system on a purely economic basis.

The space solar heating system, with an estimated initial investment of $90,000 was determined to have a payback period of about 16 years. This did not seem to be a particularly attractive investment, and we did not recommend this alternative on its financial merits. However, implementation of this alternative would save 90% of the annual amount of energy used for space heating at this facility. Therefore, based on the amount of energy displaced, approximately 19,200 therms (20,256 MJ) per year could be saved. In light of possible natural gas curtailments in the Pasadena area, we felt that this alternative deserved serious consideration.

The methods discussed in this chapter are fairly typical, and the laboratory facility in question serves as a fine example of a normal structure which might seem an appropriate candidate for an integrated combination of solar space and hot water heating technologies. Yet, in the final analysis, such a combination proved unsuitable, although new, low-cost solar technologies may make similar projects more viable in the years to come.

22. A School Solar Retrofit Case Study*

Harry T. Gordon, AIA.

Burt Hill Kosar Rittelmann Associates
Principal, Washington, D.C. Office

	Page
I. Introduction	350
II. System Description	350
III. Preparation of Construction Documents	353
IV. Bidding	353
V. Additional Project Aspects	354
VI. Conclusions	354

I. INTRODUCTION

The Solar Energy Committee of the St. Charles High School District No. 303, Kane County, Illinois retained the firms of Unteed, Scaggs, Fritch and Nelson, Buchanan, Bellows & Associates, and Burt Hill Kosar Rittelmann Associates to respond to the Department of Energy Program Opportunity Notice DSE762. Work was done with the cooperation of the Illinois Development Board, the state agency responsible for the construction of elementary and secondary schools. A proposal was submitted in which cost sharing was provided by both the school district and the Development Board.

With the receipt of the grant from the Department of Energy, work was done in the preparation of contract documents during the latter portion of 1977 and early 1978. Bids were received in the spring of 1978, and the solar energy system was completed thereafter.

II. SYSTEM DESCRIPTION

Analysis conducted during the preparation of the proposal indicated that a solar energy system combined with a heat recovery reciprocating chiller was the most effective method of providing space conditioning in the project. The building had been carefully designed for the future addition of solar energy, including the provision of solar storage tanks during the original building construction, and the selection of a heating distribution system sized to achieve the design heating requirements with low temperature water. This choice benefits both the solar energy utilization and the heat recovery operations. In addition, the structural system of the school was designed to accommodate the future addition of solar collector panels and the accompanying support structure. These early choices on the part of the St. Charles School District were important considerations, since the project was under construction during the time the solar energy proposal was being prepared.

Figure 22-1 shows a mechanical system schematic in one of the seven operational modes:

---

*This chapter, including all figures, is reprinted from *The Preprinted Papers of the Second Solar Heating and Cooling Commercial Demonstration Program Contractors' Review,* San Diego, California, December 13–15, 1978, sponsored by the U.S. Department of Energy, prepared by Johnson Environmental and Energy Center, University of Alabama, Huntsville, Alabama, November, 1978.

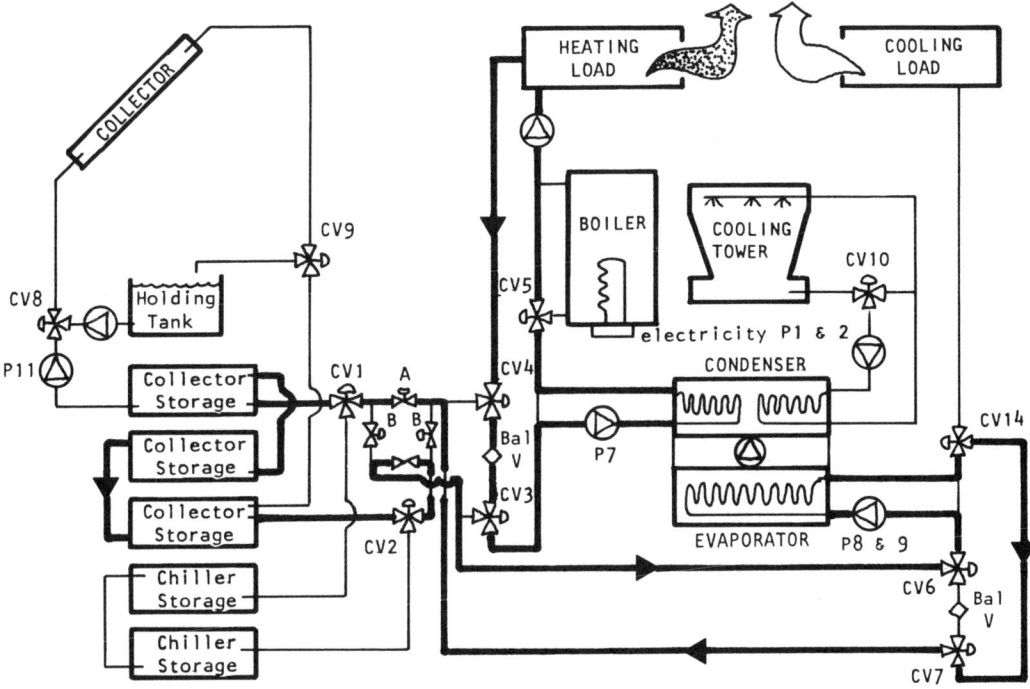

Figure 22-1. Mode E—heating (heat recovery from solar storage).

1. Space cooling with heat recovery
2. Solar collection and heat storage
3. Direct heating from solar storage
4. Direct heating from heat recovery (chiller storage)
5. Heating to the heat pump from the solar storage tanks (shown in Figure 23-1)
6. Heating with the heat pump from heat recovery tanks
7. Heating with an auxiliary electric boiler

An analysis was conducted of the solar mechanical system operation by studying an average day in each month of the heating season, September through May. A dynamic computer program was used to analyze the building heating and cooling requirements and solar collector production on an hourly basis. An example of the results of this analysis is shown in Figure 22-2, a comparison of building thermal requirements and solar production on an hourly basis through the month of February. Note that the building operation was analyzed for both occupied and unoccupied days.

Based upon the results of the foregoing hour-by-hour analysis, and calculation of the resulting storage tank temperatures, the operational modes of the system could be determined for each hour in a manner identical to that in which the system is actually controlled. In this manner, the energy requirements necessary to accomplish each mode of heating or cooling could be determined and a thorough analysis of the energy savings obtained. This analysis was conducted for both the solar assisted heat pump system and a conventional (nonsolar) heat pump system for occupied and unoccupied days for each of the nine months of the heating season.

It should be noted that in the preparation of the proposal the system was designed to provide 15,000 ft^2 (1,393.5 m^2) of aperture area in the solar system. The system, as constructed, will comprise 13,120 net ft^2 (1,218.8 m^2).

The configuration of the solar collection subsystem consists of seven parallel sawtoothed rows, each of which contain 98 solar collectors at 18.677 ft^2 (1.704 m^2) per panel. In the proposal, eight rows of solar collectors were indicated; one row was identified in the

352 7/SOLAR RETROFITTING POTENTIALS

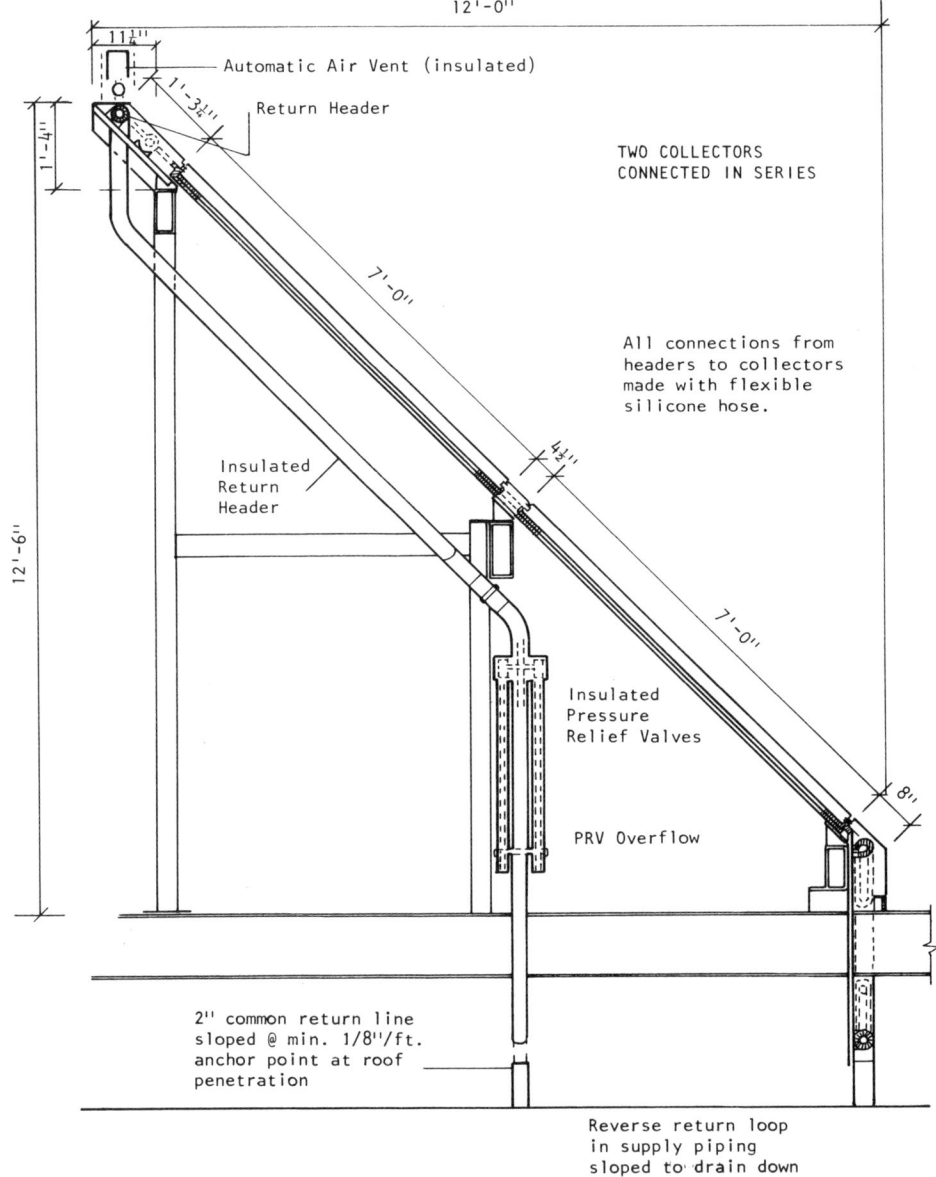

Figure 22-2. Two collectors connected in series.

bidding as an alternate to allow cost control of the total system cost.

Each of the seven rows of solar collectors consists of 49 pairs of collectors connected vertically in series. The piping connections occur at the center of the collector array in a reverse return configuration. Thus, there is a maximum of 25 parallel circuits in any one loop of the piping system. This approach was taken to minimize the difference in flow rate between the collectors in any given loop.

Freeze protection for the system is provided by draining the contents of the solar collectors and the above-roof piping into a separate drain back storage container. This is accomplished by the use of automatic control valves that isolate the pressurized distribution system of the building from the solar collection loop when it is desirable to fill or drain the collectors. There are extensive safeguards employed to provide system protection in the event of failures in the operation of individual pumps or valves, or in the event of an incorrect signal to the components from the control subsystem. Consideration was given to the employment of an antifreeze method of freeze protection, but this was rejected owing to the extensive maintenance requirements as well as the undesirable results of adding additional heat exchange requirements to the solar system.

III. PREPARATION OF CONSTRUCTION DOCUMENTS

Upon receipt of the award, the three professional corporations employed to prepare the proposal began the preparation of contract documents for the bidding phase. These contract documents consist of drawings and specifications that fully describe the solar energy system and its operation.

An important step in the preparation of contract documents was the identification of the collector manufacturer for the project. This was accomplished by conducting a separate bidding procedure prior to the initiation of the contract documents. A specification was developed to permit competitive bidding on the solar collectors by several manufacturers identified in advance. Using this technique, Sunworks, Inc. was identified as the apparent low bidder, and the preparation of the contract documents was accomplished based upon the characteristics of the Sunworks collector. All of the manufacturers were required to submit evidence of compliance with testing procedures conducted in accordance with ASHRAE Standard 92-77, and to submit evidence of the capability of their units to undergo stagnation temperatures without substantial depreciation.

An additional stage in the preparation of contract documents was discussion with the reciprocating chiller manufacturer, Dunham Bush Inc. Modifications to the in-place chillers permit more compatible operation with the characteristic temperatures produced in the solar collection system. These modifications were identified as an alternate in the bidding of the system for control purposes.

IV. BIDDING

Early in 1978, bids were received on the specifications to supply solar collectors to the St. Charles School District for installation by others. The bids were received in the form of a base bid for 13,118 ft^2 (1,218.67 m^2) of solar collector and an alternate bid for 1,458 ft^2 (135.4 m^2) of collector. In addition, a stipulation of unit price was identified for the addition or deletion of collectors from either the base or alternate bids. This technique provided a method for collector manufacturers who provide panels of various sizes to compete equally. In addition, it allowed some modification to the system configuration during the detailed design and layout of the panels.

The successful bidder was Sunworks, Inc. with a price of $10.10 per ft^2 ($108.72 per m^2) of net absorber area. The next lowest bid received was approximately $10.30 per ft^2 ($110.87 per m^2) of net absorber. The other two bids were substantially higher.

After the notice of intent to award had been made to Sunworks, Inc., the contract documents were prepared on the basis of the collector module that Sunworks, Inc. was to provide. Bidding on the construction of the solar system was conducted in early April, 1978. The results of the bidding are shown in the table. It should be noted that the base bid consists of the installation of 12,813 ft^2 (1,190 m^2) of solar panel, and the alternate consists of the installation of the remaining 1,830 ft^2 (170 m^2) of panels.

The overall project budget was approximately $566,000. It can be seen that the tech-

Bidding Results ($ in thousands)

	ESTIMATE	LOW BID RECEIVED
Structural base bid	126	143
Structural alternate bid	18	29
Mechanical base bid	300	296
Mechanical alternate bid	30	26

nique of identifying one row of solar collectors as an alternate was critical in the successful bidding of the project. The total bids received for collectors ($129,366), structure ($143,774), and mechanical installation ($296,000) exceeded the project budget. The St. Charles School District agreed to fund this amount. It can also be seen that the total bidding was within 4% of the estimated cost. The overall cost for the installation of the solar energy system totals $44.42 per ft^2 ($478.15 per m^2) of net absorber area.

It should be noted that a major advantage was achieved in prebidding the solar collector panels, from two aspects. First, it eliminated the pass-through markup that would have been applied by the installer had the collectors been bid in a more usual manner. Second, it permitted the preparation of contract documents that were exactly descriptive of the panel that would be used. This provides a substantial advantage in the bidding of the installation of the system.

V. ADDITIONAL PROJECT ASPECTS

An additional aspect of the project that was not included in the proposal is the provision of solar energy to provide swimming pool heating in an adjacent recreation center that is not directly identified with the project. The installation of piping to allow this transfer of heat will not be conducted using government funding, but the use of solar collectors during those portions of the year when excess solar energy is produced will further benefit the overall life cycle cost of the system.

A portion of the proposal that was disallowed by the reviewers was for data-recording equipment to verify the actual operation of the system. Funding is being sought separately to allow the installation of monitoring and data-gathering equipment to increase the educational aspects of the system for both the students and the solar community.

VI. CONCLUSIONS

The project team feels that the overall project has been a substantial success to the present, and anticipates relatively few difficulties in the completion of the installation. Several important lessons were learned, particularly in the bidding of solar collector systems. Although the confidence level of installing contractors is improving, there is still wide variation in bids received, making the estimation of systems costs rather difficult. It should also be noted that the rather prolonged process of submitting a proposal, having it evaluated, receiving a contract, and designing and completing the system, has produced no major difficulties in this project, but must be taken into account in those projects where sequencing of operations in a fixed time schedule is important.

Part 8

Reconciling Actual Retrofit Performance

23. Reconciling Actual Retrofit Performance with Projections

William S. Fleming
and
Michael S. Goodman

*W. S. Fleming and Associates, Inc.
Syracuse, New York*

		Page
I.	Introduction	357
II.	Types of Energy Audits	357
III.	Retrofit Performance Evaluation	358
IV.	Case Studies	360
V.	Conclusion	363

I. INTRODUCTION

Our nation is experiencing an energy crisis that we must live with for at least 30 years. Fully one-third of U.S. energy usage, the equivalent of 13.8 million barrels of oil per day, is consumed in our present building population of 74 million residential buildings and 1.5 million commercial buildings. The methodology used to design those buildings was developed when energy was inexpensive and abundant, and first cost was a primary consideration. Consequently, many of our buildings waste energy needlessly and incorporate few, if any, conservation or reclamation features.

As engineering consultants, we owe it to our clients and the nation to incorporate energy efficiency into our evaluations for all buildings. We also must be able to present to our clients reasonable projections for potential savings that can be incurred through retrofit construction and operational procedures. Life cycle costing and comparative analysis can be used in evaluations and in selecting retrofit items, and to minimize the costs of energy consumption and ownership of buildings. Marketplace leverage will ensure that manufacturers and vendors of building systems provide to the designer products that are continually improved to maximize efficiency and ease of maintenance. This is beginning to happen already, and better products are being introduced almost daily.

II. TYPES OF ENERGY AUDITS

To project the performance of an item being considered for retrofit, data must be gathered that will allow the engineer to understand how the building operates. The gathering of data is called an energy audit. There are three types of audits, labeled as Class A, B, and C audits.

Class A energy audits consist of:

1. An on-site visit by the energy auditor. This visit may or may not include instrument readings of the building to provide a detailed analysis of the energy consumption, on a system-by-system basis.
2. Evaluation by the auditor of building energy consumption and energy systems.

This includes analysis of energy and economic savings likely to result from selected modifications, and information necessary to calculate actual savings resulting from the modifications. This information is usually furnished to the building owner/operator in a report, manual, or handbook.

Class B energy audits consist of:

1. A questionnaire filled out by the building owner/operator.
2. An analysis, by the energy auditor, similar to that of a Class A audit, but based on information supplied by the building owner/operator, plus known factors, such as climate conditions and fuel costs.

Class C energy audits consist of:

1. A workbook that enables a building owner/operator to evaluate specific modifications by himself.
2. Information necessary to perform or contract for the modifications that prove economically justifiable.

The degree of complexity for gathering data depends upon the accuracy desired for estimating the potential of retrofit recommendations for saving energy. The Class A audit will provide the best information, but it is the most expensive type to perform. Both the Class A and B audits can be analyzed with the aid of a computer program or manual calculations.

The analysis of a building for energy conservation requires a professional who is familiar with energy conservation techniques. These techniques evaluate both the building envelope and the building systems. There are four basic methods of reducing energy consumption in a building:

1. Eliminate unnecessary systems or functions.
2. Reduce consumption of overenergized systems.
3. Improve the efficiency of systems and the building envelope.
4. Reduce the consumption of necessary systems.

Reducing energy consumption should be done systematically. An energy audit should be performed before modifications. Items having little or no effect on working conditions should be performed first, and the first energy audit need not necessarily be the last.

III. RETROFIT PERFORMANCE EVALUATION

Many building modifications that may be implemented as a result of recommendations made after the completion of an initial energy audit usually can be performed at little or no cost, and result in energy savings. These modifications do not require detailed energy and economic analysis and are generally the first to be implemented. Of course, any modification that requires large capital expenditures should have energy and economic analyses performed to indicate that the modification will be cost-effective.

The energy conservation program should not end with the implementation of operational or design modifications. The operating characteristics of a building change with time. Building remodeling and additions, replacement of large pieces of equipment, energy conservation modifications, replacement of systems, and changes in building personnel all contribute to this change. It is desirable for the building owner to have energy audits performed every few years, especially if a large sum of money has been spent on retrofit construction for energy conservation. It is advantageous to the building owner/operator to keep monthly energy records to evaluate the success of the building changes. In addition, a sudden jump in energy usage can be noticed immediately, and the situation can be corrected. Intensified maintenance programs and other human factors should be checked periodically, or lapses will occur. Such checks enable the owner to determine if the investment is performing as predicted, and also serve as indicators to determine if further action is needed.

The evaluation of a retrofit item does not

begin after it has been implemented, but before. In order accurately to evaluate a retrofit item, data representative of the operation of the building before retrofit should be readily accessible. An energy audit is the easiest way to obtain these data.

Some items that may be addressed both before and after the retrofit are:

1. Contacting the local utility company for a demand survey.
2. Instrumenting and observing the building over a period of several days.

Energy bills can only provide mean daily energy consumption. A great deal can be learned from an hourly energy consumption profile for occupied and unoccupied days. This is especially true when energy systems with highly variable energy consumption rates are present. To conduct this type of survey, universal recording watt meters are connected to the major electrical systems. In addition, electric, gas, and water meters are read hourly. Building occupancy by the hour is also noted. This type of survey can reveal information not normally obtained, such as:

1. Time of peak demand.
2. Building systems that operate inefficiently. (This may occur only in certain situations.)
3. If a building is used for more than one function, the one that is the most energy-intensive.
4. If energy reclamation systems are under consideration, whether storage will be required, and if so, how much.
5. Length, in time, of morning pull-up or pull-down load.
6. Operation of systems at times when they are not required.

The biggest mistake an energy auditor can make is to assume that a building operates as designed. Even if this is true on the day the building opens, it has been found that buildings tend to settle into an operating condition somewhat different from the design condition. Equipment performance can change with age, and the quality and frequency of maintenance. Air dampers leak, insulation or equipment covers may not be properly replaced after repairs have been made, lights become dirty, and weather stripping cracks. Air pollution changes the air flows around a building. Control systems may or may not control the building, even when in perfect working order. Addition or removal of interior partitions can have a considerable effect on building energy consumption. Redecoration of offices can change the amount of light absorbed or reflected by interior surfaces. Desks or filing cabinets may block heat registers.

Since these factors, plus innumerable others, can affect building performance, it is wise to investigate as much as possible and assume *as little as possible.* If performance is estimated, the estimation may be based on the assumption of an efficient operation, and this may not be the case.

A recent study, completed for the Department of Energy ("Energy Conservation in Office Buildings," Phase I, DOE contract # EY-76-C-02-2799.000), addressing energy conservation in existing office buildings, reinforces the notion that operational data both before and after the implementation of energy conservation programs should be available. This study indicated that:

> most building owners or their representatives do not have ready access to or knowledge of the information required for an adequate evaluation of their building's energy consumption. Generally they do not know how much their energy consumption changes from year to year, the quantitative benefits from changes they have implemented or will introduce, and how they compare with similar buildings. For the most part, their decisions with respect to energy conservation appear to be intuitive or based on qualitative assessment....

The study found through questioning various owners that only 10% had monitored their building's energy consumption.

IV. CASE STUDIES

The firm of W. S. Fleming and Associates, Inc., contacted two of its clients who had had detailed and instrumented Class A energy audits performed within the last few years. The clients were asked if they had quantitatively evaluated the operational, envelope, and HVAC modifications that the firm recommended. Both clients had kept records of the performance of their buildings after recommendations were implemented. The following cases explain the data gathering process and the results of implementation.

Case 1. A detailed energy audit was performed by our firm on the Agway Corporate Headquarters building in DeWitt, New York. A variety of methods and resources were utilized to evaluate many alternatives and to make recommendations for retrofit. Hourly and daily energy use profiles were obtained by measurement and observation. Mechanical system operational familiarity came through interviews and discussion with Agway's plant engineering staff, as well as through observation. Long-term energy use data needed to place results in historical perspective were obtained from Agway personnel, who had kept remarkably complete records for many years. All these data were sifted through, plotted, analyzed, and used to modify and validate a computer simulation model. With this valuable tool, alternative mechanical systems and structure modifications were evaluated, both in yearly energy use and in economic terms.

Of the many ways to reduce energy consumption in the Agway building, only a precious few were practical from the standpoint of cost, ROI, installation, and human comfort. Any energy conservation option must meet the above criteria if it is to be a realistic option.

Input for option selection was obtained from the Agway personnel most closely associated with building operation and maintenance. In addition, a mechanical contracting firm was brought to the site to aid in the selection process as well as to provide realistic material and installation prices for energy conservation equipment.

To enhance the accuracy of the audit, a computer simulation was performed. The utility of simulating a building depends upon the end-use requirements of a particular client. A design/build contractor, for instance, may only be interested in the relative performance of building materials or mechanical systems in a hypothetical building. In this case, the program's user is not overly concerned with the accuracy of input data other than for the thermal/mechanical systems he wishes to study.

A building owner's needs are different. The computer simulation is desired to provide management with a decision-making tool to use on a *real* building. In this case the building was a 15-year-old corporate headquarters building.

In our application of a computer simulation, the accuracy of input data was extremely important. Each building is unique in its daily operation, mechanical systems, and equipment quirks. It is the responsibility of those performing the simulation to identify what is unique and different, and accurately input this information into the program.

The ultimate purpose of these computer simulations was to measure the impact of alternatives to the current double-duct variable volume HVAC system in terms of energy consumption and economics. Our strategy to save money on operating costs was:

1. Perform a detailed energy audit to determine accurately the current state of total energy use within the existing building.
2. Correlate the results of the energy audit with long-term data that were available to gauge the consistency of energy use on a yearly basis.
3. Input the data into our program to generate a validation of the energy studies. This validation process is the key to studying alternate systems and the economics thereof.
4. Decide what alternate operational procedures, envelope characteristics, or systems are *practical* from both a retrofit and a new construction standpoint. Input for this step comes from (a) operating personnel, (b) the firm's past experience and research, and (c) consulting mechan-

ical contractors. Because we are an independent firm, we were not tied to any particular manufacturers, utility companies, etc. This allowed us to explore freely any and all system alternatives, including those that required *no* purchase of equipment, or architectural/engineering design changes.

5. Assemble all data pertaining to systems performance and economics and compare the results of the simulation versus the real case validation run. This was done in confidence because of (a) accuracy of the energy audit, (b) accuracy of the validation run, and (c) correlation of long-term data to the above.

The results of the energy audit, and analysis of long-term data, resulted in the definition of an "average" year for many parameters. After examining five years' worth of energy use at the building, an average year was constructed for total heating gas, total electric consumption, kilowatt demand, and hot water consumption. The results of the validation are indicated in the table.

Validation Comparison

PARAMETER	COMPUTER PROGRAM	REAL DATA	ACCURACY
Total MBtu for space heating	10053.4	10842.4	92%
Total gas (CCF) for hot water	14490	15130	96%
Total kWh	4696145	4987328	94%
Total kWh/ft²	28.13	30.57	92%

After the computer model had been validated against the actual operating data, various alternatives were evaluated. In making a choice of retrofit alternatives, several basic factors were recognized and considered:

1. Building's purpose
   (a) Life expectancy of building
   (b) Type of tenant(s) and activities
   (c) Anticipated change in use
2. Geographical location
   (a) Type of climate—heating, cooling, or both required
3. Tenant comfort
   (a) Air temperature control (minimum variation)
   (b) Humidity control
   (c) Air movement control—air changes, ventilation, exhaust
   (d) Sound level
4. Aesthetics—appearance of equipment, inside and ouside of building
5. Space limitations for equipment (piping takes up less space than ducting)
6. Costs
   (a) Installed costs
   (b) Operating and maintenance costs
7. Reliability
   (a) Frequency of required service
   (b) Inconvenience caused during service

A variety of alternatives were considered and many recommendations made.

Those recommendations that have been implemented since the completion of the energy audit consist of:

1. Modification to the operating procedures of the building plant and systems
2. Modification of the control system of the variable volume air-handling unit
3. Installation of an energy management control system.

The energy saving that could be realized from the implementation of the above recommendations was estimated to be approximately 18%, with a payback period of five years.

The building manager has kept records of the building's energy consumption since the implementation program was completed. Three full years have elapsed since the program was completed, the manager verifies that the annual energy savings have been approximately 20% or $52,000. Given the professional cost of the energy study and the capital expenditure for the purchase of products, the payback period to Agway, Inc., in present value terms, shall be less than five years.

Case 2. The Marriott Corporation, which had detailed instrumented energy audits performed by our firm for their restaurant division on three of their restaurants (Hot Shoppe, Joshua Tree, and Roy Rogers), has kept data that indicate the differential in energy consumption after certain energy-conserving recommendations had been implemented.

The energy study was divided into two main categories: (1) electric demand and (2) natural gas consumption.

Under electrical consumption, subcategories were established and instrumented for:

1. Electrical consumption
2. Refrigeration
3. Kitchen equipment
4. Exhaust fans

Under natural gas consumption, the subcategories investigated were:

1. Domestic hot water
2. Food preparation

Each of the above subcategories was subjected to a series of measurements to allow for an *energy consumption by hour* analysis. This, in turn, was related to customer count and energy costs.

The data-collecting procedures involved both observation/recording and instrumentation/recording. For example, the domestic hot water (DHW) cycles were measured with a running time meter to obtain the duty cycle (average time on per hour). The quantity of gas consumed was observed and recorded by the hour and for selected one-minute intervals. The electrical load was recorded using a series of six probes for continuous recording plus running time meters and reading the electric meter every hour. Air flow was recorded by an air flow meter for exhaust air volume and infiltration, and water usage was taken directly from the water meter. These and other measurements/observations were then correlated by time to provide the electric, gas, and water budgets.

After the initial data collection was complete, an energy and economic analysis computer program was utilized to determine the effect various retrofit alternatives had on energy consumption and the economic feasibility of each alternative. Input to the computer program consisted of building envelope description, lighting and equipment energy consumption, operating schedules, weather data, energy costs and escalation rates, estimated equipment costs, etc. The computer simulation was first performed to model the building as it actually existed and validated to within ±10% of the building's actual consumption. Once the actual model was complete, each of the alternatives was simulated. After the energy consumption of each alternative was determined, the economic analysis was performed to identify what the payback period was, if any.

The following items were recommended for retrofit, based on a low payback period and high return on investment:

1. Ceiling insulation—payback period of three years, return on investment 42%
2. Set back–set up thermostat system—payback period of two years, return on investment 72%
3. Electrical demand limiter—payback period of four years, return on investment 33%

There were other alternatives that were recommended; however, the above alternatives were the first to be implemented in the restaurants. The estimated savings in energy consumption for the above items was 8%.

Also implemented in the restaurants, but not included in the retrofit analysis, were recommendations determined from operational observances:

1. Promote a preventive maintenance schedule for all energy-consuming equipment.
2. Promote an energy conservation program for employees.
3. Eliminate unnecessary lighting.
4. Prepare a time on–time off schedule for equipment.

The total estimated savings for both retrofit and operational recommendations was 15 to

20% of annual energy consumption. The Marriott Corporation had collected data for the year prior to that when the recommendations were implemented and the year when the recommendations were implemented. The range in savings ran from a low of 10% to a high of 28%. The Marriott Corporation is very pleased with the results achieved so far and plans to implement these recommendations in as many restaurants as their yearly budget will allow.

V. CONCLUSION

In summary, the retrofit evaluation program should not end when all the options have been implemented. It must become a permanent part of building operations. Comprehensive maintenance programs and compliance with guidelines by building personnel must be periodically checked. In addition, changes in building occupancy, remodeling, and/or repartitioning may require changes in an energy reduction plan.

Energy consumption and cost should be monitored monthly. This will enable building operators to keep track of the success of their programs. Since energy consumption is weather-dependent, monthly degree-days should be charted as well. After several years, energy consumption variations with weather conditions will become predictable. It is also useful to monitor the energy usage of the large building systems.

The application of energy conservation options will alter the way a building operates. The energy audit should be repeated periodically to determine whether new methods of operation are amenable to the other retrofit options. New types of energy-conserving equipment are constantly being developed. These should be screened for possible building retrofit. This is especially true when existing building systems wear out and must be replaced. Because of the large number of products entering the market, cooperation between building operators and manufacturers will be required.

Part 9

Retrofitting and Power Generation

24. Cogeneration and Retrofit Opportunities

Eugene E. Cooper, Ph.D.

Mechanical Systems Division
Mechanical and Electrical Engineering Department
Civil Engineering Laboratory
Port Hueneme, California

	Page
I. The Principle of Cogeneration	367
II. Factors Impacting Cogeneration	368
III. Cogeneration System Options—Capacity, Type, and Ownership/Operation	370
IV. Economic/Financial Considerations	372
V. Performance Curves and Characteristics of Cogeneration Plants	374
VI. Energy/Environmental Legislation and Regulations	380
VII. Institutional Constraints	381
VIII. Implementing Cogeneration	382
References	384

I. THE PRINCIPLE OF COGENERATION

Cogeneration can be an effective means of conserving energy where both thermal energy and electrical or shaft energy are needed. Figure 24-1 illustrates the benefit of cogeneration by comparing energy balances for a conventional system and for an ideal cogeneration system serving a load center simultaneously requiring 200 units of thermal energy and 100 units of electrical energy. In the conventional system, where the thermal energy is supplied by an on-site boiler or a heater with an efficiency of 75%, an input energy of 267 units is required to meet the thermal load. The electrical load of 100 units is met by purchasing power from the utility company at a net efficiency of approximately 33%, so that 300 units of energy are input at the central power plant. The conventional system has an overall, or "universal," efficiency of 53%. By comparison, the depicted cogeneration system is able to meet the same thermal and electrical loads at an overall efficiency of 75% by capturing exhaust heat from an on-site electrical generation process to provide the thermal energy. The relative thermal-to-electrical energy ratio of 2 to 1 in this ideal example is typical of cogeneration systems using combustion turbines. Other cogeneration systems have different characteristic thermal-to-electric ratios, as illustrated in Figure 24-2.

Retrofitting a facility to utilize cogeneration potential usually involves the acquisition of a considerable amount of new equipment for generation of electrical power and capture or utilization of normally "wasted" heat. In fact, it is generally the case that "retrofitting" to accomplish cogeneration is actually a replacement of significant portions of the old systems for supplying electrical and thermal energy, plus adding more capability. Cogeneration systems typically require carefully considered capital expenditures, therefore, that will be re-

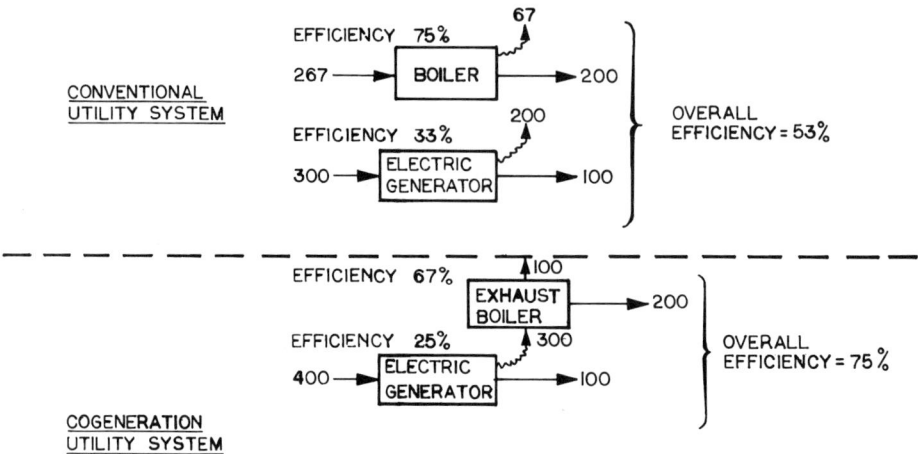

Figure 24-1. Energy balance comparison. Example: Need 200 units of energy as steam and 100 units of energy as electricity. (From Ref. 4.)

covered only if sufficient cost savings in fuel or purchase of electrical power result.

II. FACTORS IMPACTING COGENERATION

The process of deciding whether or not to implement cogeneration, and, if so, which system to select, can be a complex one during which numerous factors must usually be considered. There are probably more nontechnical than technical factors that influence the outcome of an assessment of cogeneration for a particular site. Many technically sound cogeneration system projects have failed to reach fruition because the nontechnical constraints were not anticipated or could not be overcome when encountered. Thus it is important that there be an awareness from the outset of the need to identify both the technical and nontechnical issues that apply to a specific candidate cogen-

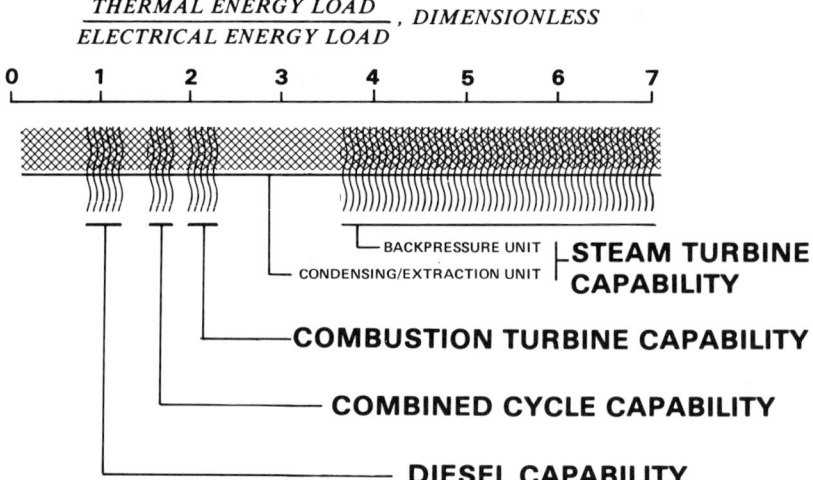

Figure 24-2. Thermal/electrical capabilities of cogeneration systems. (From Ref. 4.)

eration site. They must each be given the level of attention needed to adequately deal with them.

Factors impacting cogeneration may be placed into the following categories: (1) technical, (2) economic/financial, (3) legislative/regulatory, and (4) institutional. In some cases, a particular issue or factor relevant to the evaluation of cogeneration may affect two or more categories; for example, the type of fuel to be used may be restricted by environmental and energy legislation, and the type of fuel affects the technical design of the equipment as well as the capital costs and life cycle economics.

Although this is not an exhaustive list, some examples of issues that might arise during the examination of cogeneration are:

1. *Technical:* load patterns (profiles, magnitudes, driving functions); choices of fuels; plant site location relative to loads and condition of distribution systems; performance of candidate systems.
2. *Economic/financial:* magnitude of capital costs; impact of fuel choice and other parameters on capital costs, operating costs, ROI; available tax incentives; marketability of excess power; sources of capital, outside versus self-generated; utility company standby rates.
3. *Legislative/regulatory:* local air emissions standards and attainment status; waste disposal restrictions; jeopardy of any environmentally protected areas; national fuel use and energy legislation; state PUC facilities regulations regarding cogeneration; status of PUC jurisdiction over regulation of cogeneration facilities; zoning or siting restrictions.
4. *Institutional:* utility company policies toward cogeneration; availability of personnel and skills for operation and maintenance of system; assessment of impact on facility mission; assessment of impact on special requirements, e.g., security; impact on community; impression on community.

Some of the more important factors often encountered are discussed herein. Guidance is also provided to sources of information and data for a number of questions that may arise.

The key requirement for successful cogeneration applications is that the recovered "waste" heat be used beneficially. Heat recovery can be accomplished on engines driving electrical generating equipment or on engines producing mechanical shaft power; but whichever is the case, the incorporation of heat recovery equipment is uneconomical unless a substantial portion of the heat is used.

A good energy survey of a facility provides the basic information needed to determine if load conditions are favorable for cogeneration. An energy survey should examine the thermal loads, electrical loads, and mechanical shaft loads. It is recommended that an energy survey conducted as a preliminary step in the assessment of cogeneration include determination of the following information about the various loads.

THERMAL ENERGY

Present thermal energy usage profiles and demand levels, including extremes.
Existing thermal supply system characteristics/capacity, pressure, temperature, including purchases.
Condition and coverage of thermal distribution system.
Major thermal loads and thermal state of energy required.
Opportunities for conserving and reducing thermal energy usage.

ELECTRICAL/MECHANICAL SHAFT ENERGY

Present electrical energy usage profiles and demand levels, including extremes.
Identification of any on-site electrical generating units and/or large motors or engines providing mechanical shaft power, including the output levels and profiles.
Opportunities for conserving and reducing electrical energy usage.

In addition, it is recommended that an energy audit include consideration of:

1. Planned changes in operations or functions at the facility that would impact load profiles, demand levels, or capability to meet loads.
2. Opportunities to alter loads to benefit cogeneration potential.

The data and information accumulated during the energy survey should be applied in the following way in preparation for the cogeneration assessment:

1. Estimate the loads *after* implementation of energy conservation measures.
2. Determine whether the resulting thermal load is high enough to warrant cogeneration. Reference 1 recommends having a process heat load of at least 200–300 million Btu/hr, at which level diesels or combustion turbines with exhaust heat boilers would normally be used, or even higher for use of steam turbines.
3. Establish representative profiles that account for diurnal and seasonal effects, or other significant effects such as process variations. In general, the potential for cogeneration is enhanced when electrical and thermal load patterns are similar and in phase.

III. COGENERATION SYSTEM OPTIONS— CAPACITY, TYPE, AND OWNERSHIP/ OPERATION

There are usually numerous cogeneration system alternatives for consideration at a site, stemming from these basic questions: What types of power plant should be used? What capacity should the system have?

Commercial equipment is readily available for four generic types of cogeneration systems. Figure 24-3 illustrates the available systems, termed diesel engine systems, combustion turbine systems, steam turbine systems, and combined cycle systems. Other options, such as the organic Rankine bottoming cycle, are on the threshold of commercialization, but the sparse data base and experience level with them precludes their inclusion in this report.

The selection of a system for a particular installation depends on many site-specific factors that ultimately affect cost. One very important factor is how the system ratio of thermal/electrical output, addressed in Figure 24-2, matches that of the loads. Other important factors include such things as types of fuel available, environmental restrictions, utility rate structures, etc.

The size, or capacity, of a cogeneration system obviously depends upon the thermal and electrical loads to be served. As seen below, the electrical load may be considered "infinite" if the system is allowed to transfer power to the grid. Then the system can operate at its most efficient or economical point and can sell power to the grid if the system electrical output exceeds the facility's electrical load. As a rule of thumb, the best overall efficiency for meeting the thermal and electrical loads will be obtained if the cogeneration system is sized to meet the typical or average thermal load. Fluctuations in loads cause the system to almost always operate at off-design points. The load swings and other factors result in compromises when systems capacity is being determined.

Figure 24-4 depicts cogeneration systems in a generalized sense and illustrates that backup capability or additional capacity to meet the thermal and electrical loads may be provided by fired auxiliary boilers and utility grid connections, respectively. The connection to the utility company grid is a very important, and in most cases very beneficial, feature for a cogeneration system. First, because the grid can carry part of the electrical load, the option exists to size the cogeneration system with respect to the thermal load, if that appears to yield the best economy or fuel efficiency. Second, because the grid "backs up" the cogeneration system, standby and emergency electrical generating capacity do not have to be installed as part of the system; this reduces capital costs. Third, the utility company or customers connected to the grid become potential markets for power generated in excess of that needed to meet the on-site electrical load. And fourth, the capacity of a reliable cogeneration system in actuality adds to the total capacity that the utility company can count on to

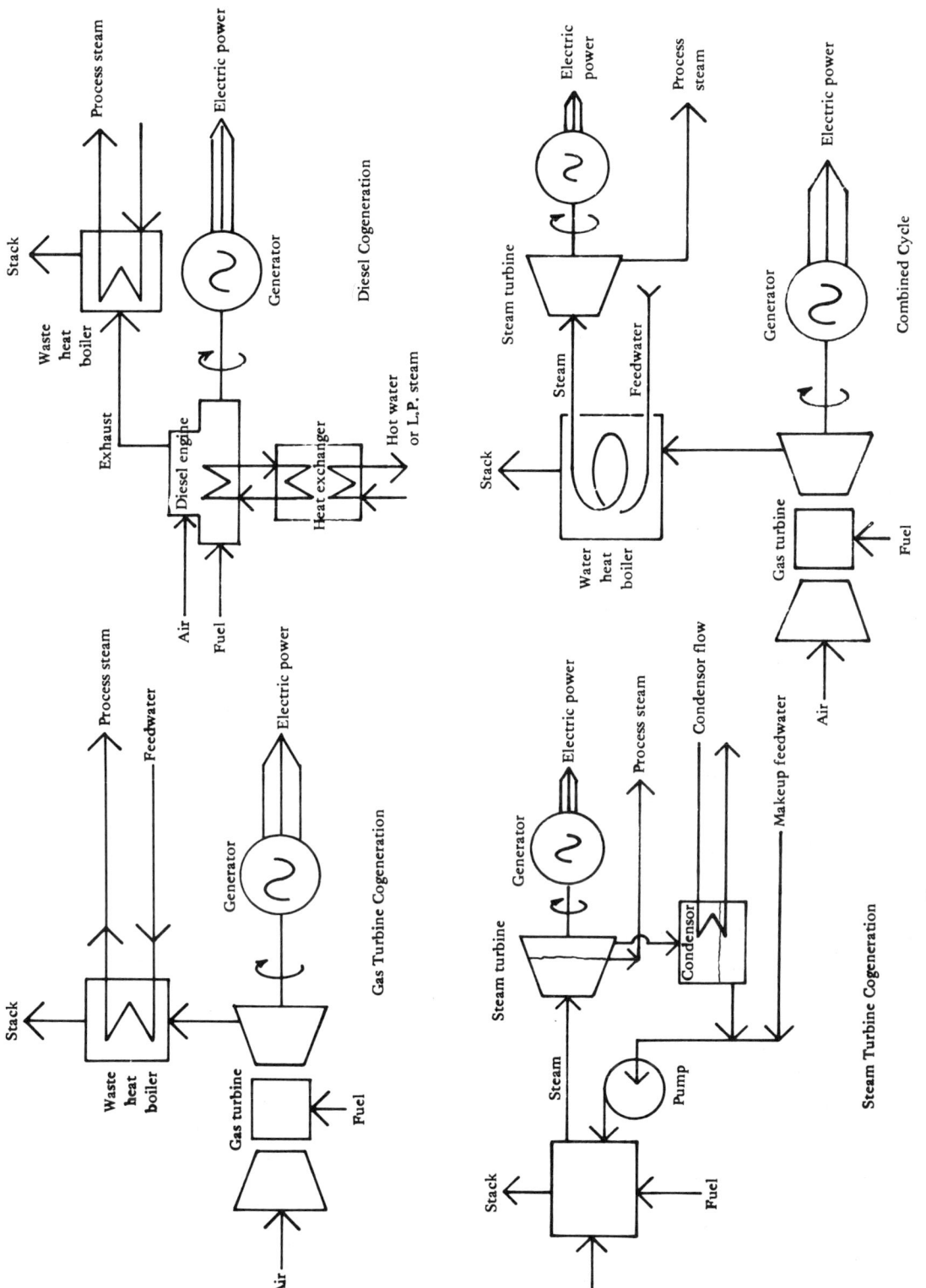

Figure 24-3. Examples of cogeneration concept. (From Ref. 4.)

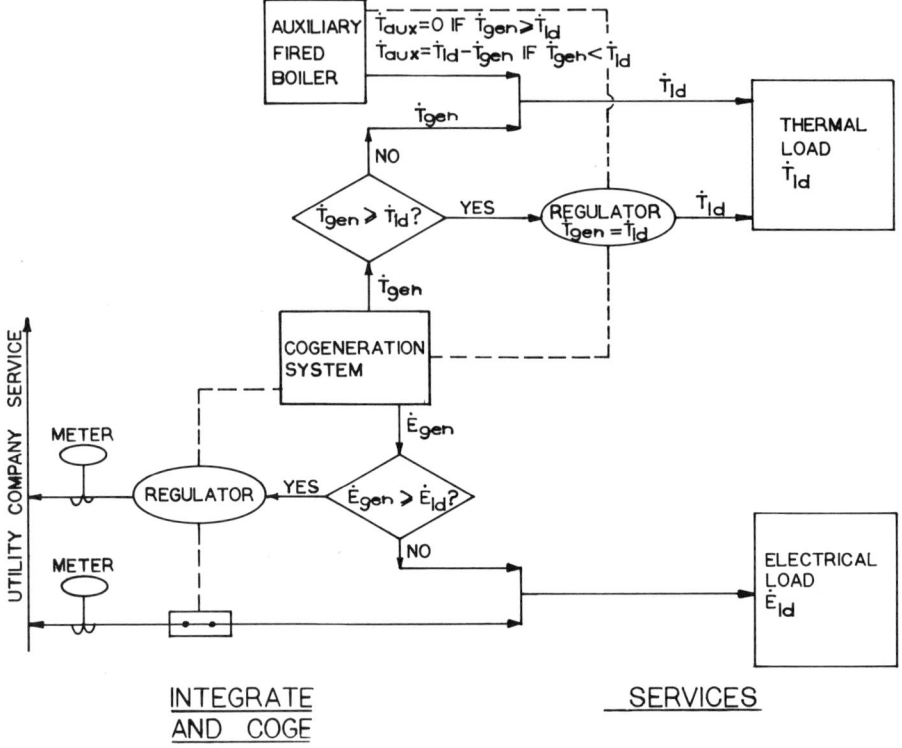

Figure 24-4. Integrated utility company service and cogeneration plant. (From Ref. 4.)

meet the grid-connected loads, so there becomes an incentive for the utility company to consider partial or total ownership and/or operation of the cogeneration system.

There are five basic types of cogeneration system arrangements that have progressive degrees of utility company involvement [1]. They are:

1. Industry ownership and operation of an under-capacity plant. The utility sells power to the facility to make up the difference between the electrical load and the capacity of the cogeneration plant.
2. Industry ownership and operation of an over-capacity plant. The utility purchases power from the facility in excess of that needed to meet the load.
3. Industry ownership with utility operation of the plant.
4. Utility/industry joint venture in a cogeneration plant. This involves a mutually acceptable sharing of capital costs and operating responsibilities. A joint venture may also be arranged between industries.
5. Utility ownership and operation of a cogeneration plant. The utility markets both thermal and electrical energy to the facility.

The degree to which a utility is willing to participate depends upon its needs and policies and varies from one utility to another. There is growing acceptance of cogeneration system connections to the grid and a trend toward cooperative participation by utilities as a result of demand growth, difficulties in providing new central plant capacity, and legislative or regulatory measures favoring cogeneration.

IV. ECONOMIC/FINANCIAL CONSIDERATIONS

An investor considering the financing of a cogeneration system is concerned with numerous factors and their impacts on the economic projections for the system. The total annual cost of providing thermal plus electrical utility service to a facility may be expressed as the summation of several basic contributing factors:

$$TC_j = CC_j(y) + F_j(y) + OM_j(y) + P_j(y) - R_j(y)$$

where: (y) indicates that the annual costs are a function of time, i.e., year; j indicates the costs are those occurring if alternative "j" is chosen to supply utility services; TC = total cost for thermal plus electrical services; CC = capital cost expenditure; F = fuel costs; OM = operation and maintenance costs; P = cost of energy purchased from outside, electrical or thermal; and R = any revenues resulting from operation of equipment at the facility.

Cost estimates for future years necessitate estimating escalation rates for fuel, electricity, O&M, etc. Since the "crystal ball" is uncertain, escalation rates are sometimes varied parametrically, and the sensitivity of the economic projections to the variations is determined.

Different alternatives for providing the necessary utility services are compared by the investor. One alternative may reduce fuel costs but increase capital costs. Another may involve a larger plant that increases both fuel and capital costs but results in revenues through sale of excess power to offset the increases. In the economic analyses of alternative options, the term "alternative" indicates not only different design options, such as steam turbine versus combustion turbines, but it also indicates different ownership/operation options or even different means of financing the construction costs.

The cost components making up TC_j provide the information needed by an investor to decide upon the economic viability of an option. Principally the investor is concerned with the return on investment (ROI), with the magnitude and means of handling the capital expenditure, and with each total life cycle savings associated with each alternative.

Return on investment is a common measure of judging the economic viability of candidate investments. A minimum ROI must be exceeded to gain approval by the investor. The minimum acceptable is set by the investor himself, and depends on the type of investment being made. The acceptable ROI is a function of the economic life for the investment, which is also established by the investor. The expression from which ROI is determined is:

$$\sum_{y=0}^{y=N} \frac{CP_j(y)}{(1 + ROI_j)^y} = \sum_{y=N+1}^{y=EL+N} \frac{S_j(y)}{(1 + ROI_j)^y}$$

where CP_j = annual construction payments made during the construction period;
N = number of years from the beginning of construction financing to system startup;
EL = economic life of the system;
and S_j = annual savings resulting from operation of the system.

It is seen from the definition that ROI may be considered a discount rate that converts future savings or expenditures into present value. The above equation must be solved iteratively for ROI.

Determination of ROI_j requires knowledge of construction costs, length of construction period, and the factors F_j, OM_j, P_j, and R_j contributing to the savings, S_j. Data provided in references 1 through 4 should be helpful in making rough planning estimates of construction costs, construction schedules, and O&M costs for various types of cogeneration plants. The information there is typical of mid-1978 costs, so projections of construction and O&M costs for future years should be made on the basis of an appropriate construction cost escalation factor and labor rate escalation factor, respectively.

Over the operating life of a cogeneration system, fuel is often the largest contributor to the ownership and operation of the system. For systems burning oil or natural gas, fuel will typically constitute 65 to 90% of the total life cycle cost for the system, and will be a significant portion of the annual total utility cost, TC_j. For economic reasons, therefore, it is advisable to consider cogeneration plant designs

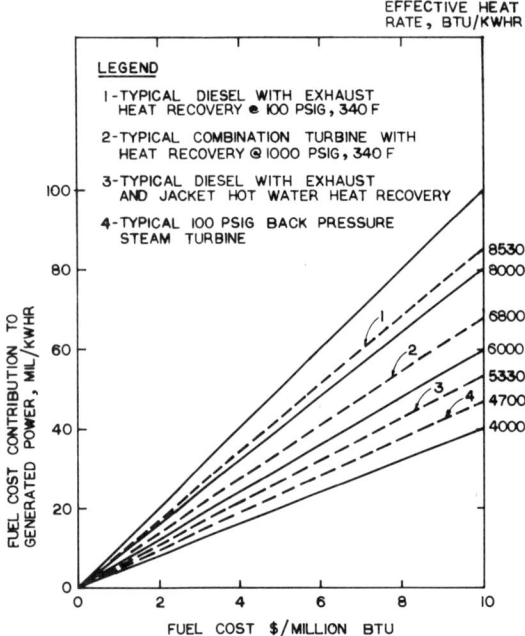

Figure 24-5. Fuel cost contribution to power costs for cogeneration systems. (From Ref. 4.)

cal power from a cogeneration system. Figure 24-5 shows how fuel cost impacts the cost of power generated on site. In a cogeneration system, additional fuel is required over the amount needed to produce only the steam. The effective heat rate for power production from the cogeneration system is:

$$HR_{EFF} = \frac{\dot{F}_{COGEN} - \dot{F}_{SO}}{\dot{E}}$$

where

HR_{EFF} = effective heat rate, Btu/kWh
$\dot{F}_{COGEN}$ = fuel flow rate to the cogeneration system, Btu/hr
$\dot{F}_{SO}$ = fuel flow rate that would be used by boiler producing only steam, Btu/hr
$\dot{E}$ = power production from the cogeneration system, kW

The fuel cost contribution to generated power is:

$$FPC = (HR_{EFF})(CF)(1/10^3)$$

where

FPC = fuel contribution to power costs, mils/kWh
CF = cost of fuel, $/million Btu

For power from the cogeneration system to be economically attractive to the site owner, the fuel cost contribution must be sufficiently less than the cost of purchased power to allow for capital recovery and O&M. For the power to be economically attractive to a utility company, the fuel cost contribution must compare favorably with costs the utility company experiences or anticipates in its system.

V. PERFORMANCE CURVES AND CHARACTERISTICS OF COGENERATION PLANTS

Table 24-1 presents and compares characteristics of diesel, combustion turbine, and steam

that burn less expensive fuels and have the flexibility to handle various fuels where it appears that two or more fuels may be competitive. The high cost of petroleum fuels and the likelihood of further rapid price escalations are making oil-fired systems difficult to justify economically in many installations. Natural gas may maintain some cost advantages over oil, but it also is a premium fuel, likely to become in short supply and undergo rapid price increases. Coal-fired systems require a larger capital expenditure for installation, but the anticipated lower fuel costs will often more than offset the greater initial outlay. Flexibility to burn other fuels can usually be incorporated into the design of a coal-fired plant at relatively low cost. For example, solid waste might be substituted for a portion of the coal if the furnace volume is slightly increased and additional storage capacity is provided. Or oil- or natural-gas-firing capability can be added to a coal-firing facility at minimum cost.

Since fuel cost is such an important contributor to the economic feasibility of cogeneration systems, it is beneficial to determine the contribution that fuel makes to the cost of electri-

Table 24-1. Characteristics of Cogeneration Systems.

	DIESEL	COMBUSTION TURBINE	STEAM TURBINE
Heat recovery source	1) Exhaust gas 2) Jacket water, plus oil cooler & turbocharger cooler (if equipped with turbocharger)	Exhaust gas	Steam extraction between intermediate turbine stages (extraction turbine) or after final stage (backpressure turbine)
State of recovered heat	1) Through heat exchangers, exhaust provides hot water or saturated steam, usually $\leq 350°F$. 2) Jacket water can provide hot water at $< 250°F$ or saturated steam at ≤ 15 psig. 3) Exhaust can heat organic fluid through heat exchanger.	1) Exhaust gas may be used directly for thermal content and O_2 content. 2) Through heat exchanger, exhaust provides hot water or steam. Steam is usually saturated at pressures 10–200 psig. Possible to produce limited superheated steam up to $\approx 600°F$. 3) Exhaust can heat organic fluid through heat exchanger.	Extracted or backpressure steam is usually at a slightly superheated state. Boiler/turbine combinations can be selected to provide steam at desired pressures from atmosphere to 600+ psig, although pressures ≤ 200 psig are most common.
Fuels	1) Distillate petroleum 2) Natural gas	1) Distillate petroleum 2) Natural gas	1) Coal 2) Residual petroleum 3) Distillate petroleum 4) Natural gas 5) Biomass 6) Solid waste 7) Liquid or gaseous waste
Capacity of units	Almost continuous sizes up to 2 MWe. Frequent sizes to 4 MWe. Discrete sizes even larger.	Limited choices available. Units available at nominal ratings of 0.5, 0.75, 2.5, 4.75, 6.8, 7.5, 10.0, 19.4, 24.7, 28.3, and larger discrete sizes.	Standard NEMA ratings are 500, 700, 1000, 1500, 2000, 2500, 3000, 4000, 5000, 6000, and 7500 kW. AIEE-ASME Preferred Standard Large 3600-rpm, 3-phase, 60 cycle condensing unit ratings are 11.5, 15, 20, 30, 40, and 60 MWe.
Typical applications	Where the thermal load: 1) Only requires hot water or low pressure steam. 2) Is near the system. 3) Is roughly equal to the electrical load.	Where the thermal load: 1) Requires hot water or moderate (up to 200 psig) steam. 2) May be remotely located, requiring thermal distribution. 3) Is generally 1 to 3 times the electrical load.	Where the thermal load: 1) Is the essential load that would require reliable boilers anyway. 2) Is generally large.

Source: Ref. 4.

turbine types of cogeneration systems, for which components and equipment are readily commercially available. Because the combined cycle system is basically a combination of a combustion turbine and a steam turbine, its characteristics can be inferred from the information presented. The table indicates that the steam turbine systems have a distinct advantage in fuel flexibility and potential for multifuel capability. The choice of fuels has a most significant impact on the economics of cogeneration and the compliance of the system with environmental and energy regulations. Diesel and combustion turbine systems are typically limited to the use of the premium fuels, petroleum or natural gas. There are exceptions; for example, some engines, not necessarily operating in the cogeneration mode, have been set up to burn waste gases or synthetic fuels such as sewer gas where a nearby cost-effective source exists. A limited number of installations have burned residual petroleum, but special facilities to "wash" the fuel, heat it, and inject additives were usually required. But, typically, the diesel and combustion turbines require premium fuels. Table 24-1 provides guidance, in the form of "Typical applications," for the use of the various types of systems.

Figure 24-6 generalizes the performance of diesel cycle engines and presents the data in normalized form [5]. The figure shows that just under one-third of the fuel energy is converted into shaft energy to drive the electrical generator (the shaft energy curve is the engine efficiency, η). Roughly one-third of the fuel energy converts to heat that is carried out in the exhaust gas; approximately 30% of the fuel energy converts to heat that is transferred to the jacket water and lubricating oil plus to air flow in the turbocharger (if the engine has a turbocharger). About 5 to 10% of the fuel energy is irretrievably lost from the engine structure. Practically all of the heat in the oil cooler, turbocharger aftercooler, and jacket water can be recovered as hot water or even low pressure (≤ 15 psig) steam. A temperature limit is set by the requirement to keep jacket water below 250°F on most engines. Figure 24-6 also shows the portion of the fuel energy recoverable from the exhaust gas stream at different temperatures. The normalized curves can be dimensionalized by assuming a full load generator output, $\dot{E}_{100}$(kW). At any fraction of the full load output, the fuel energy input is:

$$\dot{F}\left(\frac{\text{Btu}}{\text{hr}}\right) = \left(\frac{\dot{E}}{\dot{E}_{100}}\right) \cdot \frac{\dot{E}_{100}}{\eta} \cdot \frac{3,413 \text{ Btu}}{\text{kWh}}$$

Other quantities are related according to the data from Figure 24-6.

Figure 24-7 is a similarly normalized performance curve for combustion turbines [5]. Combustion turbines have no cooling jacket, and heat recovery potential from the lubricants is insignificant; so heat is only recoverable from the exhaust. It is seen that a significant portion of the fuel energy can be recovered to generate steam at 100 psig or higher, which is suitable for distribution over relatively long distances. The curves of Figure 24-7 are representative of many single-shaft, simple cycle combustion turbines. Better part load efficiencies are obtained with dual-shaft engines. Efficiency improvements may also be anticipated from future designs incorporating higher turbine inlet temperatures and recuperation heat exchangers.

Steam turbines do not lend themselves to a "universal" normalized curve of engine performance as the diesels and combustion turbines do. There are too many variables possible with steam turbine systems. Instead, a performance "map" of throttle steam flow rate versus electrical output, with extraction steam flow rate as a parameter, is illustrated in Figure 24-8. The specific generator design output throttle pressure, extraction pressure, and condenser pressure for which the curve is applicable are shown. The map is bounded by five essentially straight lines explained in the figure. It is noted that the "map" does not extend down to $\dot{E} = 0$. The curves become nonlinear at the low end of the scale; so it is better to design the equipment for operation at half load or better in order to map the performance accurately [2, 6].

An expression that relatively accurately re-

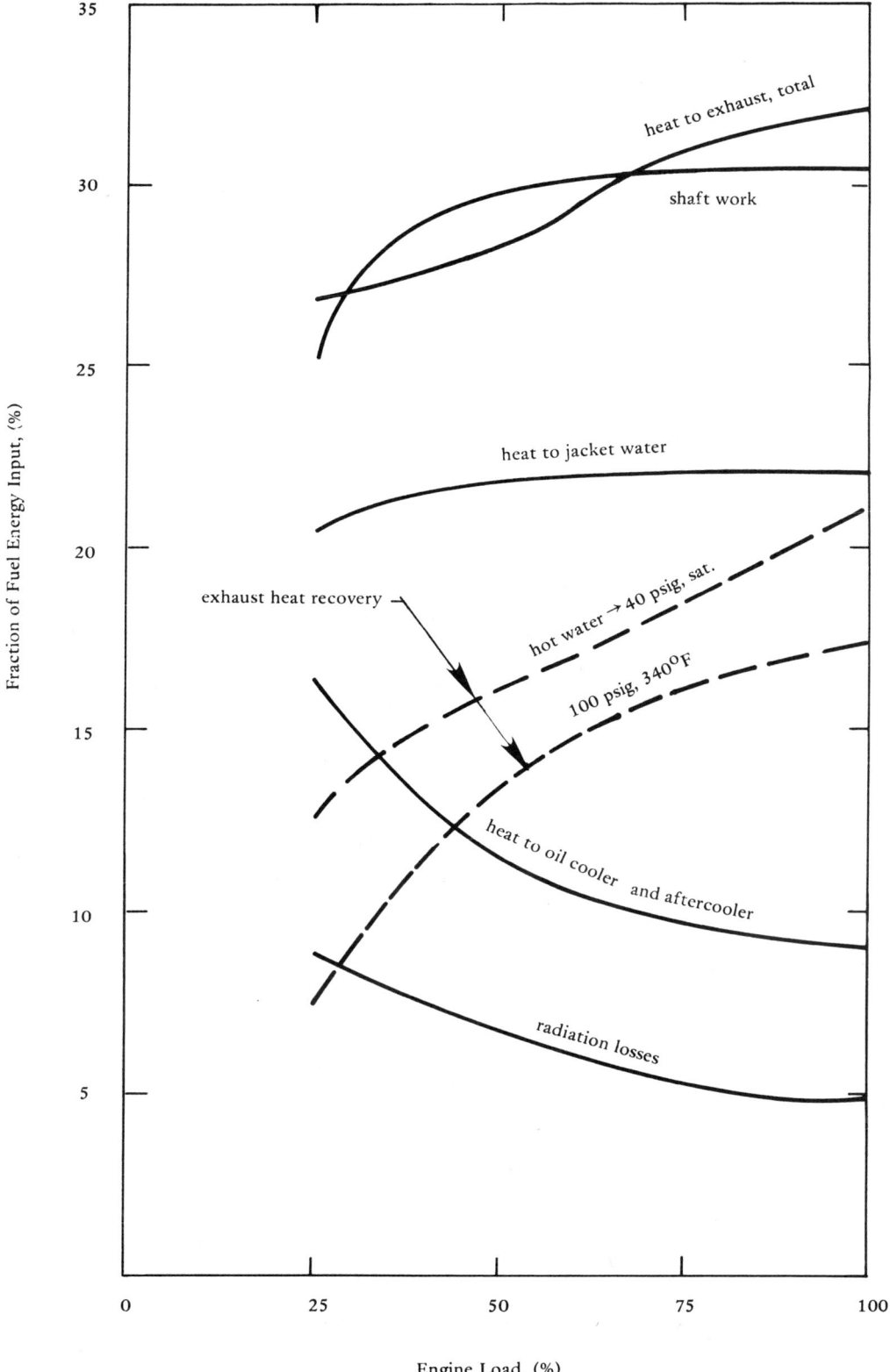

Figure 24-6. Heat balance for diesel engine in heat recovery application, relative output. (From Ref. 4.)

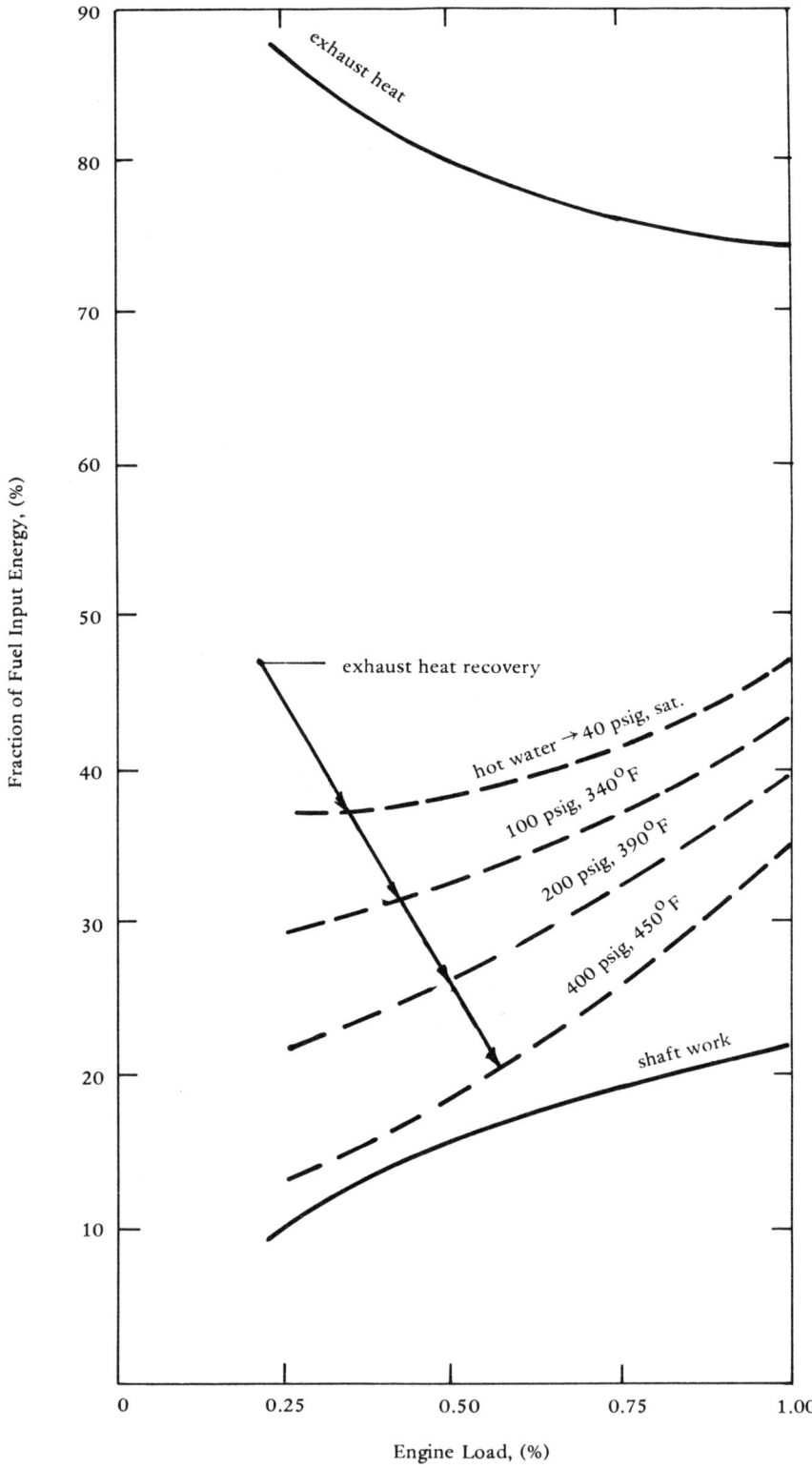

Figure 24-7. Heat balance for combustion turbine in heat recovery application, relative output. (From Ref. 4.)

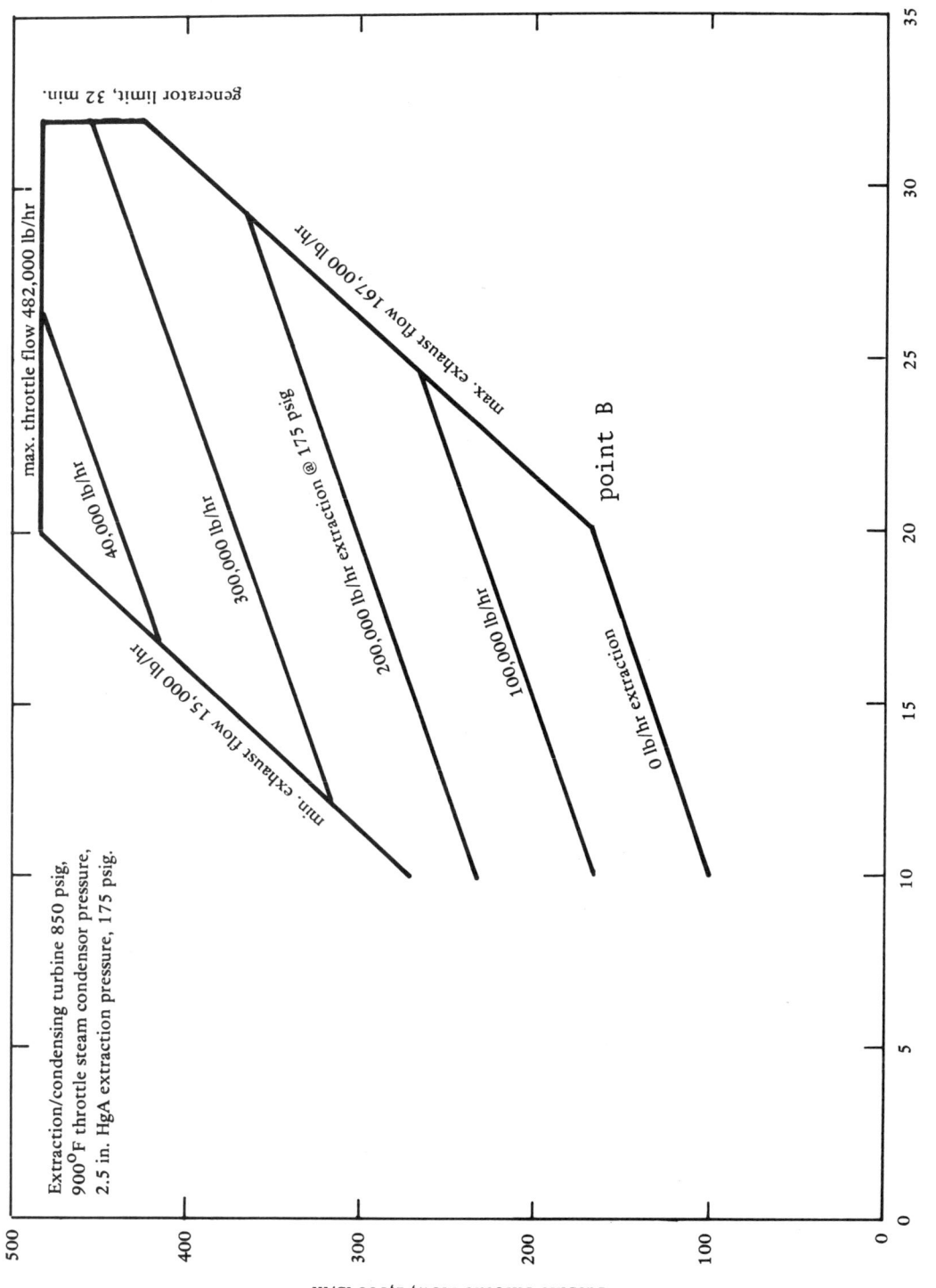

Figure 24-8. Performance map for an extraction/condensing steam turbine-generator set. (From Ref. 4.)

lates the throttle flow rate to the electrical generation and the extraction flow rate is:

$$M_{THR} = \left[\frac{\dot{E} - \frac{\dot{E}_B}{2}}{\frac{\dot{E}_B}{2}} \right] (\dot{M}_{THR,B} - \dot{M}_{THR,1/2B}) + \dot{M}_{THR,1/2B} + \dot{M}_{EXT} \left[1 - \frac{C(H_{THR} - H_{EXT})}{(H_{THR} - H_{EXH})} \right]$$

where

$\dot{M}_{THR}$ = throttle flow rate, lb/hr
$\dot{M}_{THR,B}$ = throttle flow rate at point B
$\dot{M}_{THR,1/2B}$
  = throttle flow rate at $\dot{E} = \frac{1}{2} \dot{E}_B$
$\dot{E}$ = electrical generation, kW
$\dot{E}_B$ = electrical generation at point B
$\dot{M}_{EXT}$ = extraction flow rate, lb/hr
H_{THR} = enthalpy of steam at throttle valve, Btu/lb
H_{EXT} = enthalpy of steam at extraction valve
H_{EXH} = enthalpy of steam leaving last turbine stage
C = dimensionless empirical factor,
  ≈ 0.857 when exhaust pressure ≤ 1 atm
  ≈ 0.902 when exhaust pressure ≥ 1 atm

For the special case of a backpressure turbine with no extraction, $\dot{M}_{EXT} = 0$, and $\dot{M}_{THR}$ becomes a single-line function of $\dot{E}$.

The fuel energy required for a steam turbine system is:

$$\dot{F} \text{ (Btu/hr)} = \frac{\dot{M}_{THR}(H_{THR} - H_{FW})}{\eta_{BLR}}$$

where

H_{FW} = enthalpy of feedwater (mixed makeup plus condensate return) entering water treatment, Btu/lb

η_{BLR} = overall boiler efficiency, dimensionless

VI. ENERGY/ENVIRONMENTAL LEGISLATION AND REGULATIONS

Legislation and regulations are subject to changes and revisions. It will be necessary, therefore, to verify compliance of the cogeneration system design with appropriate legislation and regulations in effect at the time and in the jurisdiction in which it will be located. The energy and environmental mandates in effect or recommended as of 1979 provide some insight, however, into the types of legislative/regulatory matters that are likely to apply to cogeneration, now or in the future [7]:

1. The Powerplant Industrial Fuel Use Act provides that new powerplants or fuel-burning installations of a single unit having a design fuel heat input of 100×10^6 Btu/hr or greater, or that result in two or more units at the same site having a combined design fuel heat input rate of 250×10^6 Btu/hr or greater, are prohibited from burning natural gas or petroleum, unless an exemption is provided by the secretary of energy. However, the secretary of energy is specifically authorized to exempt cogeneration facilities from the prohibition if the benefits of cogeneration are otherwise unobtainable. (Guidelines for exemption have been formulated.) As points of reference, 100×10^6 Btu/hr corresponds roughly to a 7,500 kWe combustion turbine or a 4,500 kWe steam turbine system. It is seen that rather small installations, which would be likely to involve natural-gas- or petroleum-fired diesels or combustion turbines, are not restricted in choice of fuels under the Fuel Use Act.

2. The Public Utility Regulatory Policies Act of 1978 directs the Federal Energy Regulatory Commission (FERC) of the Department of Energy to develop regulations for encouraging cogeneration and small power-production facilities that produce electric energy solely by the use, as a primary energy source, of biomass, waste, renewable resources, or any combination thereof; and has a power-production capacity that, together with any other facilities located at the same site, is not

greater than 80 MWe. The regulations developed by FERC must include provisions to ensure that utilities buy or sell power from these types of facilities at equitable prices. Fossil fuels (coal, petroleum, and natural gas) are not considered renewable resources.

3. The Clean Air Act of 1977 requires that each state submit documentation to EPA of the attainment status of its air-quality-control regions for each of six pollutants for which national ambient air-quality standards have been set. Areas with air quality better than the standards would be designated as areas of prevention of significant deterioration (PSD), while an area where the air quality does not meet the standards would be termed a nonattainment area (NA). EPA has proposed new source performance standards for new or modified steam-electric units capable of combusting more than a heat input of 250×10^6 Btu/hr of fossil fuel. Performance standards for new sources would apply to modified or reconstructed facilities also, where the cost is 50% or more of the cost of replacing the existing powerplant. Another provision is that addition of pollutants to the atmosphere from new sources in NA regions must be more than offset by the further removal of pollutants from nearby existing sources by means of shutdowns, process changes, or additional pollution abatement equipment. All existing facilities owned by the company commissioning the new plant must be in compliance with applicable emission limits and standards.

4. Under the Resource Conservation and Recovery Act, EPA is empowered to identify and regulate hazardous wastes. Currently, powerplant wastes (flyash, bottom ash, scrubber sludge) are called "special wastes" and "problem wastes," and as such can be disposed of in sanitary landfills rather than in hazardous-waste facilities. The characteristics of solid wastes from powerplants are still being examined, however.

5. Discharges of heavy metals and toxic pollutants are controlled under the Clean Water Act. A powerplant must have a national pollutant discharge elimination system (NPDES) permit to discharge treated wastewater directly into a navigable waterway, and discharges to municipal sewers usually have to be in compliance with some wastewater quality guidelines established by the local sewage treatment agency. Some potential impacts of the Clean Water Act may be future regulations for: (a) a tank or cover over coal piles to prevent rainwater from picking up pollutants from the coal and flowing into streams, lakes, or sewers; (b) containment dikes to prevent coal-pile runoff; (c) ash pond linings to prevent seepage of pollutants into the ground water; and (d) prescribed or controlled cooling tower biocide-treatment practices to reduce toxic substances.

6. EPA has proposed emission standards for stationary engine/generator sets, including rather small units used in cogenerating facilities. Separate standards are proposed for combustion turbines and diesels. NO_x reduction is a primary goal. For combustion turbines, water injection during the combustion process appears to be the means of meeting the requirements. It is basically the responsibility of new engine manufacturers to design their products to comply with standards that are adopted. Some future modifications of existing noncompliant engines may be required if they are used in a cogeneration system.

7. In addition to the federal legislation described above, each state maintains its jurisdiction over utility services through a public utilities commission (PUC), or comparable body. A PUC may establish regulations or issue rulings affecting potential cogeneration applications. States may also have special tax incentives or other legislation favorable to cogeneration, of which the PUC would be aware. Since specifics differ from one state to another, the cognizant PUC would have to be contacted for relevant information regarding state legislation and products of its own authority regarding cogeneration in general, or for comments on a specific application of cogeneration.

VII. INSTITUTIONAL CONSTRAINTS

Human preferences and opinions can have a strong influence on the consideration being given to cogeneration system alternatives. Those factors that are not solidly supported on technical or economic grounds, or mandated by legislation or regulations, are termed "institutional" factors. Some examples of matters that often have to be dealt with in order to implement cogeneration are:

1. Corporate resistance due to concern that the plant will become regulated by the PUC.
2. Restrictive utility company policies on standby charges or ownership/operation options in order to obtain connection to the grid.
3. Corporate policy "not to get into the utility business."
4. Unavailability of skilled personnel to operate the new facility.
5. Reluctance of operating and maintenance personnel to assume new duties and responsibilities.
6. Concern over job losses if the system is operated by the utility company.
7. Community resistance to placement of the generating facility.

Institutional factors are often very difficult to deal with. They are not generally quantifiable. Their root cause is often difficult or impossible to determine. They may even be contradictory. The best approach is to be vigilant for the existence of institutional barriers and to express the nature of the barriers in the clearest possible terms when they are encountered. Clear definition of a barrier is necessary for open examination, which is a big step toward solution.

VIII. IMPLEMENTING COGENERATION

The preceding discussions have pointed out that implementation of cogeneration can be a complex process involving technical, economic/financial, legislative/regulatory, and institutional factors. Guidance relevant to these factors has been outlined, and the need for a great deal of information has been discussed. This final section will describe a likely sequence of tasks leading from the seeds of thought regarding cogeneration to its implementation. Also in this final section, likely sources for much of the required information will be listed.

First, look at typical tasks involved in proceeding from concept to hardware. For a particular site, of course, circumstances may preclude some of the steps from occurring, or change the order, or even extend the process. However, these are basically the steps to be expected:

1. Question whether the existing thermal and electrical utility service is best for the facility, and whether cogeneration holds potential for achieving energy or cost savings. If changes such as an expansion or conversion to a different fuel are planned anyway, the consideration of incorporating cogeneration is often in order.
2. Conduct an energy audit or survey. Gather and analyze information on existing and anticipated loads, costs, and alternatives for conservation.
3. Formulate concepts of utility system alternatives. At this point, an approach somewhat like brainstorming is beneficial. Be open to various cogeneration options, including different types of equipment capacities, and ownership/operation arrangements.
4. Determine constraints applicable to the various options and eliminate infeasible concepts from consideration. This is really a critical step. It is easy to eliminate a feasible approach by perceiving something as a constraint that really is not. For example, it is easy to rationalize that no utility company would be interested in participating in the cogeneration project under consideration. By exploratory discussions, however, it may be learned otherwise. Make sure that constraints used to eliminate various approaches are well founded, and be open to reconsidering the constraints and their impact later if conditions change.
5. Perform elementary performance estimates, environmental assessments, and cost estimates on promising alterna-

tives, and thereby make a preliminary ranking.
6. Discuss acceptance of alternatives with management, utility companies, permitting agencies, fuel suppliers, and other potentially involved parties. From discussions, define items requiring negotiation and further clarification.
7. Reassess the alternatives and rankings, as necessary.
8. For the preferred alternatives, identify any contractual arrangements that are critical to the success of the alternative. Conduct discussions with involved parties, arriving at "agreements in principle" on the critical items.
9. Select an alternative for the design. Proceed with the design to an intermediate design review point. Formulate an environmental impact statement (EIS), if necessary.
10. Review the intermediate design. Submit the EIS and permit requests.
11. Complete contract negotiations with involved parties.
12. Complete the plant design.
13. Proceed with construction of the plant.

Potential involved parties and sources of information for various aspects of cogeneration are indicated in Table 24-2. In many cases there is more than one source for particular pieces of information or data.

Quite often the talents and insight of consultants and designers experienced or specializing in the cogeneration field are necessary in order effectively to address factors arising in deciding upon, designing, and successfully deploying a system. Assistance in recognizing and dealing with the tradeoffs between technical performance, economics, and obstacles to implementation can be most beneficial. The interest in cogeneration of numerous industrial firms, commercial concerns, and institutions as a means of achieving energy and cost savings, the encouragement by federal agencies and

Table 24-2. Sources of Information for Decision on Cogeneration.

Source of Information	Total thermal loads, magnitude and profile	Total electrical loads, magnitude and profile	Cooling loads	Major load centers & energy consumers	Anticipated load changes; mission or function changes	Waste heat sources	Waste fuel sources	Complementary off-site loads	Present electrical energy costs & rate formula	Projected electrical energy rate structure	Policies toward parallel generation	Ownership/operation policies & preferences	Cogenerator rate structure: Standby charges, reliability requirements, payment for power to grid	Environmental & siting constraint overview	Tax & investment incentives	Regulations relative to cogeneration	Present fuel costs	Projected fuel costs	Projected fuel availability	Fuel characteristics & properties	Performance data: Design & off-design conditions	Fuel consumption	Fuel flexibility	Emissions data & specifications	Air emissions regulations	Other emissions regulations	Fiscal policies	Funding sources	Cost and conditions of financing	Siting restrictions
Energy audit	x	x	x	x	x	x	x	x						x																
Utility company									x	x	x	x	x		x	x	x		x	x										
Fuel suppliers																	x	x	x	x										
Oil, NG, coal																														
Equipment suppliers																					x	x	x	x	x	x				
Management					x																			x	x					
Public Utility Comm.											x	x															x			
Air quality district																									x					x
Sanitation & water quality districts																										x				x
Special authorities (e.g., zoning, airport, coastal)																														x
Financial institutions																												x	x	

Source: Ref. 4.

some states, and the growing acceptance by utility companies are all helpful in removing the obstacles to cogeneration.

REFERENCES

1. Resource Planning Associates, "Guidelines for Developing State Cooperation Policies, April 1979," Report No. HCP/M 8688-01, prepared for the U.S. Department of Energy Office of Industrial Programs under contract No. EC-77-01-8688, Washington, D.C., April, 1979.
2. Bechtel National, Inc., "Coal-fired Boilers at Navy Bases, Navy Energy Guidance Study Phases II and III," Report No. CR 79.012, prepared for the U.S. Navy Civil Engineering Laboratory under contract no. N68305-77-C-0003, Port Hueneme, Calif., May, 1979.
3. Bulletin 3382270 from The Cummins Congeneration Company, Suite 1134, Empire State Bldg., New York.
4. E. E. Cooper, "Cogeneration Systems," Technical Report R879, U.S. Navy Civil Engineering Laboratory, Port Hueneme, Calif., June, 1980.
5. Booz-Allen & Hamilton, "Procedures for Feasibility Analysis and Preliminary Design of Total Energy Systems at Military Facilities," Report No. BA 9005-454, prepared for the U.S. Navy Civil Engineering Laboratory under contract no. N62399-73-C-0029, Port Hueneme, Calif., July, 1975.
6. A. G. Christie and W. Trinks, *Kent's Mechanical Engineers' Handbook, Power Volume* (J. Kenneth Salisbury, ed.), 12th ed., John Wiley & Sons, New York, 1950, Section 8.
7. J. E. Levin, "Know Key Energy, Environmental Laws," *Power, 1979 Generation Planbook,* McGraw-Hill, N.Y., N.Y., 1979, pp. 165–183.

25. Employing Solar Thermal Hybrid HVAC/Power Generation Systems in Buildings and Community Energy Systems*

Milton Meckler
President, The Meckler Group
Encino, California

	Page
I. Introduction	385
II. Range of Solar Variability	386
III. Availability: The Key	387
IV. Building Design Parameters	387
V. Utility/Solar Perspectives	388
VI. Rankine Cycle	389
VII. Rotating Magnetic Fields	390
VIII. Storing Community/On-Site Power	393
IX. Rankine Drive Train	396
X. Summary	397
References	397

I. INTRODUCTION

Current studies suggest a multimillion-dollar domestic solar heating and cooling industry will be created, with substantial impacts in solar power contribution to U.S., Mediterranean, and Middle East energy needs, "during 1980–90 and in broad terms through 2000." Although so-called solar applications also include windpower, bioconversion, ocean thermal gradients, and photovoltaic (electric) power, one important distinction is that the solar heating and cooling category is probably the most advanced in terms of practical demonstration.

Here we will address the problem of integration of promising solar augmentation concepts with conventional building and community electrical power distribution systems. In particular, we will discuss the unusual opportunities that exist at the utility interface for reducing demand, growth, and associated environmental problems.

A major consideration in the engineering and economic analysis of any solar thermal power generation system is the optimum sizing of system components to minimize life cycle costs. In achieving the optimum, it is often necessary to introduce changes in the building envelope (i.e., reduce glass ratios or improve shading, insulation, or building mass). The problem becomes one of balancing two opposing criteria, namely: the solar powered system cost, which is treated as a system input, versus the performance of the output of the system,

---

*Portions of this chapter, and all figures and tables, are reprinted from "Employing Solar Thermal Hybrid HVAC Power Generation Systems in Buildings and Community Energy Systems," conference of the International Congress on Building Energy Management, May 14, 1980, published by Pergamon Press Ltd, Oxford, England.

Table 25-1. Characteristics of Principal Concentrating Solar Collector.

MANUFACTURER	CONCENTRATING METHOD	ESTIMATED OPTICAL CONCENTRATION MAXIMUM	RECEIVER TYPE
A	Front reflecting	8–10	Integral tube structure behind concentrator
B	Fixed, tilted, reflective segments on movable cylindrical shell—back reflecting	30+	Movable gas-filled tube
C	Reflecting segments on movable cylindrical shell—back reflecting	30+	Flat plate + tubes on movable axis
D	Parabolic trough—back reflecting	30+	Coated tube surrounded by evacuated space and glass outer tube
E	Separately movable mirror segments—back reflecting	8–10	Reflective secondary concentrator with glass cover
F	Fresnel lens, front reflector	8–10	Flattened tube behind concentrator
G	Separately movable mirror segments—back reflecting	30+	Glass "vee" at 60°F in front of absorber tubes

including the building heating, ventilating, and air conditioning (HVAC) systems. This process inherently recognizes that undersizing of such systems implies that each dollar invested is not gaining its full return. Similarly, an oversized system requires the building owner to pay excessively (versus purchased utilities) for the combined utilities and electrical energy supplied under desired matching load conditions.

Careful analysis of the building design parameters on the building mix ultimately selected for study is essential for practical results. Optimization techniques exist for determining minimum cost with respect to solar collectors required for a given system and desired peak power level. Also important are the matters of selective coating, collector tilt, collector size, and thermal storage volume. Representative concentrator type collectors suitable for solar thermal power generation are listed in Table 25-1.

Consideration can be given to the degree to which modifications to building design can be employed to optimize a given solar system configuration capable of producing heating, cooling, and power generation. Solar systems that exhibit specific advantages for particular building types and/or climatic regions can be clearly identified. For each of the building design combinations established, ranges of parameter variations can be described to cover arbitrary design choices. However, introduction of passive means (i.e., building mass) or other design parameters into the picture can materially shift the balance among a building's heating, cooling, and power needs [1]. When considering the use of solar powered thermal engines, one should recognize that the special engine and useful work to be obtained are subject directly to variations in the prevailing solar flux, which varies with the time of day and season of year.

II. RANGE OF SOLAR VARIABILITY

Obviously, the economic feasibility of any solar powered building and community energy system will vary according to its geographic location. On a typical day in June, for example, anywhere from 20.9 to 29.3 MJ (megajoules) of solar flux will impinge on most parts of the United States, whereas in December, when the days are shorter and the sky generally cloudier, only 4.18 to 12.6 MJ can be expected. Furthermore, in June, both Saskatoon, Canada and Tampa, Florida receive about 25.1 MJ of solar flux daily. In December, however, the amount of solar flux received in Saskatoon drops to a mere 3.14 MJ a day (roughly 14% of the June value), while Tampa receives in excess of 12.6 MJ or roughly half of the amount received in June.

Solar collector costs typically vary in proportion to the collector area. However, the amount of solar energy that can be captured by a given collector type of known area varies less predictably. This is particularly true with the concentrator types that would normally be required to deliver the necessary solar heated fluid temperatures essential to efficient operations. Thus, in December a solar collector in Saskatoon would require four times the area needed in Tampa to supply the equivalent amount of energy. Consequently, the cost per unit of solar energy delivered may be considerably greater in Saskatoon than it is in Tampa [2].

Utilization of direct solar energy capture is relatively favorable in December in the so-called Sun Belt areas, which are comparable in latitude to Mediterranean and Middle East countries. Solar energy is also plentiful in most of the northern populated regions in June. However, applying current economic realities, direct use of captured solar energy can be expected to be only marginally economic north of some particular latitude. Indirect use of captured solar energy (i.e., also as a heat source for heat pumps) can significantly alter the economics for most northern U.S. and European areas.

III. AVAILABILITY: THE KEY

To reduce worldwide demand for fuels consumed by buildings and community energy systems, we must improve the efficiency of solar energy use. To do so we must concentrate on Second Law effects. A very large fraction of the fuel consumption for any size building goes to provide low temperature heat and uses, such as domestic hot water heating in the range of 43 to 49°C or space heating in the range of 32 to 43°C. Consuming a high quality fuel at 982°C or more to obtain low temperature heat would be fundamentally wasteful even if all of the heat liberated by combustion were to be transferred without any loss (reversibly). Heating hot water by direct fuel combustion (and assuming a 70% heat transfer efficiency) requires approximately eight times as much fuel as a reversible heat engine or a Rankine cycle expander driving a heat pump [2].

In short, work is intrinsically a more valuable form of energy than its equivalent Btu content as heat. The measure of the value of fuel depends upon the extent to which its energy content can be converted to work. Once work can be obtained from it, it may be used to provide low temperature heat. The availability principle can be used to identify inefficient practices employed in the transfer of solar energy to building HVAC systems. For example, in conventional oil- or gas-fired space heating boilers, the ratio of the thermodynamic availability required to that consumed would be less than 0.10. For electrically driven refrigeration equipment the same ratio would be less than 0.12. By means of concentrating solar collectors of the type listed in Table 25-1, we are able to utilize solar energy at higher temperature levels to do useful work (i.e., to drive a heat recovery chiller, or heat pump) and to stage the remaining thermal energy sequentially at progressively lower temperature levels for building space and domestic hot water heating needs.

IV. BUILDING DESIGN PARAMETERS

In evaluating the compatibility of solar heating and cooling systems within large buildings, key parameters include construction materials, massing, orientation, fenestration, functional use, occupancy cycle, etc. Low-rise buildings with greater roof/floor area ratios are usually better candidates than high-rise buildings packaging the same floor space. Institutional building owners employing life cycle costing can often secure low-interest loans and are thus often better than comparable candidates in the private building sector. Buildings with the greatest total hours of use maximize the productivity of solar systems. Buildings designed for low heating and cooling density, which can satisfy most of the annual energy requirements, also require less supplementary fuel or electricity. In short, well-planned buildings designed with energy conservation in mind provide the most attractive profile for solar heating and cooling systems. It is also advantageous to explore all types of energy storage, including thermal and electrical.

Thermal energy storage can be provided as hot or cold thermal mass. Chilled water storage can be employed to avoid high electrical demand charges. A lower system first cost will result from reducing installed chiller or boiler capacity by allowing it to operate over a longer period of time and to accumulate thermal storage for use during peak HVAC system demand periods. Running such equipment at higher percentages of full load also results in greater equipment operating efficiency [3]. In increasing the thermal mass of the building envelope, care must be taken to avoid applications where build-up of generated heat (e.g., from computers) would benefit more from a lighter envelope, thus negating the benefits of thermal mixing.

Mechanical storage, volumetric, and weight requirements must be checked to ensure a maximum thermal capacity within a minimum volume for required HVAC system peak demand periods. Employing night setback or system shutdown can result in peak heating (or cooling) demand for several hours and bring on excessive use of supplementary energy systems. Furthermore, it can deplete thermal storage that would be better reserved for peak shaving. Solar systems should be located close to primary HVAC or domestic water systems to minimize distribution losses. Another positive effect of thermal storage is its use in peak shaving as well as shifting the time of day that such peak demands occur.

V. UTILITY/SOLAR PERSPECTIVES

The cost of electricity as reflected to utility customers, both in the United States and elsewhere, is the average of all utility costs plus profits [4]. The public utility as manager of the nation's electrical energy distribution system must deal constantly with the true and marginal replacement cost of energy sources. By and large the solar industry itself appears deeply divided on the degree of utility involvement. Some manufacturers see both electric and gas utilities as an opportunity to channel a large market for their solar products; whereas others, particularly smaller manufacturers and installers, seem threatened by what they perceive to be an unfair competitive advantage. This fear stems from a prior history of utility involvement—which, however, has many attractive advantages, for example, as a credible conduit for accelerated marketing and readily available maintenance quite apart from any involvement in manufacturing or installation.

Recent dramatic rises in the cost of fuel and problems with nuclear and environmental concerns find the once tranquil U.S. investor-owned electric utilities caught in a web of uncertainty in dealing with energy growth (or lack of it), regulations, inflation, investor capital, bank financing, etc. Even with the soaring cost of fuel and financing, the utilities must face unsympathetic public utility commissions, now awarding on the average 50 to 60% of their rate increase requests [5]. Although some U.S. utilities now operate at growth rates below pre-1973–74 Arab oil embargo levels, many others face the same or higher growth rates yet cannot afford to build new nuclear or coal-fired plants. Therefore, in spite of today's adequate high reserve margins, some regions of the United States face the possibility of brownouts in the 1980s. As a result, pressure will be on the U.S. government to construct new oil-fired plants requiring far less lead time and an estimated 400 million barrels of oil each year over current projections. In spite of the fact that utilities now face a vicious circle of customer anger, investor disenchantment, and a more hostile regulatory climate than ever, they still are expected to maintain a clear sense of their "public responsibility" in providing a reliable distribution system capable of anticipating and providing for increased demand. Yet, many utility executives feel that curtailments are inevitable unless something can be done to alter the self-accelerating downward spiral caused by the traditional approach to the supply–demand utility generation mix.

In coming to grips with structuring new electric power rates that reflect the real costs of generating electricity, utilities should encourage solar opportunities that improve load management. System peaks should not be encouraged if we are to minimize social costs in terms of capital and natural and human resources. We must find a more effective means of providing the equivalent of standby electrical reserve than to build bigger utility plants with higher pass-through costs to the customers.

Perhaps what is needed is a clearer awareness on both sides of the opportunities and concerns of the building/electric utility interface. Inherent in the use of solar powered Rankine cycle systems is a recognition of the realities of base and peak load utility costs and the solar influence on such utility system peaks, so as to capitalize on inherent solar energy/load management opportunities when tied to passive building elements and/or active thermal storage systems [6].

VI. RANKINE CYCLE

For the purposes of our discussion we will define the solar driven thermal engine as characterized by a Rankine cycle expander. In its simplest form, the Rankine cycle engine consists of a working fluid, a boiler, a condenser, an expander, and a feed pump. Heat is supplied at the boiler, which provides saturated (or superheated) vapor to the expander. Power is extracted in the expander, after which saturated liquid is supplied to the feed pump where it is repressurized and returned to the boiler, thereby completing the cycle. Heat is normally rejected from the working fluid at the condenser. Expander shaft work transmitted through a speed-reducing gearbox can be used to drive a refrigeration compressor or generator to produce electrical power.

Figure 25-1 illustrates the pressure enthalpy diagram for the Rankine cycle, in which a regenerative heat exchanger is also shown, as is initial cycle expansion. The initial cycle expansion is indicated as starting at state 3 on the saturated vapor line and proceeding at constant entropy to state 4 in the superheat region. The regenerator transfers the enthalpy quantity $h_4 - h_5$ from the vapor to the cooler pres-

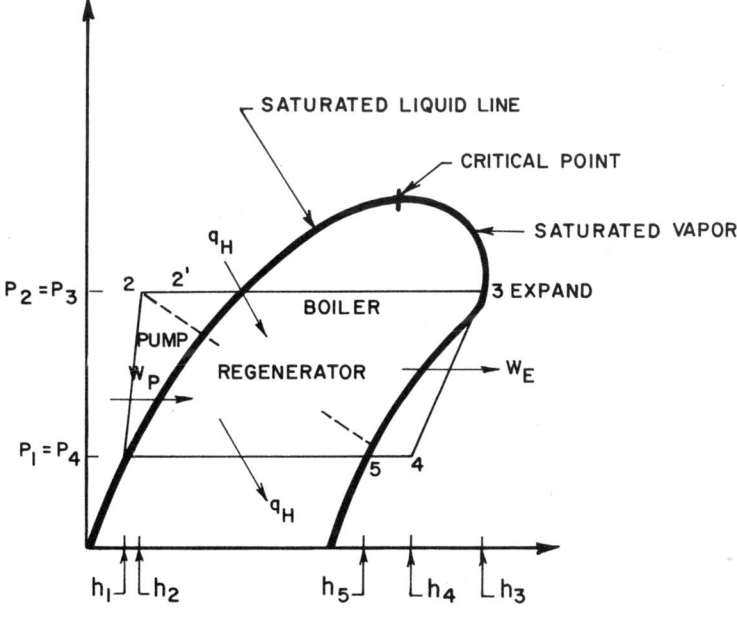

Figure 25-1. Pressure enthalpy diagram for the Rankine cycle.

surized liquid, raising the liquid enthalpy from h_2 to h_3. However, the primary factors governing the performance characteristics of a Rankine cycle are the working fluid and the operating temperatures and pressures. Of these, temperature has the strongest effect on cycle performance. In particular, for a given fluid, the choice of the fluid state at the beginning of the expansion process has a strong influence on the cycle efficiency. It can be shown that the efficiency of the conventional regenerative cycle η_e can be expressed as:

$$\eta_e = \frac{W_E - W_P}{q_h} = \frac{(h_3 - h_4) - (h_2 - h_1)}{(h_3 - h_2) - (h_4 - h_5)}$$

But why bother to get perhaps only the equivalent of 15 to 20% useful shaft work when we could use *all* of the solar energy captured as heat, which can also be used to obtain cooling through, for example, an absorption machine? For one thing, a recent study has shown that solar powered heat pumps are technically preferable to direct heat exchangers for heating applications of the absorption cycle heat pump. The principal reason for this conclusion has to do with the principle of availability, which, until the recent energy crisis, was taken seriously only in academic engineering circles.

The lowest-first-cost, most reliable, and most maintenance-free energy recovery device that can supply electrical power is the induction motor-generator, provided we have available an external power system and a source of unused energy.

VII. ROTATING MAGNETIC FIELDS

The invention of the polyphase electric system and the rotating magnetic field of the polyphase motor have made commutation of a magnetic field possible without a need for brushes or commutator bars. The three-phase polyphase system most commonly used for electric power generation and distribution consists of three single-phase systems interconnected and placed in such a position on the generator stator that the voltages induced in each phase of the system follow the preceding phase voltage by a time phase displacement of 120 electrical degrees. When this voltage is applied to a second set of windings that are arranged about a stator core in the same spacing of 120 electrical degrees per pole, the currents created will produce magnetic fields that rise

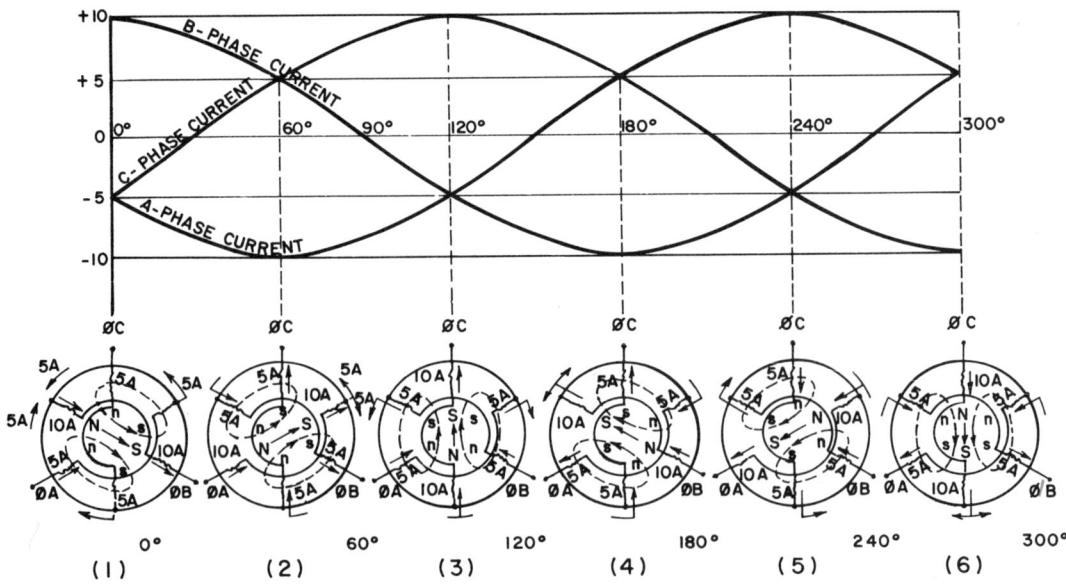

Figure 25-2. Development of a rotating field.

and fall in value in such a sequence that the north and south magnetic poles will alternately succeed each other and as a result appear to progress or rotate around the stator field of the motor. This synthesized rotating field thus becomes the means by which alternating current power can be converted to electrical power without the use of any switching or commutating devices, as shown in Figure 25-2.

Within the motor stator of a three-phase motor there must be a phase winding for each pole of the motor. The windings themselves are laid into slots in a distributed type stator that limits the loss of magnetic flux and allows for a uniform transfer of the magnetic polarities from phase to phase and pole to pole.

Once the rotating magnetic field is established, the placing of a rotor made of bars of copper or aluminum in the form of a squirrel-cage within the stator core allows a current to be induced into the rotor. The reaction between the magnetic flux of the rotating field and the flux from the current induced in the rotor bars causes an attraction between the two magnetic fields and the rotor then follows after the rotating field.

Both polyphase and single-phase induction motors have similar operating principles. The second reacting magnetic field that causes rotating torque is induced in a secondary circuit of squirrel-cage like rotor bars by the alternating current in a primary winding in both types of motors. Generally the squirrel-cage rotors of both classes of motors are similar in design and construction. The major differences between the two winding types are in the primary winding of the stator and the methods by which the rotating magnetic field is created and maintained.

Where ruggedness and simplicity are necessary, the squirrel cage motor is without doubt the principal choice for most constant speed drive requirements. The wound motor induction finds its principal application in situations where variable speed, highly controlled torques, and/or low starting torques are desired. Its starter windings can be arranged in multiples of two to provide any reasonable number of poles. If we look at its full load torque versus speed in the manner shown in Figure 25-2, we find that, assuming a starting torque of 120% of full load torque (F.L.T.), as the motor approaches synchronous speed, it rises to 245% of F.L.T. at 85% speed, followed by a drop to F.L.T. at 98% speed (i.e., representing 2% slip). As one would expect, at synchronous speed we develop no torque. However, as we drive the rotor on its synchronous speed in the same direction, the motor continues to draw a magnetizing component from the line but now delivers a power component. Thus it can be made to operate as an induction generator delivering rated load (i.e., at a negative 2% slip and at normal lagging power factor).

In essence, the squirrel cage motor operates as a transformer with a shaft, and its locked rotor torque, full load slip, and locked rotor KVA principally depend upon the resistance and reactance of the (squirrel cage) winding. Even if the rotor cage resistance is changed to maximum torque, it remains essentially the same, since, as in a transformer, a high-resistance secondary winding causes a relatively low locked rotor KVA.

Now one adds to the induction motor a prime mover and an overspeed protection device, and the motor will be ready to generate. In this device one can bring a load to speed and drive it, to regulate the speed of another prime mover or produce electricity. Such equipment is commercially available with a minimum of controls. System voltage and frequency are established at the induction motor-generator terminals. The speed of the generator then determines the amount of generated current and its corresponding power factor. The generator has its design rated output in the same way the induction motor has its name-plate loading.

The induction motor-generator automatically operates as a generator in the overspeed range (i.e., above its synchronous speed) for reasons to be seen later. It can perform several duties in a string of equipment to which it is interconnected by means of a common shaft. At times when the engine expander is unable to extract available energy or is down for repairs, it operates as a motor to direct drive equipment and can also serve as a speed controller.

Typically, the induction motor-generator

392 9/RETROFITTING AND POWER GENERATION

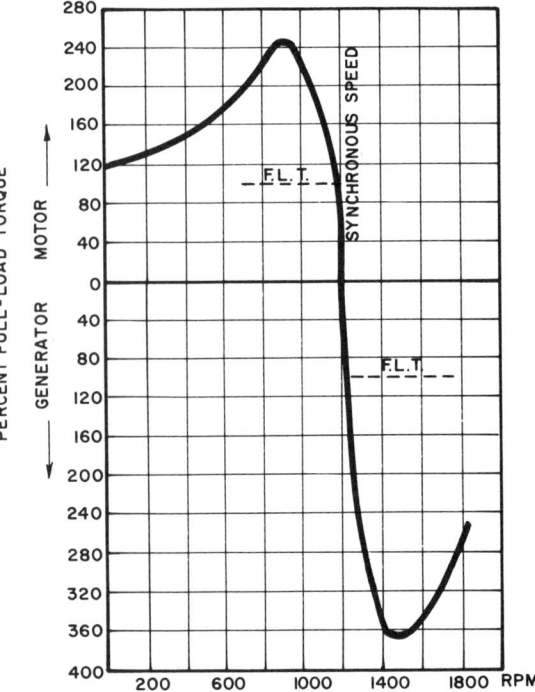

Figure 25-3. Typical torque curve of induction motor/generator versus rpm.

should, for part of the time, produce up to twice its rated torque or generate up to twice its rated kilowatts, thus providing (or consuming) torque and keeping its equipment train within ±2½% of its rated synchronous speed. The motor-generator is inherently a load matching device. The performance curves of a typical induction motor-generator are shown in Figure 25-3. Its efficiency normally remains high (i.e., about the same as a synchronous generator). From rated load to one quarter load, the power factors drop off gradually from near rated load.

Several firms now manufacture expander/induction motor-generator sets for waste energy recovery rated from 100 to 3,500 hp [5–7]. Basic criteria in the selection of a working fluid for a solar powered Rankine cycle [8] are the temperature level of the energy available from the solar collector and the condensing temperature. Chemical stability considerations generally limit maximum cycle temperatures for organic fluids to approximately the 204 to 427°C range, depending on the fluid. Use of organic fluids in the Rankine cycles has some advantages over water. Some typical fluids are listed in Table 25-2.

Although Rankine cycle generation of electrical power from solar energy has been demonstrated, cycle efficiencies have remained low. Application of a solar powered Rankine cycle incorporating the induction motor-generator and a means to utilize the energy normally wasted at the condenser can permit a significant improvement in cycle efficiency when combined with heat recovery chiller thermal storage and operated as a selective power system for use or storage of excess electricity. Solar generated excess power can be used as a positive factor in providing meaningful energy management at the utility interface. Studies of large building HVAC systems suggest that a reduction of 50% in compressor size and in peak power requirements and 6% in the total energy used by the conventional HVAC system can be achieved for a continuously operating system.

Table 25-2. Typical Fluids Suitable for Use in Rankine Cycles.

	APPROXIMATE THERMAL STABILITY LIMIT (°C)	WET VAPOR PRESSURE AT (32.2°C.) (kN/m²)	MOLECULAR WEIGHT	SATURATED VAPOR LINE ON T.S. DIAGRAM	NORMAL BOILING TEMP. (°C.)	CRITICAL POINT (°C/ MN/m²)
Thiophene (CP-34)	288	14.5	84.1	(1)	84	307/5.4
Monochlorobenzene (MCB)	315	4.1	112.6	Positive	132	358/4.5
Water	—	4.8	18.0	Negative	100	374/22.1
FC-76	315	1.9	536.0	Positive	102	227/1.6
Refrigerant-114	177	268.	170.9	Positive	4	145/3.3
Refrigerant-113	177	813.	187.4	Positive	48	214/3.4
Dowtherm A	399	3412.	166.0	Postive	257	497/3.1
Flourinal-85	315	1144.	88.0	Negative	74	240/6.4

(1) Changes from negative to positive as temperature rises through 150°F (66°C).

It is also possible to employ the organic cycle Rankine engine [9] with powered bottoming cycle systems operating in conjunction with a conventional diesel-fueled prime mover power generator. Solar energy can also be utilized in the manner shown to provide a flexible cogeneration alternate that permits decoupling the chiller requirements somewhat from power peaking or base load generation features for a more advantageous match [10] with on-site conditions and to serve utility needs, as we will see later.

Recognizing the marked increase in interest in cogeneration systems among small industries, institutions, and owners of commercial office buildings in the 500 kW to 10 MW market, such hybrid or stand-alone solar powered systems should find increasing use among consultants. If we couple the rapidly escalating costs for central power plant generation with a steeper rate of use of waste heat since 1973, a variety of cogeneration schemes can now be considered back on the boards. For example, applying New York City utility rates of up to 10 cents per kWh, payouts of from 2½ to 5 years are not uncommon, recognizing that employing the high-speed diesels' net operating cost of 3 to 3½ cents per kWh for efficient cogeneration facilities allows a comfortable 1½ to 2 cents per kWh margin for amortization of the capital investment. The use of hybrid solar bottoming cycles provides some hedge against concerns that fuel supplies may become critical.

Current U.S. energy legislation contains language that could protect owners of cogeneration systems from bans in their use of gas and oil. Furthermore, tax incentives and loan guarantees for investment in cogeneration equipment are increasingly cited as major administration strategies likely to be used to encourage the development of small cogeneration systems.

VIII. STORING COMMUNITY/ON-SITE POWER

Actually, the flywheel is one of the oldest of human inventions. Recent developments in materials and mechanical design suggest that the flywheel can help solve the steady increase in the use of energy and decrease the capacity of energy-consuming equipment and the impact of that use on the environment. Specifically, the flywheel offers the prospect of providing an efficient means of storing energy [11] on a large scale to help electric utilities handle peak loads and compact units to store comparable electrical energy in Rankine solar power in site conservation systems. A simple flywheel can take the form of a circular rim or hoop connected by thin spokes to a hub. The amount of energy stored in a flywheel depends on the mass of the rim and on how fast the wheel is spinning; the storage varies as the square of the rotation speed. The limit to the amount of energy stored is ultimately set by the tensile strength of the material from which the rim is made.

In particular the tensile strength must be sufficient to withstand the "hoop stress" resulting from centrifugal forces; otherwise the wheel would fly apart. As with the energy stored, these forces are proportional to the mass of the rim and increase as the square of its rotation speed. One therefore sees that the two properties of the material determine the amount of energy that can be stored in a flywheel: mass density, which provides kinetic energy, and tensile strength, which resists centrifugal forces.

In quantitative terms, the limiting amount of energy that can be stored per unit weight of flywheel material is equal to half the tensile stress at the breaking point divided by the density. In metric units, this relation is expressed in ergs per gram. These values are upper limits; practically realizable values will be from 40 to 60% of the upper limits.

Flywheels offer a means for coping with peak electrical loads. The problem the power companies face is that demands are growing while environmental concerns, in the United States at least, have delayed the moves that companies might make to expand capacity by building new "base load" plants. These companies have tended to deal with the problem by installing peaking units near the areas of demand. The units, which are commonly gas-turbine-driven generators, are turned on only during the hours when demand is heaviest. Capital costs for such units tend to be high (because of low utilization), and fuel costs are also high because gas turbines must burn fuel oil, which is not only expensive but also now in increasingly short supply.

Peak period pricing provides the most attractive means of overcoming the deficiencies in most utility systems' current rate structures. Furthermore, it is most consistent with the long-run energy conservation goals and basic principles, especially the objective of encouraging efficient use of energy without unduly penalizing inefficient but necessary consumption. Both time-of-day rates and seasonal rates are being considered, but given the relatively greater cost differentials between various periods of the day than between seasons, the time-of-day rates appear to be more important.

In California and elsewhere in the United States, privately owned electric utilities have been ordered by their respective state public utility commissions (PUC), to apply time-of-day rates to their largest customers immediately and to install appropriate meters that would permit implementing time-of-day electricity rates for all customers with demand over 500 kW. The utilities have been ordered to study the advisability of extending time-of-day rates to other customers in the future. The Wisconsin Public Service Commission issued a similar ruling in 1974 for the Madison Gas and Electric Company. It called for a study of the feasibility of time-of-day rates for its customers. Subsequent rulings for other utilities have followed a similar line. Two utilities in Vermont and one in Connecticut now offer optional time-of-day electricity rates to all residential customers.

The specific details of a time-of-day tariff will depend on the specific circumstances of each electric utility and may change over time as the pattern of energy use and the equipment available for generation change. In the Los Angeles DWP system, the pattern of costs is such that a single peak period rate and a single off-peak rate have been recently recommended as an initial tariff. The following rationale has been advised [12]:

The on-peak charge for energy should reflect the marginal cost of fuel to produce energy during the peak period of the day as well as a more than proportionate share of the allocation of actual generation and transmission costs. The off-peak charge should reflect the fuel cost of operating the most efficient configuration of generating facilities. A small charge based on individual maximum kilowatt demand may be made to recover customer-specific marginal capital costs of the transformer and distribution system immediately connected to the maximum kilowatt demand that the customer is ever expected to impose on the system, as well as any special metering equipment associated with his service.

This form of tariff has the advantage of promoting efficiency in energy use. Those customers who impose above-average costs on the system will see those costs reflected in their bills. Customers imposing below-average costs on the system (by consuming electricity at other than times of peak system demand for the system or by shifting a greater proportion of their use to non-peak hours) will see those savings reflected in their bills.

Specifically, with a time-of-day tariff, the most precise representation of system costs to the customer will be achieved if most of the cost difference is reflected in the kilowatt-hour charge. By minimizing the importance of the kilowatt charge, this rate conveys to the customer the principal message of system costs—namely, that it is not the momentary kilowatt demand placed on the system by the individual customer but rather the number of expensive kilowatt-hours consumed by the customer during the system peak period that is important. Figure 25-4 illustrates the importance of this distinction. For example, the figure shows the amount of electrical demand over a 24-hour period for three customers—A, B, and C.

As shown in Figure 25-4, all customers consume the same total number of kilowatt-hours over the entire day and have the same peak kilowatt demand. The *system* peak demand occurs from 12 noon until about 6:00 P.M. Customer A uses electricity at a very constant rate except for a brief period each afternoon. Customer B varies considerably in his pattern of

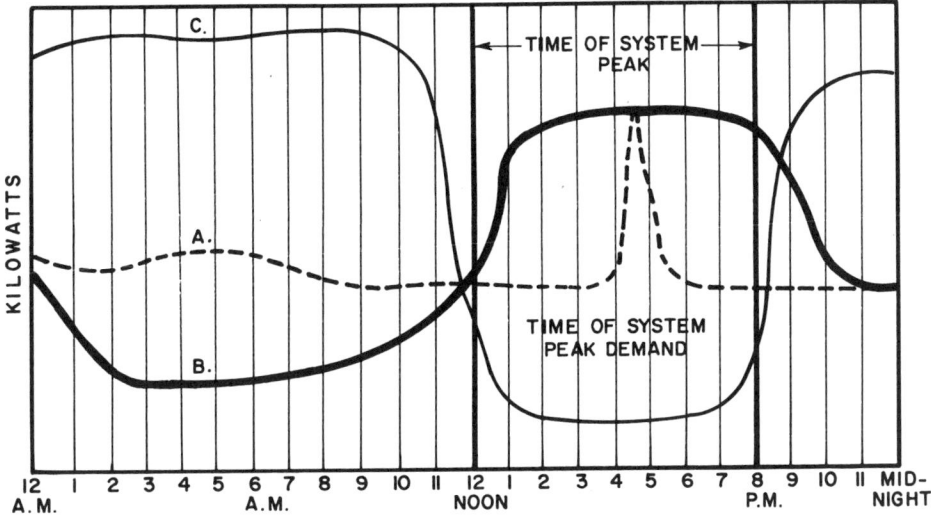

Figure 25-4. Representative variations for time-of-day rates.

use, with more than half of his consumption during the period of system peak. Under the current DWP rate structure, both customers pay identical amounts for their electricity use; however, customer B imposes significantly greater costs on the system than customer A. A tariff based on a time-differentiated kilowatt-hour charge would bill customer B a greater amount of time than customer A, thereby more accurately reflecting the greater costs B imposes on the system.

The electricity consumption of a third type of customer is also illustrated in Figure 25-4. Customer C has the same kilowatt-hour energy use and kilowatt demand as customers A and B, but this customer consumes a small fraction of his total energy during the system peak period and almost all of his energy—including his individual peak amount—outside the system peak period. This customer is the least costly of the three customers to serve, yet current DWP rates charge him exactly the same amount as customers A and B.

One way to overcome the problem of charging identical amounts to customers who impose very different system costs is to have a time-of-day tariff. Differences in energy costs and capital burden should be reflected in the kilowatt-hour charge at peak and off-periods. Electrical demand can then be monitored by utility companies to establish the maximum rate of energy usage and charges paid by their customers. The utilities reason that, since they must provide generators, transformers, conductors, etc., sized to accommodate the maximum power rate, the rate can be used to "equitably" distribute applicable capital costs.

Yet, from the user's standpoint, the average rate of usage is usually much less than his maximum. Furthermore, those maximum demand intervals are found to be of relatively short duration. Therefore, the user can either shed the peak loads through disconnection of prioritized, nonessential electrical loads approaching some predetermined demand limit, or produce the difference in electrical power on site.

Applying the above load shed strategy, let us assume that our demand limiting system through programming of appropriate algorithms has shed the demand during the peak period and redistributed it to some other time period. On this basis, the daily energy consumption in kilowatt-hours would remain the same, yet user electrical costs would be substantially lower for total electrical building usage under the proposed time-of-day utility billing procedures. The demand deficit can also be made up by on-site power generation employing a solar powered Rankine cycle system of the type described below.

396 9/RETROFITTING AND POWER GENERATION

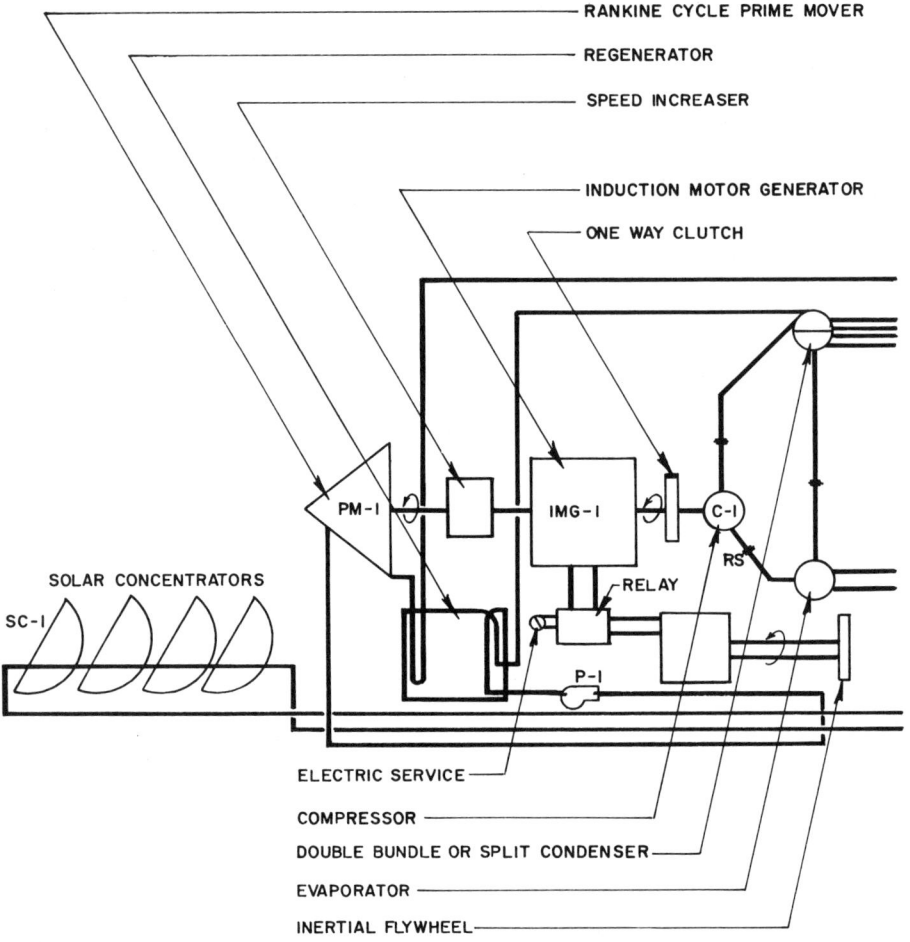

Figure 25-5. Solar power train for SPHRACS cycle.

IX. RANKINE DRIVE TRAIN

Refer to Figure 25-5. Notice that Rankine engine drive PM-1 drives CH-1 chiller compressor C-1 by means of a common drive also interconnected with induction motor-generator IMG-1 as shown. This diagram presents the power train section of the newly proposed Meckler System Group (MSG) Solar Powered HVAC/Cogeneration System, which we are currently developing. Motor-generator sets IMF-1 and MG-1 comprise induction type squirrel cage motors that operate at a speed greater than synchronous when functioning as a generator, for reasons described earlier. The drive train functions to transmit power from PM-1 to chiller compressor C-1 when net cooling is demanded, or to store the excess electrical energy through an auxiliary electrical energy shortage system comprising an inertial flywheel IF-1, which is mechanical coupled with motor-generator MG-1. The latter unit is electrically coupled with IMG-1 and may function as a motor during (HVAC system) off-peak periods to accelerate flywheel IF-1 and thereby store any excess electrical energy simultaneously generated by PM-1 through IMF-1 (i.e., operating as a generator). When operating as a generator [i.e., during (HVAC system) off-peak hours], IMG-1 can generate electrical energy that can be fed back to IMF-1 when operated as a motor. IMF-1 automatically provides speed control to the common drive train and senses when supplemental power is required by compressor C-1, causing IMF-1 to operate as a motor and drawing power from building electrical service. However, when excess solar energy is available to satisfy compressor C-1 needs, the remainder can be used to generate electricity through IMF-1 operating as a generator, through the auxiliary electrical energy storage system described earlier. Induction motor-generator IMF-1, when operated as a motor assist (or replaced PM-1), functions through operation of a one-way mechanical clutch assembly. Although excess power from solar drive Rankine cycle prime mover PM-1 is shown to be stored mechanically (i.e., by means of inertial flywheel capacity), inductive or electromechanical (i.e., batteries) options may also be employed, where competitive.

X. SUMMARY

On-site generation systems, known in the United States during the sixties and seventies principally as "total energy" systems, suffered from intense competition between the gas and electric utilities and foundered on basic plant operation and maintenance problems [13]. Performance often failed to meet projections, which often were not precise because of insufficient measurement and understanding of use patterns. In those days, the principal marketing areas centered on institutional users, shopping centers, and a number of industrial areas with high thermal load requirements.

Fortunately, a different climate exists today, and we have *more* sophisticated data on use patterns available to consultants than in earlier years [14, 15]. It is believed that the Rankine power system market will continue to develop in response to on-site power or cogeneration needs, and as a family of working fluids becomes available for power generation (i.e., in a manner similar to the evolution of refrigerants in response to a growing HVAC market), the use of solar energy will be accelerated in a variety of building applications where thermal benefits alone are not sufficient for reasonable amortization.

Although the basic design of the polyphase and split-phase motor windings has not changed to any great extent since the early 1900s, there have been improvements in rotor design, fuse or circuit protective devices, and distribution systems. These modifications have made the present alternating current motor exceptionally rugged and dependable. The most recent changes have been in the types of insulation that can withstand higher temperatures, the use of sealed roller bearings, and the reduction of the weight per horsepower ratio. The last change in some cases has proved to be a false economy as the weight reduction has been achieved by a reduction in the amount of steel used in the stator and rotor core of the motor. This reduction of the amount of steel in the magnetic circuit leaves the motor with very little reserve power output capabilities, and as a result the more recent lightweight motors often show a higher rate of failure than motors made 25 or 30 years previously when subjected to severe loading conditions. Some manufacturers are now reverting to earlier design features that provide higher efficiencies, a higher power factor, and more ample power performance reserves than any other motor made. Use of solar thermal means for power application represents a key opportunity area for developing countries and community energy systems to explore [16]. Furthermore, electric utilities on a worldwide basis will come to understand that, in the long term, use of cogeneration systems will enable customer-owned on-site power generation to retire the costly and inefficient utility peaking plants [17]. This can be done without taxing their efficient base generation plants, which will respond to system growth needs in the years ahead.

REFERENCES

1. William Ruddy and Fernando Duran, "Effect of Building Envelope Parameters, on Annual Heating/Cooling Level," *ASHRAE Journal,* July, 1975.
2. Milton Meckler, "Cost Effective Solar Augmented Heat Pump/Powered Building Systems," *Institute of Environmental Proceedings,* 1977.
3. Milton Meckler, "Solar Energy and Large Building HVAC Systems: Are They Compatible?" *ASHRAE Journal,* November, 1977.
4. Milton Meckler, "Integrating Solar Powered Heat Recovery/Cogenerator Chillers with Building and Thermal Storage Characteristics," *Changing Energy Use Future,* Vol. III, Pergamon Press, Elmsford, N.Y., 1979, 2nd International Conference on Energy

Use Management, Los Angeles, October 22–26, 1979.
5. "A Dark Future for Utilities," *Business Week,* May 28, 1979, pp. 108–111.
6. Paul Sullivan, "Solar and Utilities/Getting Along Together," *Solar Age,* III, No. 12, December, 1978.
7. Judson S. Swearingen, "Turboexpanders and Processes That Use Them," *Chemical Engineering Process,* 1972.
8. Frank R. Biancardi, Maurice D. Meader, and Elliot R. Schulman, "Feasibility Demonstration of Solar-Powered Turbocompressor Air Conditioning and Heating System," R75-951923-4, Oct., 1975, East Hartford, Conn. Distribution Category UC-59, United Technologies Research Center.
9. A. F. Veneruso, "Simulation and Operation of a Solar Powered Organic Rankine Cycle Turbine," *The American Society of Mechanical Engineers,* 76-WA/Sol-18.
10. "Feasibility Study of Solar Energy Utilization in Modular Integrated Utility Systems," June 30, 1975, prepared for the National Aeronautics and Space Administration, Lyndon B. Johnson Space Center, submitted by Arthur D. Little, Inc., Cambridge, Mass.
11. L. A. Simpson, I. E. Adaker, and J. Stermscheg, "Kinetic Energy Storage of Off-Peak Electricity," Report No. AECL-5116, Whiteshell Nuclear Research Establishment, Canada, September, 1976.
12. Mayor's Blue Ribbon Committee Report on DWP Rate Structure, Electric Rate Structure Report, Los Angeles, April, 1977.
13. Milton Meckler, "On-Site Energy Systems Stage a Comeback," *Electrical Consultant,* March, 1975.
14. Milton Meckler, "How Energy Costs in Commercial Buildings Can Be Cut 20 to 40%," *Electrical Consultant,* July, 1975.
15. Milton Meckler, "Operation and Maintenance of On-Site Energy Systems," *Maintenance Engineering,* November, 1975.
16. Ann E. Petty, "Community Energy System," *Journal of Housing* (Official Publication of the National Association of Housing and Redevelopment Officials, Washington, D.C.), Vol. 35, No. 1, January, 1978.
17. Milton Meckler, "Options for On-Site Power," *Power Magazine,* March, 1976.

26. Industrial Cogeneration

James T. Brodie

*President, Creative Power Management
Encino, California*

		Page
I.	Introduction	399
II.	Project Evaluation	400
III.	Waste Heat Steam Systems	401
IV.	Steam Topping	402
V.	Combustion Turbines	402
VI.	Waste Wood Cogeneration Facilities	403
VII.	Heavy Oil Gasification	403
VIII.	Gas Compression	404
IX.	Summary	404

I. INTRODUCTION

Nationally, we continue to have a deep and vital concern for the optimum utilization of energy forms in our facilities. Industrial companies realize energy use must be maximized before energy scarcity becomes critical enough to cause rationing and before it can damage our economic base. One way to conserve our natural resources is to increase our efficiency and effectiveness and thereby reduce the overall use and costs of fuel and power.

One conservation method is commonly called cogeneration, the simultaneous production of heat and electrical power (more commonly referred to as the utilization of wasted energy). The attitudes of electric utilities toward cogeneration have softened somewhat, to the apparent encouragement we see in some quarters today.

Cogeneration in itself is not new. In the past, a host of industrial concerns employed it in the industrial generation mode. What is new, however, is parallel operation by an industrial concern and a utility, as well as an alternative ownership (third party) option of part of, or the entire cogeneration facility by an owner other than the electric utility or the industry itself.

Industrial use of money to finance internal product-line changes is generally subject to the expected market life of the facility, and the risk associated with the venture, plus other factors. The industrial way of life generally limits forecast time to five years or less, and returns on investments to between 15 and 20%. This same formula in many cases is applied to the industrial generation project. When it is, the project never gets past the investigation stage, because the capital intensiveness of cogeneration systems does not lend itself to a five-year payback, especially when a return on investment is also applied.

Electric utilities, on the other hand, are concerned that cogeneration will affect their rate base, and operationally do not really want a whole host of small generating plants scattered across their system. Utility investment criteria are quite different from those of industry; rates of return are between 6 and 13%, and capital recovery periods are usually 20 to 25 years.

The incentive to conserve is not really enhanced in the regulated utility. Applying systems to each different industry to cogenerate would perhaps require the utility to overstaff. There are also franchise and other institutional regulations that limit the utility from seeking this form of generation.

When an industry negotiates with a utility, the industry probably has done some homework and is looking at a project from the five-year point of view. Utilities, on the other hand, start talking about ±20 years guaranteed. Obviously then, the two parties are years apart. The two get further apart when discussions start regarding who will finance, maintain, operate, construct, and design the system. Then the utility may proclaim that costs will be tied to their "average system production cost," and that standby charges, guarantees, etc., most assuredly prove that it may not be economical for the industry to generate, or it may be so marginally economical that it isn't worth the effort to the industry. These negotiations consume so much time that the original technology associated with a project may in fact be evolving so as to further stretch the time.

What is needed to help speed cogeneration into reality? Extensive investigations have revealed that the third party concept of cogeneration will be one way industries will choose. Disposing of the energy in a more favorable manner also will expedite projects.

Basically, the following steps need to be taken to complete a project:

Initial investigations
Detailed investigations
Design engineering
Financial arrangements
Construction—construction management
Utility negotiations
Disposition of energy
Maintenance and operations
Third party or utility ownership
Joint ownership

Some of the cogeneration areas that have been investigated are well known to all. However, as utility rates increase and/or power becomes in short supply, marginal, less familiar cogeneration facilities become viable.

II. PROJECT EVALUATION

In proving feasibility and financial viability of a generation project, various major areas need to be explored and evaluated.

The importance of the original analysis cannot be overstated. If a flaw exists that cannot be corrected, then the entire project need not proceed; study and negotiations should be terminated.

The following areas are briefly proposed as the major hurdles that must have conservative, satisfactory answers before we can proceed toward final negotiations:

1. *Technical:* The formula for evaluating technical risk should be quite conservative. No project should be attempted that does not have at least a ten-year history of success in installations, with the results available for technical review. The entire history should be inserted into the major areas of this risk analysis. Of course, there are exceptions; solar, photovoltaic, geothermal, wind, and other exotic systems should be left for those discussing developmental projects.

2. *Environmental:* The restraints by federal, state, and local air resource boards upon generation systems are not the only ones that need to be evaluated. Pollution by the industry itself—be it air or water—must also be considered because the financial success of the project is dependent upon that industry. Industries that lead their field in their compliance with emission standards should be considered first. More important, are the expected environmental problems solvable?

3. *Political:* Trends and policies of major governmental agencies need to be constantly monitored. Each project should, in its design, be sensitive to this obstacle.

4. *Institutional:* Industries have bona fide fears of generation that must be ferreted out of the myriad of reasons why they do not wish to proceed. These reasons are as diverse as the companies, and are sometimes subjective. Knowledge of these reasons may make a project successful, depending upon the solution offered.

5. *Regulatory:* Different combinations of ownership, lease arrangements, etc., can be structured to ensure that the project operation,

maintenance, and energy use will go forward. All arrangements should be structured around existing laws, with an eye on future legislative changes. If the project is structured in such a way that the owner is judged a utility, the owner will reevaluate the project.

6. *Legal:* An intensive search by consultants, with valuable assistance from an outside specialist, should be obtained to ensure that a legal flaw does not exist in the program. A particular industry's legal history may help the industry deal effectively with the agencies.

7. *Financial:* The financial history and potential ten-year market of a particular industry's product may lead to termination of negotiations as to third party owners or lease arrangements.

Many combinations of industry–utility–financial institution involvement exist that must be applied creatively to each case.

8. *Contractual:* Agreements must supply something for everyone involved, and greed must be avoided. Many arrangements have failed or stalled over the greed issue. Very definite goals are usually asserted by industry and the utility. If the goals are irreversibly greedy, the project has reached a fatal flaw and need not proceed.

9. *Engineering/construction:* A high level of expertise must be employed to accomplish a project. Letting projects out solely upon price may not be in the best interest of all parties. The success of a project will only add to cogeneration credibility in the years to come.

Reliable, experienced, qualified designers and builders should be employed so that project profitability starts as soon as the project is complete.

10. *Operation:* Industry is reluctant in many cases to operate a generation facility. Plant location is often very remote from operating contractors, so the value of this service must be placed into the operation economics of a project.

If the services are to be supplied by others, the proper cost figure should be applied. Such services could prove not to be available at all, in which case the project is not viable.

11. *Maintenance:* The same problems exist here as in item 10. Added emphasis should be placed on major overhauls.

12. *On-going relations:* Good communication will be required for planners to keep track of the project, solve its small problems, and stay abreast of the industry's expected use of the generation system. Close coordination with the utility is critical to a successful project.

13. *Competition:* To date, third party or joint owners have been virtually nonexistent except in the equipment lease areas.

Future encouragement is anticipated to come in the form of third party or joint owners offering long-term leases and a share in the savings created to the industry.

I have investigated various industries, applying the same basic financial formula to each area of cogeneration. The results are contained herein as an aid in defining a project's viability.

III. WASTE HEAT STEAM SYSTEMS

Originally, I thought that these types of systems were the most practical and profitable of all. The projects listed in the table reflect some of the experience to date.

Sale of steam to neighboring industries is a potentially profitable alternative, but it is not included in this evaluation.

	GLASS	CEMENT	CALCINING
Capital investment (1978)	$2,800,000	$7,000,000	$40,000,000
Boiler size	—	—	—
Generator size	2,100 kW	6,500 kW	40,000 kW
Capital recovery (10 yr)	4,200,000	10,430,000	59,600,000
Savings (10 yr revenue)	2,600,000	9,400,000	48,400,000
G & A overhead	−320,000	−2,175,000	−9,600,000
Profit (10 yr)	2,280,000	7,225,000	38,800,000

Advantages of these systems are:

1. The fuel supply is already present.
2. The systems help to reduce industry's investment in air pollution control by reducing the volume of discharge gas that has to be cleaned.
3. Usually they make a negative declaration on the environmental impact statement.
4. Easy retrofit is possible.

Disadvantages are:

1. There is total dependence upon the industry's having fuel to burn.
2. Systems could be ordered to cut back operation during heavy smog periods or could be completely closed owing to other emission standards being exceeded.
3. Factors beyond industry control could reduce operating levels to below the profitability range.

IV. STEAM TOPPING

Many variations exist in this area of generation or cogeneration, ranging from the simple installation of a steam turbine to an entire boiler and generation plant.

Several facilities have been examined. The four listed in the table are probably the best mix of the different types available to date.

	AUTOMOTIVE	BUILDING PRODUCTS	TIRE RUBBER	HOSPITAL
Capital investment (1978)	$2,900,000	$4,000,000	$5,000,000	$2,500,000
		(complete replacement of facilities)		(complete replacement)
Generator sizes	8,000 kW	4,000 kW	10,000 kW	1,200 kW
Capital recovery (10 yr)	4,300,000	5,960,000	7,500,000	4,040,000
Savings (10 yr)	5,600,000	6,800,000	8,600,000	
G & A overhead	−700,000	−800,000	−2,100,000	
Profit (10 yr)	4,900,000	6,000,000	6,500,000	1,350,000

Advantages are:

1. These facilities employ the most tried and proven method of generation.
2. There is probably the least amount of project risk.

Disadvantages are:

1. The systems are subject to steam usage changes by industry.
2. They provide the least amount of electrical output per pound of steam owing to extraction steam pressure.

V. COMBUSTION TURBINES

I have submitted preliminary studies on potential cogeneration facilities utilizing combustion turbines with waste heat boilers in an effort to enhance the profitability of the projects (combined cycle).

The results of those studies are listed in the table.

	ANIMAL FEED	BUILDING PRODUCTS	FOODS
Capital investment (1978)	$5,500,000	$5,300,000	$1,900,000
Generator size	8,000 kW	6,000 kW	3,000 kW
Capital recovery (10 yr)	9,700,000	7,900,000	2,800,000
Savings (10 yr)	8,700,000	7,600,000	3,300,000
G & A overhead	−1,200,000	−1,200,000	−400,000
Profit	7,500,000	6,400,000	2,900,000

Advantages of these facilities are:

1. There is high electrical production per pound of steam produced.
2. Prepackaged equipment speeds construction time.
3. Systems are easy to relocate to another site (salvage).

Disadvantages are:

1. The maintenance factor on the combustion turbine causes industrial concern, as compared with steam topping.
2. Systems cannot be effectively used with high ambient air temperatures or in dusty conditions.
3. Frequent starts and stops reduce the anticipated service life.
4. New point source review is necessary.

VI. WASTE WOOD COGENERATION FACILITIES

I have investigated several potential waste wood applications. Various assumptions can be formulated from the data listed in the table, but consideration should be given to the viability of these types of projects if the risk for the capital investment is covered by leverage leases or no-cut take or pay contracts.

	WOOD	WOOD	WOOD
Capital investment (1978)	$55,000,000	$7,200,000	$10,000,000
Boiler size, lb/hr	428,000	88,500	150,000
Generator size	45,000 kW	6,000 kW	7,000 kW
Capital recovery (10 yr)	82,700,000	10,700,000	14,900,000
Savings (10 yr) (revenue)	54,000,000	12,200,000	10,600,000
Tons of wood/yr	705,500	99,600	211,700
Average price/ton (10 yr)	$3.83	$6.12	$2.50
G & A overhead	6,750,000	1,500,000	1,060,000
Profit (10 yr)	47,250,000	10,700,000	9,540,000

Advantages are:

1. Value of the wood is directly related to savings.
2. Disposal of waste wood is becoming increasingly difficult.

Disadvantages are:

1. Facilities are subjected directly to changing air pollution control regulations.
2. The price per ton of wood may not be attractive enough for the lumber industry.
3. Steam sales are necessary to increase profitability of many wood projects.
4. The wood industry projects only one year in advance owing to government involvement. This diminishes the guarantee of a fuel supply.

VII. HEAVY OIL GASIFICATION

The United States has some very heavy crude oil deposits (+3.0% sulfur—8° API gravity) that are technically recoverable by a steam injection system. It is estimated that the barrels per year production would increase threefold with thermal recovery techniques.

The problem with this type of oil is the limited marketplace and price of the crude recovered. A typical case might reveal that for every four barrels recovered, one barrel goes to the recovery system, which leaves three barrels for market consumption.

To date, this system is probably one of the most environmentally acceptable large power plants we have seen because it cleans the fuel of impurities before they are burned and sent to the atmosphere.

The program, as outlined for an oil company, is quite simple. The company would sell the project crude oil and, in return, would purchase steam (the price of steam tied directly to the price of oil). A known utility would lease or purchase the combustion turbines for the electrical output.

The key to making one of these systems actually happen is to be able to dispose of the energy created and to commit the oil for the project life.

Advantages of this method are:

1. It creates a market for high sulfur crude oil.
2. It makes an environmentally sweet low Btu's gas.
3. Steam use in the field is required to get the oil out.
4. It can gasify nearly all undesirable oils and refinery bottoms.
5. The concept can be utilized in many applications.

Disadvantages are:

1. Users must satisfy environmental concerns; a heavy educational effort is required.
2. Further research is required for a dual fuel system (coal–oil slurry mixture).

VIII. GAS COMPRESSION

The natural gas industry in years past has installed gas-gathering and compression stations based on economics alone. However, gas utility companies, by the structure of the regulated community, are encouraged to invest in themselves.

More customers, winter weather, or aging gas wells may require additional pumping horsepower to meet the demand. The normal course of action to solve these problems has been to install additional fuel-burning pumping capacity.

I have proposed the following alternative: waste heat exchangers installed in the hot gas discharges off the combustion turbines in an organic Rankine cycle application. This configuration can increase the existing conditions by about 30% more horsepower without any fuel use increase.

IX. SUMMARY

Industrial generation can be developed independently of the utilities. What is needed by industrial developers is a realization of all the associated difficulties and an ability to resolve these problems. Industrial planners must develop the ability to recognize when projects have reached a fatal flaw; the widespread use of this faculty will prove beneficial to the overall success of cogeneration facilities in our future. In other words, feasibility is understanding how to complete a project successfully.

Regulation by the public utility company, design engineering, construction, maintenance, operation, utility negotiation, and financing are all areas that must be understood by the industrial developer. A solid knowledge of these areas makes cogeneration projects easier to understand. It also shortens development time.

Short-term return and high return on investment considerations will prevent industries from developing their own cogeneration plans unless cogeneration becomes absolutely necessary for continued production. Long-term capital recovery, and a desire not to have cogeneration facilities, will prevent the utilities from developing many cogeneration facilities. Many people will plan cogeneration, but few installations will actually come about unless a developer provides the assistance, or proper incentives are forthcoming.

Still, gratifying financial rewards can be achieved from cogeneration projects even though cogeneration facilities normally do not produce heavy front-end profits. I believe there are companies with enough flexibility to supply services to a cogeneration plant that the industry does not want to provide for itself. This can encompass any and all phases of the project. I also believe that, by proper coordination and review, we can ensure that industrial generation can and will occur.

There are two basic questions that an industry must ask itself:

1. Does the industry have a waste or by-product energy source?
2. Does the industry wish to capture and/or capitalize on this resource?

If the answer to both of these questions is "yes," then the options and opportunities offered by industrial generation should be investigated, even though success may seem implausible at first glance. Presently, for example, there is a cogeneration project operating satisfactorily in which the steam is used approximately 600 miles from the electrical energy user.

The charts that follow illustrate some of the options that may be available on a cogeneration project. The review of each individual project should include a careful examination of a wide variety of options. All too often, a project has been halted after it has been discovered that an industry could not successfully use all the additional energy internally.

Many more options and configurations exist than have been outlined in this chapter. The few demonstrated here will, it is hoped, stimulate the imagination and result in more cogeneration projects successfully completed.

	ORDER OF PRIORITY		ORDER OF PRIORITY
BOILER/HEAT EXCHANGER		*TURBINE GENERATOR*	
1. Type	☐	1. Ownership	☐
a. Coal/RDF	☐	a. Industry	☐
b. Oil/gas	☐	b. Utility	☐
c. Wood/RDF	☐	c. Third party	☐
d. Waste Heat	☐	2. Operation	☐
2. Ownership	☐	a. Industry	☐
a. Industry	☐	b. Utility	☐
b. Utility	☐	c. Third party	☐
c. Third party	☐	3. Maintenance	☐
3. Maintenance	☐	a. Industry	☐
a. Industry	☐	b. Utility	☐
b. Utility	☐	c. Third party	☐
c. Third party	☐	4. Electrical Energy Use	☐
4. Steam Use	☐	a. Industry	☐
a. Industry 100 percent	☐	1. Standby	☐
b. Industry condense	☐	2. Emergency	☐
c. Industry others/cond.	☐	3. Normal	☐
d. None	☐	b. Utility	☐
5. Exchange/Economics of Steam	☐	1. Capacity	☐
a. Purchase	☐	2. Energy	☐
b. Credit	☐	c. Industry - Total output	☐
c. Other, specify.	☐	d. Combination	☐
		1. Industry capacity	☐
		2. Industry energy	☐
		3. Utility capacity	☐
		4. Utility energy	☐

Index

Acceletron device, 245
Agway corporate headquarters, New York, 360, 361
Air compressors (*See* Compressors)
Air conditioning, 3, 8, 12, 38, 48–50, 57, 59, 52, 67, 68, 70, 72, 74–80, 82, 83, 91, 93, 99, 100, 106–108, 111, 114, 128, 131–135, 159, 164, 171, 174, 176, 180, 185, 186, 207, 210–213, 218, 230, 232, 270, 285, 298, 299, 301–304, 306, 307, 310, 312, 315, 316, 329, 330–332, 340, 350, 351, 361, 385, 396
  See also Heating, ventilation and air conditioning (HVAC)
  condensers (*See* Condensers)
Air dampers (*See* Dampers)
Air flow meters, 51, 55
Air incinerator, 261
Air stratification, 62, 63
Alarms, 119
Allis-Chalmers Company, 298
Aluminum jacketing, 53
Aluminum scrap melting furnace (*See* Ovens)
American Air Filter, 298, 303
American Association of School Administration (AASA), 87, 90–94
American Society of Heating, Refrigeration and Air Conditioning Engineers (ASHRAE), 52, 176, 182, 295, 307, 334, 353
  Handbook (1981), 68, 70
Analog transmission, 15, 16
Anemometer, 54
Apartments, 74, 77, 82, 84
Arab oil embargo, 388
Architects, 4, 8, 22, 25, 43, 90, 98, 105, 156, 206, 207, 212, 213, 270, 271, 361
ARI standard, 182
Astrodome, 304–314
Asymmetrics, 63
Auditor certification, 144
Audits, energy, 3–5, 43–63, 87, 89–171, 176, 218, 233, 247, 248, 252, 357–363, 370, 382
  Class A type—43, 357, 358, 360
  Class B type—357, 358
  Class C type—357, 358
  Preliminary energy audit (PEA), 57, 98
  (*See also* Surveys)
Australia, 38
Australian Solar Water Heater Publication, 336
Automated energy management monitoring and control system (AEMS), 27, 28

Automotive plants, 242
AXCESS, 44

Banks, 79–81
Benefit/cost analysis, 25–28, 31, 39, 44, 84, 176, 177, 217, 249
Bin method, 68–71
Bioconversion, 385
BLAST, 44
BLDSIM program, 208
Bleed-in air, 260
Boilers, 53, 59, 62, 91, 101, 102, 123, 124, 126, 127, 131, 161, 169, 170, 175, 176, 181, 182, 190, 208, 220, 225, 240, 243, 270, 290, 293, 295, 301, 306, 310, 312, 343, 345, 346, 348, 367, 375, 380, 388, 389, 401, 402
  auxiliary electric, 351
  efficiency, 98, 219, 221, 223, 224, 232, 295
  exhaust, 368, 370
  fire auxiliary, 370
  gas-fired, 81, 387
  multiple, 121
  oil fired, 75, 223, 387
  plants, 59, 63
  steam, 220, 239, 240, 241, 269, 305, 342, 346
  tube cleaning, 53
  waste heat, 181, 182, 240, 262, 371, 402
Boundary energy, 173, 181
Brownouts, 388
Building codes, 8
Building envelope, 12, 49, 53, 55, 98, 99, 102, 150, 151, 167–171, 203, 267, 295, 299, 326, 358, 360, 362, 385, 388
Building schedules, 49, 50
Building shell, (*See* Building envelope)
Bureau of Reclamations, 32

Calculations
  hour-by-hour, 71, 72
  manual, 71, 72
California, 27, 28, 90, 394
  Air Pollution Board (*See* Pollution)
  Los Angeles, 343
  Los Angeles Department of Water and Power, 394
  Pasadena, 342, 343, 345, 349
  Pasadena Department of Water and Power, 342
  Southern California Edison, 260
  (*See also* Legislation, proposition 13)

Camera, 51
Canada, Saskatoon, 387
Capital improvements, 46
Capital investments (*See* Investments)
Cash flow, 25–27, 31
Caulking, 57, 98, 102, 113–115, 123, 167, 170
Cement plants, 244, 256, 260
Central plants, 57, 174, 175, 192, 208, 218, 219, 269, 270, 293
Chemical laboratory, 315–326
Chillers, 82, 176, 179, 182, 190, 191, 210, 211, 228–230, 353, 388, 393, 396
  absorption, 182, 229, 295
  centrifugal, 208, 229, 269, 286
  compressor, 396
  double bundle, 56, 186
  electric, 310
  heat recovery, 376, 387, 392
  load, 229
  reciprocating, 229, 287, 350, 353
  screw-type, 229
  tube cleaning, 53
Chimney effect, 295
Circular demand chart, 74
Clean Air Act (1977), 381
Clean Water Act, 381
Client Relations, 51, 52, 54
Climate (*See* Weather)
Coal, 34, 95–97, 145, 152, 164, 374, 381, 388
  coal strikes, 94
Cogeneration, 367–384, 393, 397, 399–405
  combined cycle, 370
  combustion turbine, 370, 374–376, 380, 381, 403
  diesel engine, 370
  Rankine bottoming cycle, 370
  steam turbine, 370, 371, 374–376, 380, 403, 404
  waste wood systems, 403
Coils, 49, 307
  chilled water, 318
  condenser, 135, 173, 174
  cooling, 182, 208, 270, 280
  dehumidification, 133
  electric reheat, 175, 269
  evaporator, 133, 135
  fan (*See* Fans)
  heat, 280
  heating, 270, 279
  preheat, 316, 318
  refrigeration, 138
  reheat 131, 180, 181
  run-around, 232, 233, 316
Coincident load demand, 35
Colorado, Grand Junction, 331, 332
Combustion turbine (*See* Cogeneration)
Commercial buildings, 3, 33, 36, 57, 74, 90, 205, 218, 305, 341, 357, 393
Compressor, 132, 171, 173, 185, 192, 396
  air, 59, 62, 288
  refrigeration, 135
  valves, 133

Computer, 267–269, 292, 388
  analysis, 4, 52, 73, 233
  CDC, 271
  central processing unit (CPU), 13
  controls, 56, 175
  costs, 13, 14, 17, 22
  digital, 14–17, 22, 38
  maintenance, 14, 22
  modeling, 54, 55, 176, 177, 185
  noise environment band, 16, 17
  programs, 71–73, 87, 190, 202, 254, 267, 351, 362
  rooms, 6, 8, 174, 175
  satellite processors, 19
  simulations, 189–204, 254, 360–362
  technology, 38
  See also Axcess, BLAST, Data, Micro computers, Mini computers, Multiplexed systems, TRNYS, Trace
  (*See also* Department of Energy, DOE-2)
Computer messages
  fully-connected (Hybrid), 18
  hierarchy (tree), 18
  point-to-point, 18
  ring (loop), 18
  star, 18, 19
Condensers, 59, 179, 229, 376, 389
  air conditioning, 62, 133
  evaporative, 185
  on-line brush cleaning, 230, 231
  refrigeration, 62
Congress (United States), 28, 87
Connecticut, 90, 240, 243
Contamination, 55, 56, 62, 232, 233, 241, 261, 347, 381
Contracts, 10
Cooling (*See* Air conditioning)
Costs
  capital, 5, 25–28, 85, 91, 105, 177, 246, 326, 329, 358, 369, 372, 373, 394, 395
  community, 32
  construction, 373
  future, 6
  initial, 85–87, 91, 98, 205, 326, 348, 357
  (*See also* Life cycle costing)
Council for American Private Education, 90

Dairies, 241
Dampers, 49, 102, 122, 129–131, 133, 169, 170, 175, 180, 183, 184, 234, 245, 270, 295
  air, 359
  draft, 168
  outside air, 44, 48, 128, 177, 178
  vortex, 262
Data
  acquisition, 14
  transmission, 14, 16, 17
Degree-day method, 68, 69
Delaware, 90
Department of Energy (DOE), 87, 92, 143, 150, 218, 234, 235, 350, 359, 380
  DOE-2, 44, 45

Department of Housing and Urban Development, 218
Diesel, 393
Discounted cash flow, 25–27
Domestic hot water (DHW), 29, 30, 50, 75, 175, 219, 223, 224, 229, 232, 291, 299, 300, 303, 329, 331–333, 335–338, 340, 342, 343, 345–349, 362, 387
Draft effect, 130
Drying processes (*See* Ovens)
Dry transformers, 74
Ducts, 69, 130, 132, 175, 233, 234, 245, 253
  dual duct system, 56, 150, 161, 178, 180, 190, 191, 194, 195, 202, 203, 206, 207, 211, 212, 256, 270, 280, 292, 360, 361

Economizer, 171, 207, 209-213, 220, 240
  cycle, 128, 131, 190, 191, 203, 278–280, 289, 295
  enthalpy-controlled, 190, 203, 234
  full range valve, 229
Economy cycles, 80
Educational Research Service, 92
Education Commission of the States, 90
Electric Power Research Institute (EPRI), 296
Elevators, 12, 83, 196, 200, 201, 205, 293
Energy conservation measures (ECM), 5, 6, 8–12, 45, 93, 101, 105–143, 150–152, 156, 189–204, 217–219, 234, 266, 296, 370
Energy crisis, 33, 34, 205, 206, 357, 390
Energy efficient rate (EER), 105, 107, 110, 171, 173, 182
Energy management, 3–23, 27, 35, 43–47, 57, 59–63, 89–91, 95–99, 100, 102, 103, 115, 118, 241, 267, 271, 361
  control system (EMCS), 267, 295
  master plan for energy management, 46, 47
  (*See also* Automated energy management monitoring and control system)
Energy Research and Development Administration, 29
Energy utilization index (EUI), 5–8
Enthalpy control, 128, 190, 203, 206, 207, 212, 225, 296, 380
Environmental impact statement (EIS), 383, 402
Environmental Protection Agency (EPA), 381
Equipment schedules, 50
Europe, 38, 69, 243, 244, 387
Evaporator, 133
Exhaust, 49, 98, 137, 175, 248, 249, 262, 561
  air, 161, 232, 233, 261, 303, 315, 316, 362
  fans (*See* Fans)
  gases, 55, 257, 260, 261, 376
  heat, 367
  hoods, 138, 140, 175
  systems, 48, 49, 54, 56, 62, 129, 175
  ventilation (*See* Ventilation)
Existing buildings, 3, 6, 24, 28, 67, 72–74, 194, 208, 329, 337, 359, 360
Expander, 389

Fabric transmission, 60, 61
Factories, 33, 59, 62
  (*See also* Industrial buildings)

Fans, 83, 91, 133, 139, 151, 173, 175, 178, 179, 181, 183–185, 190, 191, 194, 195, 197, 205–213, 243, 246, 260, 270, 283, 285, 312
  air, 190, 203, 289, 299, 318
  coils, 161, 206, 208, 210, 211, 213, 233, 281, 292
  condenser, 173
  discharge, 55
  double duct (*See* Ducts, dual)
  evaporator, 173, 178
  exhaust, 44, 49, 59, 62, 102, 129, 137, 138, 168–170, 284, 315, 316, 325, 362
  hoods, 138
  induced draft, 262
  induction unit, 76
  motors (*See* Motors)
  reheat, 206–212
  roof, 102
  (*See also* Variable air volume)
F-chart, 333
Feasibility study, 29
Federal Energy Administration (FEA), 93, 185, 189
Federal Energy Regulatory Commission (FERC), 380, 381
Federal government, 43
Federal register, 163
Fenestration, 56
Filters, 49, 51, 53, 137, 140, 182, 183, 267, 316
  air, 121, 130, 133, 242
Fire codes, 57
Fire protection, 14, 205, 299
  Safety codes, 299
  (*See also* Smoke doors)
Fireside soot buildup, 219, 223
Flashlight, 51
Florida, Tampa, 387
Flowmeter, 181, 182
Flow switches, 175
Flue gas analysis, 121, 125, 181, 240–243, 245, 296
Folding rule, 50
Food services, 162
Foreign energy supply, 11
Free cooling, 179, 190
Fuel cost escalation rate, 27, 28
Fuel cupolas, 241

G-Value, 60
Gasoline station, 251, 252
Geography, 34, 152, 361, 386
Georgia, 90, 167
Geothermal, 400
Glass plants, 244
Glazing, 79
Gravity ventilators, 54

Hansen, Dr. Shirley J., 90
Hawaii, Honolulu, 331, 332
Hazardous wastes, 381

INDEX

Heat exchangers, 182, 183, 259, 296, 299, 300, 333, 340, 347, 349, 353, 390, 404, 405
  air-to-air, 253, 261
  installation, 231, 232
  recuperation, 376
  steam-to-water, 342
  tube-type, 253–256, 261
Heat gain, 98, 109, 296, 305
Heat generators, 59
Heating, 8, 12, 29, 48, 49, 59, 63, 67–70, 72–75, 79, 81–83, 87, 91, 93, 98, 99, 102, 106–108, 111, 121–127, 130, 136, 137, 159, 162, 164, 171, 176, 205, 206, 208, 210–213, 218, 221, 227, 232, 233, 240–246, 249, 258, 270, 279, 282, 289, 293, 299, 300, 303, 306, 307, 312, 326, 329–338, 340, 341, 349–354, 361, 367, 387
  coils (*See* Coils)
  (*See also* Heating, ventilation and air conditioning)
Heating, ventilating and air conditioning (HVAC), 3, 4, 6, 8, 12, 14, 43, 50, 56, 57, 59, 67, 106, 109, 150, 151, 161, 164, 172–186, 189–204, 206–208, 213, 218, 228, 233, 234, 265, 274, 296, 299–301, 360
  loads, 70, 72
  (*See also* Chillers, Boilers, Compressors, Condensers, Efficiency tests, Multizone fans and Solar thermal hybrid HVAC/power generation)
Heat loads, 32
Heat loss, 53, 69, 98, 102, 115, 124, 174, 219, 223, 224, 226, 228, 232, 241, 242, 244, 246, 296, 301, 306, 338, 345, 387
Heat recovery, 49, 57, 254, 260, 261, 303, 315, 350, 351, 369
Heavy oil gasification, 403
High velocity, 195, 196, 202
High water alarms, 267
Hill-Burton Federal Guide, 57, 218
Hirt systems, 251, 252
Hospitals, 43, 56, 57, 84, 218, 220, 230, 232–235, 306
Hot spots, 53
Hotels, 3, 81–83
Houston (*See* Texas)
Human systems, 9, 106–111, 155, 157, 158
Humidity, 11, 50
Hydro energy, 83
Hydronic heating system, 102

Illinois, 90, 168
  Chicago, 83
  Development Board, 350
  Kane County, 350
Illuminating Engineering Society (IES), 269, 296
Incinerators (air), 261
Induction unit, 79, 150, 161, 391
  fan, 76
  motor generator (*See* Motors)
Industrial buildings, 56, 57, 59–63, 89, 239–248, 341, 399
Industrial waste, 274

Infiltration, 54, 60, 61, 69, 91, 99, 101, 102, 114, 115, 362
  air change method, 54
  crack method, 54
Inflation, 24, 28, 31, 34, 35, 85, 177, 348
Infrared camera, 53
Infrared scanning (*See* Thermography)
Insolation, 152, 154, 165, 330, 332, 339
Institute of Gas Technology (IGT), 296
Institutional buildings, 217–237
  (*See also* Schools)
Insulation, 53, 57, 62, 79, 123, 136, 159, 168, 170, 219, 226, 240–242, 246, 248, 301, 333, 338, 359, 385, 397
  boiler, 126
  ceiling, 362
  curve, 228
  door, 112
  duct, 132, 178, 243
  envelope, 69
  equipment, 53
  fiberglass, 245, 338
  high temperature wool, 243
  insulating fire brick (IFB), 244
  piping, 53, 125, 132, 178, 219, 227
  roof, 53, 57, 90, 115, 167, 169, 171, 298, 299
  solar panel, 344, 345, 346
  thermal, 296
  wall, 44, 53, 57, 90, 114, 167, 270, 298, 299
  window, 99, 102, 113, 298, 303
  (*See also* Insolation)
Interest rate, 25, 26, 28, 85, 177, 192, 326, 369, 387
Internal rate of return (IRR), 25–28
Investment
  capital, 45, 46, 87, 177, 191, 401, 402
  discount rate, 28
  initial, 25–27, 31, 269, 347, 349
Iron foundry cupola, 260–261

Johnson Controls Company, 202–204

Kentucky, 54
Kilns (*See* Ovens)

Laws (*See* Legislation)
Legislation, 28, 34, 330, 369, 372, 380, 381, 382, 393, 401
  proposition 13, 298
  solar, 28
  state legislators, 28
  (*See also* Solar)
Life cycle costing, 8, 24, 25, 27–29, 31, 39, 43, 52, 85–87, 105, 191, 218, 269, 329, 346–349, 357, 373, 385, 387
Life safety code, 205
Lighting, 4, 6, 8, 9, 12, 43, 44, 46, 48, 50, 52, 56, 57, 59, 67, 75, 76, 80, 83, 99, 107, 110, 120, 151, 155, 164, 175, 176, 179, 184, 207, 208, 218, 234, 235, 246, 265–267, 296, 300, 305, 306, 310, 313, 314, 359, 362
  delamping, 117, 168, 170

fluorescent, 48, 57, 91, 100, 116, 117, 160, 167–171, 235, 246, 277, 293
HID, 48
high pressure sodium, 119, 120, 269
incandescent, 48, 100, 116, 117, 120, 160, 235, 246, 269, 277, 293
mercury vapor, 57, 116, 119, 246
metal halide, 119, 120, 246
natural, 109, 118
relamping, 269
security, 119
sodium vapor, 246
thermal codes, 184
(*See also* Illumination Engineers Society)
Light cavity heat, 194
Light meters (*See* Meters)
Load calculations, 68
Lone Star Company (*See* Texas)
Los Angeles (*See* California)
Lower explosive limit (LEL), 245
Luminary (*See* Lighting)

M-B-M building cost file, 31
Madison Gas and Electric Company, 394
Magnetic energy, 172
  rotary fields, 390, 391
Magnetic tape, 74
Main frame computer, 13, 267, 268
Maintenance, 5–12, 14, 26, 45, 49, 51–53, 55, 56, 62, 89–91, 94, 96–100, 105, 107, 126, 186, 190, 205, 206, 221, 230, 240, 242, 246, 256, 260, 265, 267, 269–271, 275, 319, 330, 331, 342, 345–348, 353, 357–363, 369, 373, 382, 388, 390, 397, 400, 401, 403, 404
  Computer (*See* Computer)
  (*See also* Manual, operation and maintenance)
Manchester biphase encoding, 16
Manpower, 10, 26, 240, 267
Manual, operation and maintenance (O&M), 8, 12, 45, 106, 146
Marriott Corporation, 362, 363
Maryland, Baltimore, 86
Massachusetts, 90
MEANS, Building Construction Cost Data, 31
Mechanical rooms, 80, 165, 190, 195
Meckler, Milton, 90
Meckler Systems Group, 196
Mediterranean, 385, 387
Melting furnace (aluminum) (*See* Ovens)
Mercury vapor bulbs, 116
Metals
  heat treating, 242–243
  hot forming, 242
  melting, 241–242
Meters, 73–75, 83, 87, 176, 203, 220, 253, 266, 293, 316, 318, 319
  air flow, 362
  electric, 359, 394
  light, 8, 50
  submeters, 310
  water, 359, 361

Michigan, 90, 99, 240, 244, 245
Microcomputer, 13–23
Microprocessor, 234, 267, 339
Middle East, 385, 387
Military, 32, 51, 57
Minicomputer, 13–23, 179
Minnesota, 245
Motels, 3
Motivation, 46
Motors, 53, 91, 129, 151, 184–186, 246
  efficiency, 173, 183
  fan, 173, 174, 176, 179, 184
  generator, 392
  induction, 184, 390–392, 396, 397
  polyphase, 390, 391, 397
  squirrel cage, 391, 396
  variable speed, 129
Multiplex systems, 13–16
Multizone units, 150, 161, 178, 180, 195, 203, 270, 278, 292
  (*See also* Induction)

Nameplate data, 49, 51, 55, 151, 161, 195, 278, 279, 280, 282–285, 391
National Electrical Manufacturers Association (NEMA), 8
National Electric Code, 74
National Energy Act (1978), 34, 57
National Oceanic and Atmospheric Administration, 98
National Science Foundation, 29
Natural gas, 32, 50, 74, 75, 90, 94–98, 145, 146, 164, 193, 208, 221, 227, 228, 241, 242, 245, 249, 261, 262, 264, 266, 275, 304, 311, 347–349, 362, 373–376, 380, 381, 404
Net present value (NPV), 26–28, 31, 348
  (*See also* Present Value)
New buildings, 3, 24, 28, 67, 68, 73, 85, 208
New Hampshire, 90, 169
New York, 90, 170, 218
  DeWitt, 360
  New York City, 393
  Rochester, 326
  State Energy Office, 218
Nonattainment area (NA), 381
Nonresidential buildings, 68, 74
Nuclear energy, 34, 72, 83, 388

Occupancy, 8, 24, 28–30, 44, 50–52, 56, 57, 60, 63, 74, 76, 80, 82, 87, 98, 101–102, 107, 108, 110, 111, 119, 128, 129, 136, 157, 158, 178, 189, 202, 203, 205, 207, 245, 292, 301, 305, 315, 338, 351, 359, 387
Occupant comfort, 175, 307, 360
Ocean thermal gradient, 385
Office buildings, 3, 6, 33, 56, 57, 74, 84, 160, 178, 189, 206, 207, 218, 234, 235, 306, 359, 393
Office machines, 141, 162, 176
Ohio, 90, 242
Oil burners, 125
Oil embargo (1973), 3, 388
Oil pressure, 125

Operating costs, 3, 4, 10, 24–26, 33, 34, 91, 92, 185, 186, 190, 247, 249, 316, 361, 369, 373
Oregon, Portland, 244
Ovens
  bakery, 138, 139, 244, 245
  drying, 245, 246
  kilns, 163, 167
  melting furnace (aluminum), 249, 250
  paint, 259
  process curing, 247, 248
  rotary kilns, 244
  vertical kilns, 24
Owens-Corning, 86

Paper mill, 244
Pasadena (*See* California)
Payback, 8, 25, 27, 28, 31, 49, 62, 85, 87, 91, 98, 101, 102, 105, 177, 182, 183, 185, 189, 217, 219, 233, 266, 269, 270, 296, 326, 348, 349, 399
  simple, 85
Pennsylvania, 69, 90, 241
  Philadelphia, 75, 77, 79, 81, 242
  Pittsburgh, 242
Periodic behavior, 63
Photovoltaic cells (*See* Solar photovoltaic)
Pipes, 132, 175, 232, 270, 318, 343, 346, 361
  chilled water, 134, 270
  condensate, 342, 343
  heat, 62, 233
  high temperature, 59
  hot gas, 174
  solar, 333, 336, 338, 340, 344, 346, 351
  steam, 270, 342, 343
  water, 124, 136
Plant process loads, 49
Plaques (energy conservation), 110, 140, 167–169
Plumbing, 50, 56
  codes, 337
Politics, 32, 34–36, 400
Pollution
  air, 249, 251, 259, 261, 359, 381, 400, 402, 403
  California Air Pollution Board, 251
  National Pollutant Discharge Elimination System (NPDES), 381
  water, 381, 400
Polyphase electric system, 390, 391, 397
Pools (*See* Swimming pools)
Post office, 72
Power Plant Industrial Fuel Use Act, 380
Present value, 25, 26, 28, 348, 373
  (*See also* Net present value)
Pressure sensors, 175
Primary voltage rate, 83
Propane, 48, 50, 52, 95–97, 145, 148, 164
Psychometers, 8, 50, 296
Public address system, 14
Public care facilities, 57
Public relations, 5, 45
Public Utilities Commission (PUC), 369, 381, 382, 388, 394

Public Utility Regulatory Policies Act (PURPA), 34, 380
Pulse Amplitude Modulation (PAM), 16
Pulse Code Modulation (PCM), 16, 17
Pumps, 29, 91, 102, 123, 127, 151, 175, 176, 183, 185, 232, 312, 333, 346
  circulating, 38, 291, 299, 318, 337, 339
  feed, 389
  heat, 62, 161, 167, 296, 351, 387, 390
  hydronic, 100, 169
  rates, 169
  recirculating, 136, 169
  solar, 333–338, 351–353
  swimming pool (*See* Swimming pool)
  water, 127, 287
  water source heat, 299–303, 338

Radiation, 150, 161
Rankine cycle, 387–397
Rasbach, Roger, 90
Raw source energy, 173, 174, 181
Recuperators, 62
Refrigeration, 135, 138, 139
  condensers, 62
  (*See also* Air conditioning and HVAC)
Regenerators, 62
Regulations (*See* Legislation)
Relative importance factors (RIF), 150–154
Residential buildings, 33, 36, 57, 68, 69, 73–76, 90, 170, 338, 341, 357
Resin manufacturing, 249
Resource Conservation and Recovery Act, 381
Restaurants, 362
Return on investment (ROI), 360, 369, 373
Rhode Island, 90
Risk analysis, 400
Rotary kilns (*See* Ovens)
Rotating magnetic fields, 390, 391
Runaround coils, 62

Salvage value, 26, 28
Schoolhouse Energy Efficiency Demonstration (SEED), 89–171
Schools, 29, 31, 43, 56, 57, 74, 84, 87, 89–98, 101, 102, 105–171, 218, 298–303, 350–355
  Hand Middle School, 298–303
  Saint Charles High School District, 350, 353, 354
  (*See also* Schoolhouse Energy Efficiency Demonstration)
"Scoring system," 46
Security, 57, 80, 292, 369
  alarms, 119
  clearance, 51
  lights (*See* Lights)
  office, 292
  systems, 14
Separate direct expansion (DX), 174, 175
Shading, 54, 56, 109, 152
Skylights, 50, 100, 115
Smoke doors, 299

Social Security Administration, 86
Solar collectors, 29, 30, 154, 300, 301, 303, 330–340, 343–346, 351–354, 386, 387, 392
  double-glazed, 330
  unglazed, 330
  (*See also* Insolation)
Solar energy, 54, 85, 109, 143, 152–155, 165, 191, 218, 329–355, 385, 387–389, 392, 397, 400
  active, 329, 330, 332, 389
  passive, 69, 329, 330, 386, 389
Solar heater
  auxiliary, 29, 30
  water, 137, 336
Solar investments, 25, 28–32
Solar legislation (*See* Legislation)
Solar panels, 29, 30, 298, 343–347, 350–354
Solar photovoltaic cells, 329, 339, 385, 400
Solar pipes (*See* Pipes)
Solar powered HVAC/cogeneration system, 396
Solar pumps (*See* Pumps)
Solar radiation, 44, 331
Solar routine, 54
Solar screening devices, 270
Solar storage tank, 29, 30, 345, 346, 347, 350, 351
Solar thermal hybrid HVAC/power generation, 385–397
Solar thermosyphan system, 334, 336–338
Solid waste, 55, 374
South Carolina, Columbia, 298, 299
Southern California Edison (SCE) (*See* California)
Stairwells, 10, 82, 165, 269, 299
Steam, 74, 127, 145, 149, 161, 164, 218, 221, 229–233, 249, 262, 315, 319, 342, 343, 345, 346, 370, 373, 374
  boilers (*See* Boilers)
  distribution, 240, 241
  radiators, 124
  topping, 402
  traps, 53, 219, 224, 241, 346
  (*See also* Waste heat steam systems)
Stores (retail) (*See* Commercial buildings)
Stratification, 63
Sunbelt, 340, 342, 387
Sundry energy, 59, 61, 62
Sunworks, Inc., 353
Superheat, 241, 389
Surveys (energy), 5, 8–10, 48–58, 63, 98, 202, 217, 218, 230, 239, 240, 246, 265–267, 342, 359, 369, 370, 382
  empirical, 12
  plant, 253, 254
  presurvey organization, 48
Swimming pools, 137, 162, 329, 330, 354
  pumps, 38
  synthetic fuels, 376

Tape measures, 50
Tax, 29, 37, 85, 312
  impacts, 25
  incentives/credit, 28, 314, 369, 381, 393
  income, 26
  property, 26
  taxpayers, 89
  transfer, 36
Technical Assistance Program (TAP), 43, 57
Tenant
  complaints, 26
  cooperation, 10–12
Tenneco Inc., 90
Texas, 90, 171
  East Texas State University, 265, 271
  Houston, 178, 304, 307
  Lone Star Gas Company, 266
  Texas Power and Light Company, 266, 269
  (*See also* Astrodome)
Thermal cascading, 242–245
Thermal energy, 172, 173, 220, 297, 329, 330, 367, 369, 370, 372, 373, 375, 382
  (*See also* Solar thermal hybrid HVAC/power generation)
Thermal oxidizer, 248, 249, 259, 261
Thermography, 53
Thermometers, 50
  stack gas, 220
Thermostates, 10, 45, 62, 69, 99, 101, 103, 106–109, 122, 124, 131, 136, 167, 175, 177, 178, 182, 203, 233, 234, 298–300, 340, 362
  deadband, 179, 180, 206–208, 212
  mainzone, 180
Thermosyphon, 336, 337
Time clocks, 108, 119, 136, 178, 179, 234
Time division multiplaning (TDM), 16, 17
Torque, full load (FLT), 391
TRACE, 44, 72, 194
Traffic controls, 14
TRNSYS (Transient stimulation program), 29–31
Turbo charger, 376
Two-way radio, 51

U-Value, 61
United Kingdom, 59, 60
United States, 34, 38, 71, 91, 211, 240, 243, 332, 334, 337, 357, 385–388, 393, 394, 403
  Air Force Manual, 70
  Congress (*See* Congress)
  Department of Housing and Urban Development, 218, 333
  Drug Administration, 342, 343, 347
  Steel, 86
  (*See also* Department of Energy)
University of Rochester River, 315
Utility, 74, 83, 163, 253–263, 386, 388, 389, 399–401
  company, 55, 359, 367, 369, 370, 372, 374, 382–384, 388, 389, 393, 395, 397, 404
  consumption, 270
  cost, 32, 271, 275, 315, 388, 389, 394
  electric, 399, 400
  interface, 37–39, 385, 389, 392
  (*See also* California, and Public Utilities Commission)

Utility bills, 4, 6, 22, 43, 48, 50, 73, 74, 145, 163, 246, 266, 269, 312, 347
Utility costs, 4, 56
Utility distribution, 57, 269, 270
Utility information, 53, 55, 56
Utility loads
 demand, 38
 user-controlled, 38
 utility-controlled, 38
Utility rates, 6, 8, 33–40, 43, 73, 83, 84, 184, 304, 305, 347, 370, 389, 393, 394, 399, 400
 commodity, 36
 demand, 35–37
 historical background, 34, 56, 266, 269, 388
 interruptable, 35, 37
 marginal costing, 35
 seasonal, 394
 socially motivated, 36
 time-of-day use, 35, 37, 38, 394, 395
Utility storage, 38, 39

Vane anemometer, 51, 54
Vapor barrier, 53
Variable air volume system (VAV), 131, 150, 161, 178–181, 184, 190, 194, 203, 204, 206–213, 270, 292, 360, 361
 fans, 206–213
Velmeters, 8
Ventilation, 48, 49, 54, 60–62, 82, 91, 98–102, 125, 127–130, 137, 138, 150, 185, 207, 211, 251, 270, 282, 283, 297, 334, 361
 codes, 299
 overventing, 62
 rates, 6–9, 49, 56
 (*See also* Exhaust, Gravity ventilators, and Heating, ventilation and air conditioning)
Vents (*See* Ventilation)
Vermont, 394
Virginia, 90
Vertical kilns (*See* Ovens)
Vitreous enameling, 243, 244
Volt ammeter, 51

Washington, D.C., 207
Waste energy, 54, 61, 62, 330, 393
Waste heat steam systems, 401
Water treatment, 220, 347
Water cost, 190, 192
Waterside scale buildup, 219, 222, 223
Water towers, 165
Weather, 6, 8, 11, 14, 34, 45, 46, 52, 54, 56, 68–71, 73, 75, 77, 92, 93, 98, 114, 115, 145, 157, 177, 189, 203, 217, 232, 251, 331, 332, 333, 343, 358, 361–363, 386, 404
 macro climate, 44
 micro climate, 28, 44
 National Climatic Center, 8
 National Weather Service, 152, 165
 weatherstripping, 44, 62, 98, 100, 102, 112, 113, 123, 167, 170, 297, 359
 (*See also* Caulking)
Weather stations, 44
Weather tape, 54
West Virginia, 244
Wind, 44, 54, 69, 112, 152, 154, 165, 167, 171, 244, 343, 400
 break, 102, 171
 power, 385
 screens, 112
Windows, 45, 49, 53, 57, 62, 75, 80, 83, 91, 98, 100–102, 106, 109, 111, 113, 150, 167, 169–171, 175, 207, 269, 298–300, 303
 double-glazed, 45, 79, 81, 113, 207, 300
 insulation (*See* Insulation)
 single-glazed, 53, 79, 81, 168, 246, 298
 thermopane, 113, 233
 triple-glazed, 79
Wisconsin, 242, 243
 Public Service Commission, 394
 University of Wisconsin-Madison, 29
Wolff, Dr. Calvin M., 90

Zones, heating and cooling (U.S.), 157, 158, 202, 278